QUANTITATIVE FINANCE

QUANTITATIVE FINANCE

Editor

Orhan Petrić

scitus academics

Quantitative Finance

Edited by **Orhan Petrić**

ISBN: 9781681179667
Library of Congress Control Number: 2017949098

SCITUS Academics LLC,
616, Corporate Way, Suite 2, 4766,
Valley Cottage, NY 10989

www.scitusacademics.com

Printed in United States of America.

Preface

Quantitative finance, also known as mathematical finance, is a field of applied mathematics, concerned with financial markets. Generally, mathematical finance will derive and extend the mathematical or numerical models without necessarily establishing a link to financial theory, taking observed market prices as input. Mathematical consistency is required, not compatibility with economic theory. In particular mathematical finance has been ranked, without a doubt, as the first among many and quants, as mathematicians are known in the industry, have been blamed for developing and using esoteric models which are believed to have caused the deepening of the financial crisis. As in many fields of science, mathematics plays a relevant role in finance. The basis of financial theory is an analysis of actions of 'economic agents' on an effective use of resources under uncertainty, and as a complex procedure, it makes necessary the use of advanced methods of mathematical modeling, the application of which has a direct and considerable impact on the financial world.

This volume 'Quantitative Finance' presents perspective on how the development of finance theory has influenced and in turn been influenced by the development of mathematical finance theory. It evaluates the importance of using mathematical modeling tools in finance and discussion of its positive and negative influence on financial markets. The use of such methods in finance might take different directions regarding various aspects of financial market types, each of which might involve sophisticated analytical and numerical techniques. Therefore, with enormous fluctuations occurring frequently in the financial markets the demand for better forecasting and decision-making methods is increasing. Since organizations responsible for producing economic and financial forecasts have a huge amount of information to process and a growing variety of techniques. So, financial practitioners must not only heavily rely on them, but on their intuition and judgment. Only the proper use of mathematical models in finance with consideration of all accurate financial data, trends of financial markets, and useful variables of a system might bring a good understanding of dynamic markets and assist practitioners to make financial projections and decisions adequately.

Table of Contents

CHAPTER 1

Extended Correlations in Finance

Mark Burgin[1], Gunter Meissner[2,3,4]

[1]*UCLA Department of Mathematics, Los Angeles, CA, USA*

[2]*Derivatives Software, www.dersoft.com*

[3]*Cassandra Capital Management, www.cassandracm.com*

[4]*MathFinance at NYU-Courant, New York, USA*

ABSTRACT

Extended correlations, i.e. correlations that can take values less than −1 and/or larger than 1, occur naturally in mathematical models of financial processes. Extended correlations also occur in financial practice, especially in dispersion trading, implying arbitrage opportunities. Based on theoretical and practical emergence of extended correlations, we derive a mathematical framework for extended correlations explaining interpretations and applications. We develop a broader mathematical approach, which can model conventional as well as extended correlations.

Keywords: Extended Correlation, Pearson Correlation, Financial Process, n-Aspect Correlation Coefficient, n-Factor Correlation Coefficient, Complete Correlation Coefficient, Total Correlation Coefficient

1. INTRODUCTION

New discoveries are often met with skepticism and resistance. When negative numbers came to Europe in books of Eastern mathematicians, critics dismissed their sensibility. Many well-known European mathematicians, e.g., Jean le Rond d'Alembert (1717-1783) or Augustus De Morgan (1806-1871), rejected the sensibility of negative numbers until the 18th century and referred to them as "absurd" or "meaningless" (Kline, 1980 [1] ; Mattessich, 1998 [2]). Even in the 19th century, it was a common practice to criticize any negative results derived from equations, on the assumption that they were meaningless (Martinez, 2006 [3]). For instance, Lazare Carnot (1753- 1823) argued that "the attempt to take it [a number] away from a number less than itself is ridiculous", affirming that the idea of something being less than nothing is absurd (Mattessich, 1998 [2]). Some other outstanding mathematicians, such as William Hamilton (1805-1865) and August De Morgan (1806-1871), had similar opinions.

Likewise, irrational numbers and later imaginary numbers were firstly rejected. Today these concepts are accepted and applied in numerous scientific and practical fields, such as physics, chemistry, biology and finance.

Similarly, probabilities less than 0 and greater than 1 were long considered non-sensible. However, these extended probabilities and especially their important case-negative probabilities with values between 1 and −1, have been applied in physics for quite a while (cf. Wigner, 1932 [4] ; Dirac, 1942 [5] ; 1943 [6] ; Bartlett, 1945 [7] ; Feynman, 1987 [8] ; Mückenheim, 1986 [9] ; Krennikov, 1992 [10] and 2009 [11]). The probabilities which are negative and bigger than 1 are also applied in finance (Haug, 2004 [12] ; Székely, 2005 [13] ; Burgin and Meissner, 2010 [14] and 2012 [15] ; and 2012 [16]).

Mathematical patterns of negative probabilities were studied in Bartlett 1945 [7] and in Allen 1976 [17]. A mathematical theory of negative probabilities in p-adic fields was developed in (Krennikov, 2009) [11]. A mathematical theory of extended probabilities, which included negative probabilities, was created by Burgin and Meissner in 2010 [14] and 2012 [15] for the standard situations in physics, economics and finance, where real numbers and not p-adic numbers are used. Mathematically grounded interpretations of negative probabilities were constructed in Burgin, 2012 [18] ; Abramsky and Brandenburger, 2014 [19].

In this paper, we study correlation coefficients used in mathematical models of financial markets. By their conventional construction, correlation coefficients cannot be larger than 1 and smaller than −1. However, by exploring the theory and practice of financial markets, we have discovered emergence of correlation coefficients beyond these limits. We call them extended correlations and develop a mathematical theory for them.

The significance of our research lies on the fact that conventional mathematical theories cannot fully model all existing correlation values and processes in finance. That is why we introduce and study the concept of extended correlations, which is more general than conventional correlations including them as a special case. Hence all correlations in finance (and other sciences), i.e. conventional, and correlations that are smaller than −1 or larger than 1 can be evaluated using the concept of extended correlations.

The rest of the paper is organized as follows. In Section 2, we show that extended correlations can naturally occur in mathematical models of financial processes. In Section 3, we find extended correlations in the practice of financial markets, which imply arbitrage opportunities. In Section 4, we build mathematical foundations for extended correlations. Section 5 addresses the limitations of our concept. Section 6 concludes.

2. EXTENDED CORRELATIONS IN FINANCIAL MODELING

Financial variables such as stocks, bonds, interest rates, commodities, and volatilities are stochastic, i.e. they can be only predicted with a certain probability. Therefore, it is a good idea to model financial variables with stochastic processes as it is done in mathematical finance.

Recent events, such as the global financial crisis 2007-2009, have highlighted an important critical financial variable: correlation. In the crisis, correlations between many financial variables such as stocks, bonds, loans, and especially, sub-prime mortgage loans often securitized in a CDO, increased sharply and led to large unexpected losses. Hence, "Correlation Risk", the risk of unfavorable change in correlation, has recently been addressed in financial modeling as well as in risk management and regulation[1].

It has been suggested (see Emmerich, 2006 [20] , Ma, 2009 [21] and 2009 [22]) to model correlations with a bounded Jacobi process of the form

$$d\rho = a\left(m_\rho - \rho_t\right)dt + \sigma_\rho\sqrt{\left(h-\rho_t\right)\left(\rho_t - f\right)}\varepsilon_t\sqrt{dt} \qquad (1)$$

where:

ρ is the Pearson correlation coefficient[2],

a is the mean reversion parameter (speed, gravity) i.e. degree with which the correlation at time t,

ρ_t is pulled back to its long term mean m_ρ, $0 \le a \le 1$,

σ_ρ is the volatility of ρ; $\sigma_\rho > 0$,

m_ρ is the long term mean of the correlation ρ,

h is the upper boundary level, f: lower boundary level, i.e. $h \geq \rho \geq f$,

ε_t is the random drawing from a standard normal distribution at time t, $\varepsilon = n \sim (0,1)$.

Applying traditional Pearson correlation values, *i.e.* $-1 \leq \rho \leq +1$, the upper bound h becomes +1 and the lower bound f becomes −1 and Equation (1) reduces to

$$\mathrm{d}\rho = a\left(m_\rho - \rho_t\right)\mathrm{d}t + \sigma_\rho\sqrt{\left(1-\rho_t^2\right)}\varepsilon_t\sqrt{\mathrm{d}t} \tag{2}$$

For high values of the correlation volatility σ_ρ and low values of the mean reversion parameter "a", Equations (1) and (2) can result in correlation values ρ of < −1 and > +1, and the equations cannot be evaluated (since the terms $\sqrt{(h-\rho_t)(\rho_t - f)}$ and $\sqrt{(1-\rho_t^2)}$ cannot be evaluated). Hence we have to introduce boundary condi-

tions, which are

$$\alpha \geq \frac{\sigma^2(h-f)/2}{(m_\rho - f)} \quad \text{and} \quad \alpha \geq \frac{\sigma^2(h-f)/2}{(h-m_\rho)} \tag{3}$$

for Equation (1) and

$$\alpha \geq \frac{\sigma^2}{(m_\rho + 1)} \quad \text{and} \quad \alpha \geq \frac{\sigma^2}{(1-m_\rho)} \tag{4}$$

for Equation (2).

However, there are problematic issues with Equations (1) and (2) and its boundary conditions (3) and (4). When modeling financial correlations in practice, we have to discretize Equations (1) and (2). Equations (1) and (2) then become

$$\rho_{t+1} = \rho_t + a\left(m_\rho - \rho_t\right)\Delta t + \sigma_\rho\sqrt{\left(h-\rho_t\right)\left(\rho_t - f\right)}\varepsilon_t\sqrt{\Delta t} \tag{5}$$

$$\rho_{t+1} = \rho_t + a\left(m_\rho - \rho_t\right)\Delta t + \sigma_\rho\sqrt{\left(1-\rho_t^2\right)}\varepsilon_t\sqrt{\Delta t} \tag{6}$$

respectively.

For Equations (5) and (6) the boundary conditions (3) and (4) are invalid, i.e. even if the boundary conditions are met, it can happen that Equations (5) and (6) cannot be evaluated. This is especially the case for high correlation volatility σ_ρ and low values of the mean reversion parameter "a". There are several solutions to this problem:

1) We can introduce limits which the correlation values can take. The limits would be h and f for Equation (5) and −1 and +1 for Equation (6). We could compute Equation (5) as

$$\rho_{t+1} = \text{If}\left(\rho_t < f, f, \text{if}\left(\rho_t > h, h, \left(\rho_t + a\left(m_\rho - \rho_t\right)\Delta t + \sigma_\rho\sqrt{(h-\rho_t)(\rho_t - f)}\varepsilon_t\sqrt{\Delta t}\right)\right)\right) \tag{7}$$

Equation (7) reads: If the simulated correlation coefficient at t, ρ_t, is smaller than the lower boundary f, take the value f; if the simulated correlation coefficient at t, ρ_t, is greater than the upper boundary h, take the value h,

otherwise apply Equation (5)

$$\rho_{t+1} = \rho_t + a\left(m_\rho - \rho_t\right)\Delta t + \sigma_\rho\sqrt{(h-\rho_t)(\rho_t - f)}\varepsilon_t\sqrt{\Delta t}$$

. For Equation (6), the

lower boundary $f = -1$, and the higher boundary $h = +1$. However, the approach (7) arbitrary and model-incon- sistent.

2) We can allow extended correlations increasing the upper boundary h above +1 and decreasing the lower boundary f below −1 in Equation (5). However, increasing h and decreasing f would effectively increase the volatility of the correlation ρ since the last terms of Equations (5) and (6) are amplified.

3) A viable solution to this problem is to allow utilization of extended correlations and to describe correlation by means of a standard mean-reverting Vasicek model of the form

$$\mathrm{d}\rho = a\left(m_\rho - \rho_t\right)\mathrm{d}t + \sigma_\rho\varepsilon_t\sqrt{\mathrm{d}t} \tag{8}$$

or by its discrete version

$$\rho_{t+1} = \rho_t + a\left(m_\rho - \rho_t\right)\Delta t + \sigma_\rho\varepsilon_t\sqrt{\Delta t} \tag{9}$$

Equations (8) and (9) can be evaluated in every simulation with the standard parameter values $0 \le a \le 1$ and $\sigma_\rho > 0$.

In summary, the Jacobi process, which has been suggested to model correlation, can lead to errors when simulating discrete correlations in reality. Increasing the boundaries in a Jacobi process solves the problem, however at the cost of increased correlation volatility.a viable solution is to apply a standard Vasicek model and allow extended correlations. This will guarantee that every real-world discrete correlation simulation is executed without receiving error values.

Extended Correlations as an Input for Financial Models

In finance, the Pearson correlation coefficient or a Pearson correlation matrix serves as an input for many financial models. For example, Copulas typically apply the Pearson correlation coefficient or a Pearson correlation matrix. Due to its convenient properties, the thin-tailed Gaussian copula is often used in finance. For the bivariate Gaussian copula, the density function is

$$c_G(x,y) = \frac{1}{\sqrt{1-\rho^2}} \exp\left(\frac{\zeta_1^2 + \zeta_2^2}{2} + \frac{2\rho\zeta_1\zeta_2 - \zeta_1^2 - \zeta_2^2}{2(1-\rho^2)} \right) \tag{10}$$

where $\zeta_1 := N^{-1}(x), \zeta_2 := N^{-1}(y)$ and N^{-1} is the inverse of the standard normal distribution, and ρ is the Pearson correlation coefficient. Figure 1 and Figure 2 show the density of the bivariate copula with uniform marginals.

From Figure 1 and Figure 2 we can observe the critical impact of the correlation parameter ρ on the density function.

To derive the dependency for more than two variables, often a factorization is applied. This OFGC (one-fac- tor Gaussian copula) model is

$$x_i = \rho M + \sqrt{1-\rho^2} Z_i \tag{11}$$

where:

M is the systematic factor, which impacts all variables x_i. As ε, M is a random drawing from a standard normal distribution, $M = n \sim (0,1)$.

Z_i is the idiosyncratic factor of entity *i*. Just like *M*, Z_i is a random drawing from a standard normal distribution, $Z_i = n \sim (0,1)$.

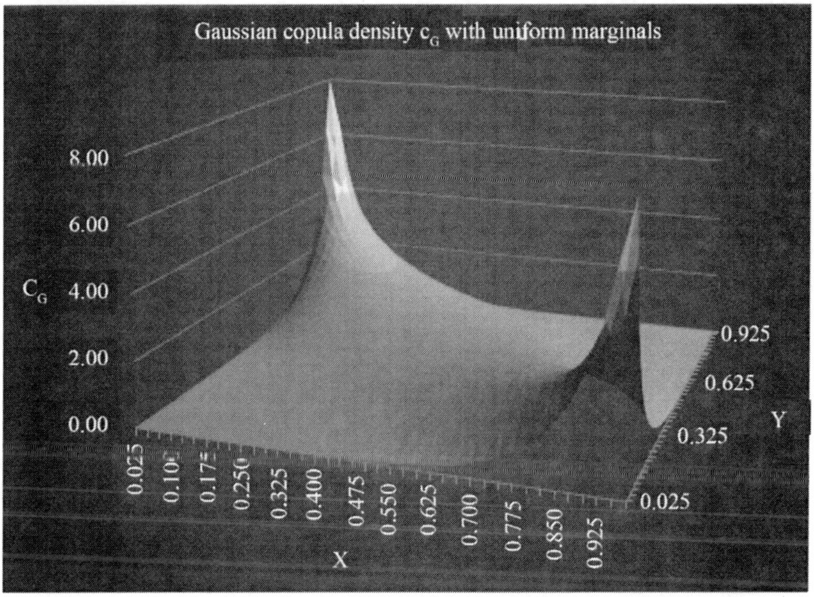

Figure 1. Gaussian copula density with ρ = −0.8.

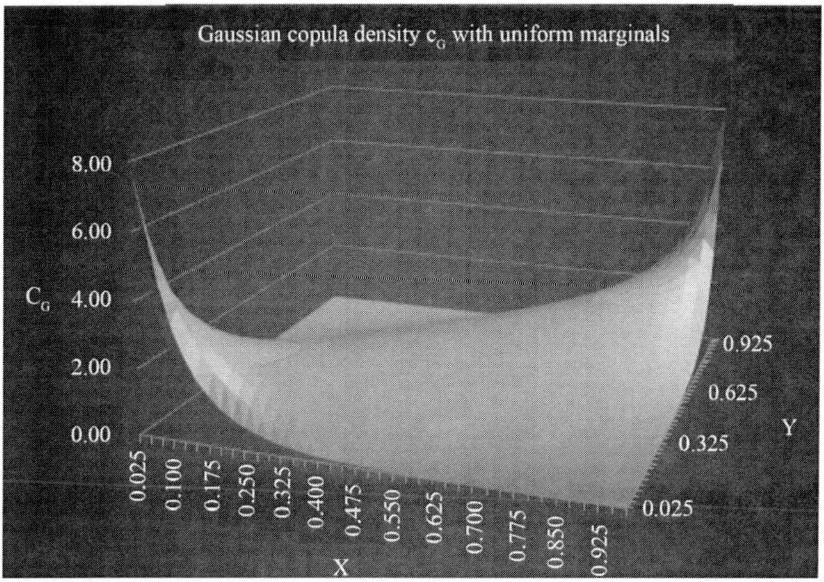

Figure 2. Gaussian copula density with ρ = +0.8.

x_i: The variables x_i for $i = 1,2,\dots,n$ are correlated with each other by "conditioning on M". From Equation (11) we observe that for $\rho = 1$, all x_i are equal to M in a certain simulation, i.e. all x_i are identical. For $\rho = 0$, all x_i are equal to their idiosyncratic factor Z_i, *i.e.* they are independent.

The seminal Heston model applies a similar model as Equation (11). It correlates two Brownian motions dz_1 and dz_2 with the Pearson correlation coefficient ρ, where $dz = \varepsilon_t dt$ and ε_t is defined as in Equation (1). The core equation is

$$\mathrm{d}z_1(t) = \rho \mathrm{d}z_2(t) + \sqrt{1-\rho^2}\,\mathrm{d}z_3(t) \tag{12}$$

Numerous extensions of the Heston (1993) model exist, see for example Hagan et al. (2002) [25] and Buraschi et al. (2006) [26].

By construction, Equations (10) to (12) limit the correlation parameter ρ to $-1 \le \rho \le 1$. Hence in order to apply extended correlations, we have to alter the equations. Changing the term $\sqrt{1-p^2}$ in Equations (11) and (12) to $1-p^2$ or $1 - \rho$ would add model flexibility. For $\rho > 1$, the dependent variable x_i or $dz_1(t)$ have a higher than 100% positive dependence on M or $dz_2(t)$ respectively, and a negative dependency on Z_i or dz_3 respectively. Vice versa, for $\rho < -1$, the dependent variable x_i or $dz_1(t)$ have a higher than 100% negative dependence on M or $dz_2(t)$ respectively, and a positive dependency on Z_i or dz_3 respectively. The higher flexibility comes at the cost of loss of standard normality for x_i and dz_1. While the mean, skewness and kurtosis of $dz_1$and x_i would still be zero, the variance is unequal to 1 for all $\rho\backslash\{0,1\}$. Equation (10) is not a good candidate for extended correlations since they can change the sign of the equation.

3. EXTENDED CORRELATIONS IN FINANCIAL PRACTICE

Extended correlations occur in finance if arbitrage opportunities exist. We will show this with the example of dispersion trading.

Dispersion trading emerged in the late 1990s from index arbitrage. In a long index arbitrage trade, the trader buys certain components (e.g. stocks) of an index (e.g. the S&P 500) and shorts the whole index. The index components are expected to outperform the index, so that $\sum_{i=1}^{n} w_i r_i > r_I$ where w_i are the component weights, r_i is the return of the index components and r_I is the return of the Index.

Dispersion trading applies the same idea, just with respect to component volatility and index volatility. The strategy can be well implemented with options. For details on dispersion trading see Willmott (2009) [27] or Meissner (2015) [28].

Let's briefly derive the core equation of dispersion trading. We start with the variance equation for two assets *i* and *j*, $Var_{ij} = Var_i + Var_j + 2Cov_{ij}\ (i, j = 1, \cdots, n\ )$. Generalizing for *n* assets, which comprise the index *I*, and using financial notation, *i.e.* $Var \equiv \sigma^2$, we derive

$$\sigma_I^2 = \sum_{i=1}^{n} w_i^2 \sigma_i^2 + 2\sum_{i=1}^{n-1}\sum_{j>i}^{n} w_i w_j \sigma_i \sigma_j \rho_{ij} \tag{13}$$

where σ^2_I is the implied variance of the Index, i.e. the variance implied by option prices on the index, and σ^2_i is the implied variance of an option on the component *i*, and w_i and w_j are weighting factors. Solving equation (13) for the average implied pairwise correlation coefficient between assets *i* and *j*, ρ_{ij}, we derive

$$\rho_{ij} = \frac{\sigma_I^2 - \sum_{i=1}^{n} w_i^2 \sigma_i^2}{2\sum_{i=1}^{n-1}\sum_{j>i}^{n} w_i w_j \sigma_i \sigma_j} \tag{14}$$

Equation (14) shows the general concept of dispersion trading. The correlation between the components *i* and *j*, ρ_{ij}, is not derived by data points in a two-dimensional coordinate system as in the Pearson model, but by the relationship between the index implied volatility σ_I and component implied volatility σ_i.

The CBOE disseminates the implied correlation index of the S&P 500 derived by Equation (14) since 2007, ticker symbol ICJ, JCJ, and KCJ, see http://www.cboe.com/micro/impliedcorrelation. So far the CBOE has reported 3 extended correlations. Not surprisingly, they occurred at the height of the global financial crisis: On November 6, 2008, implied correlation was 100.8%, on November 13, 2008, 105.93% and on November 20, 2008, 103.04% for the KCJ, the January 2009 option maturity. This confirms that extended correlations exist in financial practice.

The extended correlations imply arbitrage opportunities: If $\rho_{ij} > 1$, from Equation (14), we observe that implied index volatility σ^2_i is too high relative to the component volatility σ^2_i. *i.e.* index volatility can be sold and component volatility bought realizing a risk-free profit (assuming no transaction cost, and mid-market execution). The type of arbitrage, $\rho_{ij} > 1$, not $\rho_{ij} < -1$, is plausible. For instance, in the severe crisis of 2008, traders assumed that many stocks would decline jointly. Hence, they bought puts on the whole index I, driving up index volatility σ^2_i to levels, which generated the correlation coefficients $\rho_{ij} > 1$, i.e., they became not conventional but extended correlations.

In the following section, we derive a mathematical model for extended correlations constructing two basic types of extended correlation coefficients: complete correlation coefficients and total correlation coefficients.

4. A MATHEMATICAL MODEL FOR EXTENDED CORRELATIONS

Let us first review constructions and properties of Pearson's correlation coefficient, also called population correlation coefficient, if a whole population is modeled. For two variables X and Y, it is defined by

$$r_{X,Y} = \operatorname{cov}(X,Y)/(\sigma_X \times \sigma_Y) \tag{15}$$

where *cov*(*X*, *Y*) is the covariance of *X* and *Y*, while s_X and s_Y are the standard deviations of *X* and *Y*, respectively.

If we model random variables, it is possible to express the correlation coefficient in terms of expectations

$$\rho_{X,Y} = \left(E(XY) - E(X)E(Y)\right) \Big/ \sqrt{\left(E(X^2) - E(X)^2\right)\left(E(Y^2) - E(Y)^2\right)} \tag{16}$$

In the case when random variables *X* and *Y* are represented by samples, the sample Pearson correlation coefficient, or sample correlation coefficient, $r_{X,Y}$ is defined by

$$r_{X,Y} = \frac{\sum_{i=1}^{n}(x_i - \mu_x)(y_i - \mu_y)}{\sqrt{\sum_{i=1}^{n}(x_i - \mu_x)}\sqrt{\sum_{i=1}^{n}(y_i - \mu_y)}} \qquad (17)$$

where $\{x_1, \cdots, x_n\}$ is a sample for X with the sample mean m_x and $\{y_1, \cdots, y_n\}$ is a sample for *Y* with the sample mean ⍰$_y$.

The properties of the Pearson correlation coefficients are well-known:

1) The population correlation coefficient has the following boundaries

$$-1 \leq \rho_{X,Y} \leq 1$$

2) The sample correlation coefficient has the same boundaries

$$-1 \leq r_{X,Y} \leq 1$$

3) The population correlation coefficients symmetric, i.e., ⍰$_{X,Y}$ = ⍰$_{Y,X}$.

4) The sample correlation coefficients symmetric, i.e., $r_{X,Y} = r_{Y,X}$.

5) The population correlation coefficients invariant with respect to linear transformations of the two variables, namely, changing *X* to $a + bX$ and *Y* to $c + dY$, where a, b, c, and d are constants with $b, d > 0$, does not change the population correlation coefficient ⍰$_{X,Y}$.

6) The sample correlation coefficients invariant with respect to linear transformations of the measurement scales, i.e., of the x-y-coordinates, namely, changing the coordinate x to $a + bx$ and the coordinate *y* to $c + dy$, where a, b, c, and d are constants with b, d > 0, does not change the sample correlation coefficient ⍰$_{X,Y}$.

7) The equality ⍰$_{X,Y}$ = 1 implies that a linear equation perfectly describes the relationship between X and Y, with all data points lying on a line for which Y increases as X increases.

8) The equality ⍰$_{X,Y}$ = −1 implies that a linear equation perfectly describes the relationship between X and Y, with all data points lie on a line for which Y decreases as X increases.

9) The equality ⍰$_{X,Y}$ = 0 implies that there is no linear dependency between the variables X and Y.

However, as demonstrated above, to better model financial reality, it is necessary to use extended correlation coefficients. At first, we consider population correlation coefficients.

To understand how extended correlation coefficients emerge in a mathematical context, we delineate a set A of aspects of the random variables *X* and *Y* assuming that it is possible to numerically represent each aspect *A* from *A*, i.e., there are variables X_A and Y_A that represent the aspect A in the variables *X* and *Y*.

This allows us to define the aspect population correlation coefficient for variables X and Y as

$$\rho_{X,Y}^{A} = \operatorname{cov}\left(X_A, Y_A\right) / \left(\sigma_{X_A} \cdot \sigma_{Y_A}\right) \tag{18}$$

To find an integral correlation characteristic, it is possible to use the complete population correlation coefficient for variables X and Y, which is computed by the following formula:

$$\rho_{X,Y}^{C} = \sum_{A \in \mathbf{A}} \rho_{X,Y}^{A} \tag{19}$$

When the set A consists of n aspects, the value $\rho_{X,Y}^{C}$ is called the n-aspect population correlation coefficient.

It is also possible to use the aspect sample correlation coefficient for variables X and Y

$$r_{x,y}^{A} = \frac{\sum_{i=1}^{n}\left(x_i^A - \mu_x^A\right)\left(y_i^A - \mu_y^A\right)}{\sqrt{\sum_{i=1}^{n}\left(x_i^A - \mu_x^A\right)}\sqrt{\sum_{i=1}^{n}\left(y_i^A - \mu_y^A\right)}} \tag{20}$$

Then the complete sample correlation coefficient for variables X and Y is based on the aspect sample correlation coefficients and is computed by the following formula:

$$r_{X,Y}^{C} = \sum_{A \in \mathbf{A}} r_{X,Y}^{A} \tag{21}$$

When the set A consists of n aspects, the value $r_{X,Y}^{C}$ is called the n-aspect sample correlation coefficient.

Example 1. Let us consider a situation when a company wants to find the correlation between its profit and expenses. The company has three factories and wants to take into account all of them. To find the necessary correlation, a statistician represents the profit by the random variable X and the expenses by the random variable Y. Each factory represents a factor of these variables. Then it's possible to use the complete correlation coefficient, which better represents relations between profit and expenses than the conventional correlation coefficient.

In this case, the profit of the first factory is represented by the random variable X_1 and the expenses related to the first factory - by the random variable Y_1. The profit of the second factory is represented by the random variable X_2 and the expenses related to the second factory-by the random variable Y_2. The profit of the first factory is represented by the random variable X_3 and the expenses related to the first factory-by the random variable Y_3. Then the complete (or 3-aspect) sample correlation coefficient for variables X and Y is equal to

$$r^C_{X,Y} = r_{X_1,Y_1} + r_{X_2,Y_2} + r_{X_3,Y_3}$$

For instance $r_{X_1,Y_1} = 0.3$, $r_{X_2,Y_2} = 0.8$,, and $r_{X_3,Y_3} = 0.6$. In this case, the complete sample correlation coefficient $r^C_{X,Y} = 1.7$.

If $r_{X_1,Y_1} = 0.3$, $r_{X_2,Y_2} = -0.7$, and $r_{X_3,Y_3} = -0.8$. In this case, the complete sample correlation coefficient $r^C_{X,Y} = -1.2$.

If $r_{X_1,Y_1} = -0.5$, $r_{X_2,Y_2} = 0.3$, and $r_{X_3,Y_3} = 0.7$. In this case, the complete sample correlation coefficient $r^C_{X,Y} = 0.5$.

Example 2. Let us consider a situation when a biologist wants to find correlation between traits of fathers and sons. Then it possible to use the complete correlation coefficient taking such features as the height, weight, color of eyes, IQ and education in the role of aspects of the compared trait.

In this case, the height of the father is represented by the random variable X_1 and the height of the son—by the random variable Y_1. The weight of the father is represented by the random variable X_2 and the weight of the son—by the random variable Y_2. The IQ of the father is represented by the random variable X_3 and the IQ of the son by the random variable Y_3. In addition, it is possible to assign numerical values to the color of eyes and education, for example, assigning 0 to the case when a person does not have any education and 10 when a person has PhD. This makes possible representation of father's color of eyes by the random variable X_4 and son's color of eyes by the random variable Y_4. In a similar way, we represent father's education by the random variable X_5 and son's education by the random variable Y_5.

Then the complete sample correlation coefficient for variables X and Y is equal to

$$r^C_{X,Y} = r_{X_1,Y_1} + r_{X_2,Y_2} + r_{X_3,Y_3} + r_{X_4,Y_4} + r_{X_5,Y_5}$$

For instance, $r_{X_1,Y_1} = 0.3 r_{X_2,Y_2} = -0.5$, $r_{x3,y3}$ = -0.5, $r_{X_4,Y_4} = 0.8$, $r_{X_5,Y_5} = 0.7$, and. In this case, the complete (or 5-aspect) sample correlation coefficient $r^C_{X,Y} = 1.8$.

Some properties of complete correlation coefficients are the same as properties of conventional correlation coefficients, while other properties are different. Based on the properties 1 - 8 of conventional correlation coefficients considered above, we obtain the following results.

1) The complete (n-aspect) population correlation coefficient $\rho^C_{X,Y}$ has the following boundaries

$$-n \leq \rho^C_{X,Y} \leq n.$$

2) The complete (n-aspect) correlation coefficient $r^C_{X,Y}$ has the same boundaries

$$-n \leq r^C_{X,Y} \leq n.$$

3) The complete (n-aspect) population correlation coefficient is symmetric, i.e. $\rho^C_{X,Y} = \rho^C_{Y,X}$.

4) The complete (n-aspect) sample correlation coefficient is symmetric, i.e., $r^C_{X,Y} = r^C_{Y,X}$.

5) The complete (n-aspect) population correlation coefficient is invariant with respect to linear transformations of the two variables and their aspects, namely, changing X to $a + bX$ and Y to $c + dY$, where a, b, c, and d are constants with $b, d > 0$, does not change the complete population correlation coefficient $\rho^C_{X,Y}$.

6) The complete (n-aspect) sample correlation coefficient is invariant with respect to linear transformations of the measurement scales, i.e., of the *x-y*-coordinates, namely, changing the coordinate x to $a + bx$ and the coordinate y to $c + dy$, where a, b, c, are constants with $b, d > 0$, does not change the complete sample correlation coefficient $r^{C}_{X,Y}$.

7) The equality $\rho^{C}_{X,Y} = n$ implies that a linear equation perfectly describes the relationship between all corresponding aspects of X and Y, with all data points lying on a line for which the aspect of Y increases as the corresponding aspect of X increases.

8) The equality $\rho^{C}_{X,Y} = -n$ implies that a linear equation perfectly describes the relationship between all corresponding aspects of X and Y, with all data points lying on a line for which the aspect of Y decreases as the corresponding aspect of X increases.

Proofs of these properties are based on the properties 1 - 8 of conventional correlation coefficients.

Another way to build extended correlation coefficients is based on factors of random variables. Let us assume that we have a set F of factors of the random variables X and Y assuming that it is possible to numerically represent each factor F from F, *i.e.*, there are variables X_F and Y_F that represent the factor A in the variables X and Y.

This allows us to define the factor population correlation coefficient for variables X and Y as

$$\rho^{F}_{X,Y} = \mathrm{cov}\left(X_F, Y_F\right) / \left(\sigma_{X_F} \cdot \sigma_{Y_F}\right) \tag{22}$$

To find an integral correlation characteristic, it is possible to use the total population correlation coefficient for variables X and Y, which is computed by the following formula:

$$\rho^{T}_{X,Y} = \sum_{F \in \mathbf{F}} \rho^{F}_{X,Y} \tag{23}$$

When the set F consists of n factors, the value $\rho^{F}_{X,Y}$ is called the n-factor population correlation coefficient.

It is also possible to use the factor sample correlation coefficient for variables X and Y

$$r_{x,y}^{F} = \frac{\sum_{i=1}^{n}\left(x_i^F - \mu_x^F\right)\left(y_i^F - \mu_y^F\right)}{\sqrt{\sum_{i=1}^{n}\left(x_i^F - \mu_x^F\right)}\sqrt{\sum_{i=1}^{n}\left(y_i^F - \mu_y^F\right)}} \tag{24}$$

Then the total sample correlation coefficient for variables *X* and *Y* is based on the factor sample correlation coefficients and is computed by the following formula:

$$r_{X,Y}^{T} = \sum_{F \in \mathbf{F}} r_{X,Y}^{F} \tag{25}$$

Example 3. Let us consider a situation when an investor wants to find correlation between prices of stocks of companies A and B. Then it possible to use the total correlation coefficient, taking into account factors that influence prices of stocks, which could be the PE (price earnings ratio), EPS (earnings per share), and the dividend yield. This allows achieving better representation of dependencies than the conventional correlation coefficient.

In this case, the PE of the company *A* is represented by the random variable X_1 and the PE of the company B- by the random variable Y_1. The EPS of the company *A* is represented by the random variable X_2 and the EPS of the company B- by the random variable Y_2. The dividend yield of the company *A* is represented by the random variable X_3 and the dividend yield of the company B-by the random variable Y_3. Then the total (or 3-aspect) sample correlation coefficient for variables *X* and *Y* is equal to

$$r_{X,Y}^{T} = r_{X_1,Y_1} + r_{X_2,Y_2} + r_{X_3,Y_3}$$

For instance, $r_{X_1,Y_1} = 0.5$, $r_{X_2,Y_2} = 0.7$, and $r_{X_3,Y_3} = 0.3$. In this case, the complete sample correlation coefficient $r_{X,Y}^{T} = 1.5$.

If $r_{X_1,Y_1} = 0.5$, $r_{X_2,Y_2} = -0.9$, and $r_{X_3,Y_3} = -0.8$. In this case, the complete sample correlation coefficient $r_{X,Y}^{T} = -1.2$.

If $r_{X_1,Y_1} = 0.5$, $r_{X_2,Y_2} = -0.3$, and $r_{X_3,Y_3} = -0.7$. In this case, the complete sample correlation coefficient $r_{X,Y}^{T} = -0.5$.

Some properties of total correlation coefficients are the same as properties of conventional correlation coefficients, while other properties are different. Based on the properties 1 - 8 of conventional correlation coefficients considered above, we obtain the following results.

1) The total (n-factor) population correlation coefficient has the following boundaries

$$-n \leq \rho^{T}_{X,Y} \leq n$$.

2) The total (n-factor) sample correlation coefficient has the same boundaries

$$-n \leq r^{T}_{X,Y} \leq n$$.

3) The total (n-factor) population correlation coefficient is symmetric, i.e., $\rho^{T}_{X,Y} = \rho^{T}_{Y,X}$.

4) The total (n-factor) sample correlation coefficient is symmetric, i.e., $r^{T}_{X,Y} = r^{T}_{Y,X}$.

5) The total (n-factor) population correlation coefficient is invariant with respect to linear transformations of the two variables, namely, changing *X* to *a* + *bX* and *Y* to *c* + *dY*, where *a*, *b*, *c*, and *d* are constants with *b*, *d* > 0, does not change the total population correlation coefficient $\rho^{T}_{X,Y}$.

6) The total (n-factor) sample correlation coefficient is invariant with respect to linear transformations of the measurement scales, i.e., of the x-y-coordinates, namely, changing the coordinate *x* to *a* + *bx* and the coordinate y to *c* + *dy*, where *a*, *b*, *c*, and *d* are constants with *b*, *d* > 0, does not change the total sample correlation coefficient $r^{T}_{X,Y}$.

7) The equality $\rho^{T}_{X,Y} = n$ implies that a linear equation perfectly describes the relationship between all corresponding factors of *X* and *Y*, with all data points lying on a line for which the factor of *Y* increases as the corresponding factor of *X* increases.

8) The equality $\rho^{T}_{X,Y} = -n$ implies that a linear equation perfectly describes the relationship between all corresponding factors of *X* and *Y*, with all data points lying on a line for which the factor of *Y* decreases as the corresponding factor of *X* increases.

Proofs of these properties are based on the properties 1 - 8 of conventional correlation coefficients.

5. LIMITATIONS OF THE MODEL

The correlations, which are created in this paper, are extended from the Pearson correlation model. While the Pearson correlation model is by far the most applied correlation model in finance, it suffers from several limitations. The main limitations are

1) The Pearson can only evaluate linear associations between variables

2) As a consequence of 1), The Pearson coefficients can only be meaningfully interpreted if the data distribution is approximately elliptical.

3) Outliers can distort the correlation results

4) Different time frames can lead to very different results

5) The causality has to be determined exogenously

6) Spurious correlation can occur

For a detailed discussion on ten limitations of the Pearson model, see Meissner 2015 [14]. These limitations are inherited by the new model.

6. CONCLUSIONS

When modeling financial processes with discrete stochastic processes, extended correlations, i.e. correlations which can be <-1 and >1, naturally occur. Rather than discarding the whole model or applying arbitrary boundaries, extended correlations can be implemented to utilize the model.

Correlations often serve as an input for more complex mathematical models such as copulas or Heston models. Utilization of extended correlations in these models adds flexibility such as extending of the dependencies between variables beyond unity.

Extended correlations occur in financial practice as the 2008 global financial crisis demonstrated. Consequently, there is a need to create a sound mathematical model for extended correlations.

We derive one type of extended correlation coefficients by delineating a set A of aspects of random variables and creating aspect correlation coefficients and combining them into the complete, or n-aspect, correlation coefficient. Some properties of the complete correlation coefficient are different from the traditional Pearson correlation coefficient, for example, boundaries of the coefficient become −n and +n, while other properties, such as symmetry, remain unchanged.

Another way to build extended correlation coefficients is based on factors of random variables. Taking a factor F from a set F, which consists of n factors of considered processes, we build the quantity $\rho^F_{X,Y}$ called a factor correlation coefficient and combine factor correlation coefficients into the total, or n-factor, correlation coefficient. As with the complete correlation coefficient, the total correlation coefficient can have wider boundaries, while some of its properties are identical to the Pearson correlation coefficient.

In conclusion, conventional mathematical correlation approaches cannot fully model all existing correlation values and processes. Therefore we introduce the model of extended correlations, which is more general than the conventional correlation models. Hence all correlations in finance (and other sciences), *i.e.* conventional, and correlations that are smaller than −1 or larger than 1, can be represented by extended correlations.

REFERENCES

1. Kline, M. (1980) Mathematics: The Loss of Certainty. Oxford University Press, New York.
2. Mattessich, R. (1998) From Accounting to Negative Numbers: A Signal Contribution of Medieval India to Mathematics. Accounting Historians Journal, 25, 129-145
3. Martinez, A.A. (2005) Negative Math: How Mathematical Rules Can Be Positively Bent. Princeton University Press.
4. Wigner, E.P. (1932) On the Quantum Correction for Thermodynamic Equilibrium. Physics Review, 40, 749-759. http://dx.doi.org/10.1103/physrev.40.749
5. Dirac, P.A.M. (1942) The Physical Interpretation of Quantum Mechanics. Proceedings of the Royal Society of London, 180, 1-39. http://dx.doi.org/10.1098/rspa.1942.0023
6. Dirac, P.A.M. (1943) Quantum Electrodynamics. Communications of the Dublin Institute for Advanced Studies.
7. Bartlett, S. (1945) Negative Probability. Mathematical Proceedings of the Cambridge Philosophical Society, 41, 71-73. http://dx.doi.org/10.1017/S0305004100022398
8. Feynman, R. (1987) Negative Probability. In: Hiley, B.J. and Peat, F., Eds., Quantum Implications: Essays in Honour of David Bohm, Routledge & Kegan Paul Ltd., London and New York, 235-248.
9. Mückenheim, W., Ludwig, G., Dewdney, C., Holland, P.R., Kyprianidis, A., Vigier, J.P., et al. (1986) A Review of Extended Probabilities. Physics Reports, 133, 337-401. http://dx.doi.org/10.1016/0370-1573(86)90110-9
10. Khrennikov, A. (1992) p-Adic Probability and Statistics. Dokl. Acad. Nauk, 322, 1075-1079.
11. Khrennikov, A. (2009) Interpretations of Probability. Walter de Gruyter, Berlin/New York. http://dx.doi.org/10.1515/9783110213195

12. Haug, E.G. (2004) The Collector: Why So Negative to Negative Probabilities? Wilmott Magazine, 34-38.
13. Székely, G.J. (2005) Half of a Coin: Negative Probabilities. Wilmott Magazine, 66-68.
14. Burgin, M. and Meissner, G. (2010) Negative Probabilities in Modeling Random Phenomena. Integration: Mathematical Theory and Application, 2, 305-322.
15. Burgin, M. and Meissner, G. (2012) Larger Than One Probabilities in Mathematical and Practical Finance. Review of Economics and Finance, No. 4, 1-13.
16. Burgin, M. and Meissner, G. (2012) Negative Probabilities in Financial Modeling. Wilmott Magazine, 58, 60-65.
17. Allen, E.H. (1976) Negative Probabilities and the Uses of Signed Probability Theory. Philosophy of Science, 43, 53-70. http://dx.doi.org/10.1086/288669
18. Burgin, M. (2012) Integrating Random Properties and the Concept of Probability. Integration: Mathematical Theory and Applications, 3, 137-181.
19. Abramsky, S. and Brandenburger, A. (2014) An Operational Interpretation of Negative Probabilities and No-Signalling Models. Lecture Notes in Computer Science, 8464, 59-75. http://dx.doi.org/10.1007/978-3-319-06880-0_3
20. Emmerich, C. (2006) Modeling Correlation as a Stochastic Process. Working Paper, Bergische Universitaet, Wuppertal.
21. Ma, J. (2009) Pricing Foreign Equity Options with Stochastic Correlation and Volatility. Annals of Economics and Finance, 10, 303-327.
22. Ma, J. (2009) A Stochastic Correlation Model with Mean Reversion for Pricing Multi-Asset Options. Asia Pacific Financial Markets, 16, 97-109.
23. Basel Committee of Banking Supervision (2009) Principles of Sounds Stress Testing Practices and Supervision. BCBS, Basel.
24. Meissner, G. (2015) The Pearson Correlation Model—Work of the Devil? Working Paper.
25. Hagan, P.S., Kumar, D., Lesniewski, A.S. and Woodward, D.E. (2002) Managing Smile Risk. Wilmott Magazine, 84-108.
26. Buraschi, A., Porchia, P. and Trojani, F. (2006) Correlation Risk and Optimal Portfolio Choice. Journal of Finance, 65, 392-342. http://dx.doi.org/10.2139/ssrn.908664
27. Wilmott, P. (2009) Frequently Asked Question in Quantitative Finance. John Wiley, Hoboken.
28. Meissner, G. (2015) Correlation Trading—Opportunities and Limitations. Working Paper.

CHAPTER 2

Interest Rate Models

Alex Paseka[1,2], Theodoro Koulis[1,2], Aerambamoorthy Thavaneswaran[1,2]

[1]*Department of Accounting and Finance, University of Manitoba, Winnipeg, Canada*

[2]*Department of Statistics, University of Manitoba, Winnipeg, Canada*

ABSTRACT

In this paper, we review recent developments in modeling term structures of market yields on default-free bonds. Our discussion is restricted to continuous-time dynamic term structure models (DTSMs). We derive joint conditional moment generating functions (CMGFs) of state variables for DTSMs in which state variables follow multivariate affine diffusions and jump-diffusion processes with random intensity. As an illustration of the pricing methods, we provide special cases of the general formulations as examples. The examples span a wide cross-section of models from early one-factor models of Vasicek to more recent interest rate models with stochastic volatility, random intensity jump-diffusions and quadratic-Gaussian DTSMs. We also derive the European call option price on a zero-coupon bond for linear quadratic term structure models.

Keywords: affine process; dynamic term structure models; jump-diffusions; quadratic-Gaussian DTSMs

1. INTRODUCTION

In this review, we summarize available continuous time technology for pricing default-free term structures. Our main goal is to provide a unified approach to the pricing exercise based on well-developed application of stochastic calculus to finance problems. In our discussions, we always start from state variable processes given under a risk-neutral measure. We make no attempt to systematically present the transition between the actual measure and the risk-neutral measure. Thus, we leave out the question of reconciling time series properties of the underlying state variables (such as short rate), which are defined under the actual measure, and the properties of the market yields (or, equivalently, bond prices), which are derived under the risk-neutral measure.

We begin with affine diffusion models of interest rates. The menu of available models in this class is truly vast.[1] The most redeeming features of this class of models are the availability of closed form solutions for derivatives on the short rate and the ability of models to reproduce a variety of term structure patterns. Models in this class have closed form solutions for bond and bond option prices, which makes them particularly attractive for empirical work. Some models in this class, especially twoor three-factor models, are flexible enough to have a satisfactory fit to observed term structures. The last property is essential for pricing options on bonds. Extended versions of Vasicek and CIR (e.g., [2,5,14]) allow for timevarying coefficients in state variable dynamics. This extra feature allows the model to fit the initial term structure.

Discontinuous movements in interest rates are caused by central banks and unexpected news announcements. In order to implement a specific policy, monetary authorities often use specific rate-setting rules which result in entire yield curve shifts in line with movements in the benchmark rate. Empirical evidence suggests that interest rates exhibit substantial skewness and kurtosis, and hence jump-diffusion interest rate models are more appropriate[2]. In deriving the term structure of spot rates for the case of discontinuous interest rates, we stay within the affine jump-diffusion class of models. The bond prices are still available in near-closed form up to the solutions of Riccati ordinary differential equations for the coefficients of the conditional moment generating function of the state variables.

Market spot rates are mostly positive, and a class of quadratic interest rate models is designed to restrict spot rates to be positive without ruining the bond price tractability. Square-Gaussian models, in which the short rate is a sum of squares of Gaussian state variables, have been investigated by [28-30]. The models of Quadratic-Gaussian class were studied by [31-35]. The mathematics of these models is similar to that of affine jump diffusions and also allows closed form solutions for bond prices in the exponential quadratic form.

Previous jump-diffusion TSMs treat short rate intensity in one of the following restrictive ways:

1) A deterministic function of the short rate ([26]);

2) Constant except for FOMC meeting days and macroeconomic news announcements, in which case it is a linear function of short rate, interest rate volatility, and Fed rate target; or

3) A function of the spread between the Fed funds rate and the target.

Thus, we also consider a recent extension of a class of Linear-Quadratic term structure models (LQTSMs) to a class of jump-diffusion TSMs in which short rate jump intensity follows its own stochastic process (see [36-38] for details). We show how this model class is designed to have a closed form conditional moment generating function and bond prices. Despite strong restrictions on state variable processes that are necessary to obtain closed form bond prices, [36] shows that the inclusion of random intensity improves the model fit to both the dynamics of the term structure and that of the volatility term structure. The main mechanism of improved fit is the short term kurtosis of the short rate.

The rest of this paper is organized as follows. In Section 2, we present a general technique of solving for joint CMGF of an affine state vector and its integral. Special cases include CMGF of an affine vector and bond prices within affine model class. Section 3 covers affine jumpdiffusion DTSMs. Finally, we close with a review of quadratic Gaussian models and a more general class of linear quadratic class of DTSMs with jumps of random intensity. We also derive the price of a European call option using a generalized transform.

2. AFFINE TERM STRUCTURE MODELS

In this section we provide general results for affine models. As an illustration, we also give detailed account of some popular special cases.

Definition. Let Y be a vector of n state variables, and let r be the short rate. A model is affine if the state vector solves the following diffusion SDE:

$$\mathrm{d}Y = \kappa\left(\theta - Y\right)\mathrm{d}t + \Sigma V \mathrm{d}W, \tag{2.1}$$

$$r = \delta_0 + \delta_1' Y, \tag{2.2}$$

where $\kappa \in \mathbb{R}^{n\times n}, \theta \in \mathbb{R}^{n\times 1}, \Sigma \in \mathbb{R}^{n\times n}$ (constant matrix), $\delta_0 \in \mathbb{R}, \delta_1 \in \mathbb{R}^{n\times 1}$, V is a diagonal $\mathbb{R}^{n\times n}$ -matrix such that $VV' = \mathrm{diag}\left(\alpha_i + \beta_i' Y\right)$, and $W \in \mathbb{R}^{n\times 1}$ is a vector of standard Wiener processes. We assume that an equivalent martingale measure (EMM) exists, and all processes and expectations (unless stated otherwise) are understood as those under the EMM, e.g., the process in (2.1) is written under the EMM. The SDE (2.1) implies that the joint conditional moment generating function (CMGF) of the state vector is exponential affine:

$$\phi(u,t)=E_t\left[e^{u'Y_T}\right]=e^{u_0(\tau)+u_1(\tau)'Y_t}, \tag{2.3}$$

Where $\tau = T-t$.

The following lemma is a special case of [39] with no jumps.

Lemma 2.1. Assume that a state vector Y satisfies (2.1). Consider the following boundary value problem:

$$\frac{\partial f}{\partial t}+\left[\kappa(\theta-Y)\right]'\partial_Y f$$
$$+\frac{1}{2}tr\left[V'\Sigma'\partial_{YY'}f\Sigma V\right]-\alpha r(t,Y)f=0 \tag{2.4}$$

$$f(T,Y)=e^{u'Y} \tag{2.5}$$

This problem has a stochastic solution of the form[3]

$$f(t,Y)=E_t\left[\exp\left(-\alpha\int_t^T r(s,Y_s)\mathrm{d}s+u'Y_T\right)\right]$$
$$=e^{\varphi_0(\tau)+\varphi_1(\tau)'Y_t} \tag{2.6}$$

with the following initial conditions: $\varphi_0(0)=0$ and $\varphi_1(0)=u$.

Proof. Applying Ito's lemma to function

$$Z_w=f(w,Y_w)\exp\left(-\alpha\int_t^w r(s,Y_s)\mathrm{d}s\right),$$ we have[4]

$$\begin{aligned} Z_T &= f(T,Y_T)\exp\left(-\alpha\int_t^T r(s,Y_s)\mathrm{d}s\right) \\ &= f(t,Y_t)+\int_0^T \partial_Y f\Sigma V\mathrm{d}W_s \\ &+\int_0^T\left(\frac{\partial f}{\partial s}+[\kappa(\theta-Y)]'\partial_Y f\right. \\ &\left.+\frac{1}{2}tr\left[V'\Sigma'\partial_{YY'}f\Sigma V\right]-\alpha r(s,Y)f\right)\mathrm{d}s. \end{aligned}$$

After taking the conditional expectation of both sides and noting that the first of the two integrals on the right hand side is zero due to (1), we have

$$\begin{aligned} f(t,Y_t) &= E_t\left[\exp\left(-\alpha\int_t^T r(s,Y_s)\mathrm{d}s\right)f(T,Y_T)\right] \\ &= E_t\left[\exp\left(-\alpha\int_t^T r(s,Y_s)\mathrm{d}s+u'Y_T\right)\right] \end{aligned}$$

In order to prove the exponential affine form of the solution (2.6) to (2.4), it suffices to use it as a guess and find the equations determining n+1 coefficients $\varphi_0(\tau)$ and $\varphi_1(\tau)$. Substitution of $f(t,Y)=e^{\varphi_0(\tau)+\varphi_1(\tau)'Y_t}$ into (2.4) produces the following equation that must hold for any value of the state vector:

$$\begin{aligned} &\partial_t\varphi_0(\tau)+\partial_t\varphi_1(\tau)'Y+[\kappa(\theta-Y)]'\varphi_1(\tau) \\ &+\frac{1}{2}\underbrace{tr\left[V'\Sigma'\varphi_1(\tau)\varphi_1(\tau)'\Sigma V\right]}_{tr\left[\varphi_1(\tau)'\Sigma VV'\Sigma'\varphi_1(\tau)\right]}-\alpha(\delta_0+\delta_1'Y)=0 \end{aligned}$$

The left hand side of the above equation is an affine function in Y. Since the equation must hold for arbitrary vector Y, both coefficients of the affine function must be zero. This requirement gives two Riccati ODEs for coefficients $\varphi_0(\tau)$ and $\varphi_1(\tau)$:

$$\partial_t \varphi_0(\tau) + \kappa\theta\varphi_1(\tau) + \frac{1}{2}\varphi_1(\tau)' H_0 \varphi_1(\tau) - \alpha\delta_0 = 0 \tag{2.7}$$

$$\partial_t \varphi_1(\tau) - \kappa\varphi_1(\tau) + \frac{1}{2}\varphi_1(\tau)' H_1 \varphi_1(\tau) - \alpha\delta_1 = 0, \tag{2.8}$$

where H_0 and H_1 can be inferred from the following expression:

$$\begin{aligned} \Sigma V V' \Sigma' &= \Sigma \operatorname{diag}(\alpha_i) \Sigma' + \Sigma \operatorname{diag}(\beta_i Y) \Sigma' \\ &= H_0 + H_1 Y \end{aligned}$$

It follows from (2.5) that the initial values are $\varphi_0(0) = 0$ and $\varphi_1(0) = u$.

The expression for MGF for an affine diffusion is just a special case of the above result for a=0:[5]

$$\phi(u,t) = E_t\left[e^{u'Y_T}\right] = e^{u_0(\tau) + u_1(\tau)' Y_t} \tag{2.9}$$

The coefficients satisfy the following Riccati ODEs:

$$\partial_t u_0(\tau) + \kappa\theta u_1(\tau) + \frac{1}{2}u_1(\tau)'' H_0 u_1(\tau) = 0,$$

$$\partial_t u_1(\tau) - \kappa u_1(\tau) + \frac{1}{2}u_1(\tau)' H_1 u_1(\tau) = 0,$$

$$u_0(0) = 0, u_1(0) = u.$$

Zero coupon bond prices are a special case of the Lemma. Assuming again that the short rate process is affine in the state variables as given in (2.2), the price of a zero-coupon default-free bond with maturity T and face value of $1 is given by

$$D_t(t, Y_t) = E_t\left[\exp\left(-\int_t^T r(s, Y_s)\,ds\right)\right] \tag{2.10}$$

where the expectation is under a EMM. The expectation in (2.10) is a special case of (2.6) with a = 1 and u = 0. Given these restrictions, the bond price has exponential affine form

$$D_t(t, Y_t) = e^{\varphi_0(\tau) + \varphi_1(\tau)' Y_t}$$

and satisfies the following PDE:

$$\partial_t D + \left[\kappa(\theta - Y)\right]' \partial_Y D - r(t,Y)D$$
$$+\frac{1}{2}tr\left[V'\Sigma'\partial_{YY'}D\Sigma V\right] = 0 \tag{2.11}$$

Bond price coefficients $\varphi_0(\tau)$ and $\varphi_1(\tau)$ solve the following system of Riccati equations (a special case of (2.7)-(2.8)):

$$\partial_t\varphi_0(\tau) + \kappa\theta\varphi_1(\tau) + \frac{1}{2}\varphi_1(\tau)' H_0\varphi_1(\tau) - \delta_0 = 0 \tag{2.12}$$

$$\partial_t\varphi_1(\tau) - \kappa\varphi_1(\tau) + \frac{1}{2}\varphi_1(\tau)' H_1\varphi_1(\tau) - \delta_1 = 0 \tag{2.13}$$

$$\varphi_0(0) = \varphi_1(0) = 0 \tag{2.14}$$

The spot rate is given by

$$r_t(\tau) = -\frac{1}{\tau}\ln D_t(t, Y_t)$$
$$= -\frac{\varphi_0(\tau)}{\tau} - \frac{\varphi_1(\tau)}{\tau}Y_t \tag{2.15}$$

Taking limits as $\tau \to 0$, we obtain the short rate r_t (which is also given in (2.2))

$$r_t = \delta_0 + \delta_1' Y_t = -\partial_\tau\varphi_0(\tau)\Big|_{\tau=0} - \partial_\tau\varphi_1(\tau)\Big|_{\tau=0} Y_t$$

2.1. One-Factor Interest-Rate Models

In this section we provide closed form expressions of the default-free bond prices with details. Under the EMM, the evolution of this process is described by a diffusion Gauss-Markov process of the form:

$$\mathrm{d}r(t) = \mu(t)\mathrm{d}t + \sigma(t)\mathrm{d}W(t), \tag{2.16}$$

where $\{W(t), t \geq 0\}$ is a standard Brownian motion and $\mu(t)$, $\sigma(t)$ are suitable processes adapted to $\mathcal{F}_t = \sigma(W(u), 0 \leq u \leq t)$. The usual assumption is that $\mu(t)$ and $\sigma(t)$ are simply functions of the short rate $r(t)$ and time t, i.e., $\mu(t) = \mu(r(t), t)$ and $\sigma(t) = \sigma(r(t), t)$.

Theorem 2.2. Consider the Hull and White model of the form

$$dr(t) = (\alpha(t) - \beta(t) r(t)) dt + \sigma(t) dW(t) \tag{2.17}$$

where $\alpha(t)$, $\beta(t)$ and $\sigma(t)$ are non-random functions of t such that

$$\int_0^T (|\alpha(t)| + |\beta(t)| + \sigma^2(t)) dt < \infty. \tag{2.18}$$

The price at time t of a zero-coupon default free bond paying \$1 at time T is

$$B(t,T) = \exp(A(t,T) - r(t) C(t,T)), \tag{2.19}$$

where

$$A(t,T) = \frac{1}{2} \int_t^T \left(\int_s^T \frac{g(u)}{g(s)} \sigma(s) du \right)^2 ds$$

$$- \int_t^T \left(\int_t^u \frac{g(u)}{g(s)} \alpha(s) ds \right) du,$$

$$C(t,T) = \int_t^T \frac{g(u)}{g(t)} du \text{ and } g(t)$$

$$= \exp\left(-\int_0^t \beta(s) ds \right).$$

Note: The Merton [43] model, Vasicek [1] model, and the Ho-Lee model [44] are special cases of (2.17).

Proof. Under these assumptions (2.18) the equation (2.17) has a unique (strong) solution

$$r(t) = g(t) \left(r_0 + \int_0^t \frac{\alpha(s)}{g(s)} ds + \frac{\sigma(s)}{g(s)} dW_s \right)$$

Since $r = (r(t))_{t \le T}$ is the Markov process, it follows that

$$B(t,T) = \mathrm{E}\left[\exp\left(-\int_t^T r(s)\,\mathrm{d}s\right)\middle| r(t)\right]$$
$$= \mathrm{E}\left[\exp(-I(t,T))\middle| r(t)\right]$$

where $I(t,T) = \int_t^T r(s)\,\mathrm{d}s$ is conditionally normal. The conditional mean, variance, and the Laplace transform are given as:

$$\mathrm{E}\left[I(t,T)\middle| r(t)\right]$$
$$= r(t)\int_t^T \frac{g(u)}{g(t)}\,\mathrm{d}u + \int_t^T\left(\int_t^u \frac{g(u)}{g(s)}\alpha(s)\,\mathrm{d}s\right)\mathrm{d}u,$$

$$\mathrm{var}\left[I(t,T)\middle| r(t)\right] = \int_t^T\left(\int_s^T \frac{g(u)}{g(s)}\sigma(s)\,\mathrm{d}u\right)^2 \mathrm{d}s,$$

and

$$\mathrm{E}\left[\exp(-I(t,T))r(t)\right]$$
$$= \exp\left(\frac{1}{2}\mathrm{Var}\left[I(t,T)\middle| r(t)\right] - \mathrm{E}\left[I(t,T)\middle| r(t)\right]\right)$$

Hence the default-free time ***t*** bond price is given by

$$B(t,T) = \exp\left(A(t,T) - r(t)C(t,T)\right).$$

Note 1. The Vasicek model is a special case with constant coefficients: $\mu(t) = \alpha(\beta - r(t))$ and $\sigma(t) = \sigma$ (*a*>0), such that $\alpha(t) = \alpha\beta, \beta(t) = \alpha, \sigma(t) = \sigma$. It follows from Theorem 2 that

$$C(t,T) = e^{\alpha t}\int_t^T e^{-\alpha y}\,\mathrm{d}y = \frac{1}{\alpha}\left(1 - e^{-\alpha(T-t)}\right) \tag{2.20}$$

$$
\begin{aligned}
&A(t,T)\\
&=\int_t^T\left(e^{\alpha v}\alpha\beta\left(\int_v^T e^{-\alpha y}\mathrm{d}y\right)-\frac{1}{2}e^{2\alpha v}\sigma^2\left(\int_v^T e^{-\alpha y}\mathrm{d}y\right)^2\right)\mathrm{d}v\\
&=\alpha\beta\int_t^T C(v,T)\mathrm{d}v-\frac{1}{2}\sigma^2\int_t^T C^2(v,T)\mathrm{d}v\\
&=-\frac{\left(C(t,T)-T+t\right)\left(\alpha^2\beta-\sigma^2/2\right)}{\alpha^2}+\frac{\sigma^2 C^2(t,T)}{4\alpha}.
\end{aligned}
\tag{2.21}
$$

Note 2. For the extended Vasicek model, $\alpha(t)=\alpha\theta(t)$, $\beta(t)=\alpha$, it follows from Theorem 2.2 that

$$
C(t,T)=e^{\alpha t}\int_t^T e^{-\alpha y}\mathrm{d}y=\frac{1}{\alpha}\left(1-e^{-\alpha(T-t)}\right)
\tag{2.22}
$$

$$
\begin{aligned}
&A(t,T)\\
&=\int_t^T\left(e^{\alpha v}\alpha\theta(v)\left(\int_v^T e^{-\alpha y}\mathrm{d}y\right)-\frac{1}{2}e^{2\alpha v}\sigma^2\left(\int_v^T e^{-\alpha y}\mathrm{d}y\right)^2\right)\mathrm{d}v\\
&=\alpha\int_t^T C(v,T)\theta(v)\mathrm{d}v-\frac{1}{2}\sigma^2\int_t^T C^2(v,T)\mathrm{d}v\\
&\quad+\frac{\sigma^2}{2\alpha}\left(\frac{C(t,T)-T+t}{\alpha}+\frac{C^2(t,T)}{2}\right).
\end{aligned}
\tag{2.23}
$$

Theorem 2.3. Consider the Cox, Ingersoll and Ross (CIR) model, the interest rate process $r(t)$ is given by

$$
\mathrm{d}r(t)=\left(\alpha-\beta r(t)\right)\mathrm{d}t+\sigma\sqrt{r(t)}\mathrm{d}W(t),
$$

where $r(t)$ is given. The price at time t of a zerocoupon bond paying \$1 at time t is

$$
B(t,T)=e^{-r(t)C(t,T)-A(t,T)}
$$

where

$$C(t,T)=\frac{\sinh(\gamma(T-t))}{\gamma\cosh(\gamma(T-t))+\frac{1}{2}\beta\sinh(\gamma(T-t))},$$

$$A(t,T)$$

$$=-\frac{2\alpha}{\sigma^2}\log\left(\frac{\gamma e^{\frac{1}{2}\beta(T-t)}}{\gamma\cosh(\gamma(T-t))+\frac{1}{2}\beta\sinh(\gamma(T-t))}\right),$$

And $\gamma=\frac{1}{2}\sqrt{\beta^2+2\sigma^2}$.

Proof. See the Appendix.

2.2. Multi-Factor TSMs

There is substantial evidence that bond yields have timevarying conditional volatilities (see, e.g., [45,46]). Even though general single-factor DTSMs do have built-in time varying bond yield volatilities (except for basic Gaussian affine and lognormal models), however, they can not match the variation observed in practice. For example, single-factor models tend to understate the volatility of long-term yields, overstate the correlation between yields of different maturities, require mean reversion properties that cannot simultaneously match cross section and time series of bond yields. There is also evidence that linear drift term in single factor models is misspecified. Multifactor models are designed to address some of these issues.

Here, we give an example of a three-factor affine family model presented in [11]. The short rate is a CIR process with long run mean (central tendency) following Ornstein-Uhlenbeck process and stochastic volatility following a CIR process all under an EMM. The state vector $Y=(r,\theta,V)^T$ is defined as follows:

$$r=\kappa(\theta-r)\mathrm{d}t+\sqrt{V}\mathrm{d}W_r$$

$$\mathrm{d}\theta=\alpha(\beta-\theta)\mathrm{d}t+\sigma_\theta dW_\theta$$

$$\mathrm{d}V=a(b-V)\mathrm{d}t+\sigma_V\sqrt{V}\mathrm{d}W_V$$

Where $E[\mathrm{d}W_r\mathrm{d}W_V]=\rho\mathrm{d}t$, and all other correlations are zero.

According to (2.11), the price of a zero coupon bond in this model satisfies the following PDE:

$$\partial_t D + \kappa(\theta - r)\partial_r D + \alpha(\beta - \theta)\partial_\theta D + a(b - V)\partial_V D$$
$$+\frac{1}{2}\left(V\partial_{rr}D + \sigma_\theta^2\partial_{\theta\theta}D + \sigma_V^2 V\partial_{VV}D + 2\rho\sigma_V V\partial_{rV}D\right) - rD = 0$$

Since the model is in the affine category, we look for a solution in the following form:

$$D_t\left(t, r, \theta, V\right) = e^{\varphi_0(\tau) + \varphi_1(\tau)' Y_t}$$

where vector $\varphi_1(\tau)$ has three components $\varphi_{11}(\tau)$, $\varphi_{12}(\tau)$ and $\varphi_{13}(\tau)$. The Riccati equations (2.12) and (2.13) along with the boundary conditions (2.14) imply

$$\partial_\tau\varphi_0 + \alpha\beta\varphi_{12} + ab\varphi_{13} + \frac{1}{2}\sigma_\theta^2\varphi_{12}^2 = 0 \tag{2.24}$$

$$\partial_\tau\varphi_{11} + \kappa\varphi_{11} + 1 = 0 \tag{2.25}$$

$$-\partial_\tau\varphi_{12} + \kappa\varphi_{11} - \alpha\varphi_{12} = 0 \tag{2.26}$$

$$-\partial_\tau\varphi_{13} + \frac{1}{2}\varphi_{11}^2 + \frac{1}{2}\sigma_V^2\varphi_{13}^2 + \left(a + \rho\sigma_V\varphi_{11}\right)\varphi_{13} = 0 \tag{2.27}$$

$$\varphi_0(0) = \varphi_1(0) = 0 \tag{2.28}$$

Solutions to equations (2.25) and (2.26) are available in closed form:

$$\varphi_{11}(\tau) = -\frac{1 - e^{-\kappa\tau}}{\kappa}$$

$$\varphi_{12}(\tau) = -\frac{1 - e^{-\kappa\tau} + \frac{\kappa}{\alpha}e^{-\alpha\tau}\left(1 - e^{-\alpha\tau}\right)}{\alpha - \kappa}.$$

Explicit solutions for $\varphi_{13}(\tau)$ and $\varphi_0(\tau)$ are not available. However, numerical solutions are easy to obtain.

3. AFFINE JUMP-DIFFUSION DTSMS

There is a growing body of empirical research showing that discontinuous behavior of interest rates is essential in correctly describing their distribution and fitting terms structures to the data. e.g., [24-26] all find that introducing jumps substantially improves the fit of the conditional distribution of short-term interest rates compared to that of nested diffusion models.[6]

The relative scarcity of empirical work on models of discontinuous interest rates is most likely related to far less tactable mathematics involved in these models. Interest rates are mean reverting. In general, the mean reversion feature combined with jumps in the interest rates makes the moments of jump-diffusion distribution dependant on the timing of the jumps. However, in jumpdiffusion models it turns out that the density of equity returns does not depend on the jump times. In order to derive the expression for the CMGF of the state variables and bond prices for the case of affine jump diffusion we generalize the Lemma 2.1 in order to include jumps. The CMGF for a jump-diffusion process turns out to be the product of CMGFs of continuous and jump parts of the processes.

Definition. Let Y be a vector of n state variables and r is the short rate. A model is affine if the state vector solves the following diffusion SDE:

$$\mathrm{d}Y = \kappa\left(\theta - Y\right)\mathrm{d}t + \Sigma V \mathrm{d}W + J\mathrm{d}N \tag{3.1}$$

Where $\kappa \in \mathbb{R}^{n\times n}, \theta \in \mathbb{R}^{n\times 1}, \Sigma \in \mathbb{R}^{n\times n}$ constant matrix, $\delta_0 \in \mathbb{R}$, $\delta_1 \in \mathbb{R}^{n\times 1}$, $V$is a diagonal $\mathbb{R}^{n\times n}$ -matrix such that $VV' = \mathrm{diag}\left(\alpha_i + \beta_i' Y\right)$, $W \in \mathbb{R}^{n\times 1}$ is a vector of standard Wiener processes, $\boldsymbol{q}$ is a Poisson counter with state-dependent intensity, a positive affine function of $\boldsymbol{Y}$, $(Y) = \lambda_0 + \lambda_1' Y$, and $\boldsymbol{J}$ is a jump size with time-homogeneous MGF $\phi_J(u) = Ee^{u'J}$. We assume that jump size and jump arrival times are uncorrelated with the diffusion part of the process. In general, jump sizes can be correlated across different state variables.

As before, (2.2) is the expression for the short rate. We again assume that an equivalent martingale measure (EMM) exists, and all processes and expectations (unless stated otherwise) are understood as those under the EMM. The SDE (3.1) implies that the joint conditional moment generating function (CMGF) of the state vector is exponential affine. To see why this is the case, we extend the lemma of section 2 to include jumps.

Lemma 3.1. Assume that a state vector Y satisfies (3.1). Then, the joint CMGF of process $\boldsymbol{Y}$ and its integral is given by the following expression

$$f(t,Y) = E_t\left[\exp\left(-\int_t^T r(s,Y_s)\mathrm{d}s + u'Y_T\right)\right]$$

$$= e^{\varphi_0(\tau)+\varphi_1(\tau)'Y_t} \tag{3.2}$$

Where $\tau = T - t$. This CMGF solves the following boundary value problem:

$$\partial_t f + Af - r(t,Y)f = 0 \tag{3.3}$$

$$f(T,Y) = e^{u'Y} \tag{3.4}$$

where

$$Af = \left[\kappa(\theta - Y)\right]'\partial_Y f + \frac{1}{2}tr\left[V'\Sigma'\partial_{YY'}f\Sigma V\right]$$
$$+\lambda(Y)\left[E_t f(t,Y+J) - f(t,Y)\right] \tag{3.5}$$

is the infinitesimal generator of state variable vector $\boldsymbol{Y}$ applied to the function f. Coefficients $\varphi_0(\tau)$ and $\varphi_1(\tau)$ satisfy the following ODEs:

$$\partial_t\varphi_0(\tau) + \left[\kappa\theta\right]'\varphi_1(\tau) + \frac{1}{2}\varphi_1(\tau)'H_0\varphi_1(\tau)$$
$$+\lambda_0\left[\phi_J(\varphi_1(\tau)) - 1\right] - \delta_0 = 0 \tag{3.6}$$

$$\partial_t\varphi_1(\tau) - \kappa'\varphi_1(\tau) + \frac{1}{2}\varphi_1(\tau)'H_1\varphi_1(\tau)$$
$$+\lambda_1\left[\phi_J(\varphi_1(\tau)) - 1\right] - \delta_1 = 0 \tag{3.7}$$

$$\varphi_0(0) = 0, \varphi_1(0) = u \tag{3.8}$$

Proof. The proof is somewhat similar to the proof of Lemma 2.1 To find equations for the coefficients $\varphi_0(\tau)$ and $\varphi_1(\tau)$, we insert exponential affine solution expression (3.2) into (3.3). The resulting expression

$$\partial_t\varphi_0(\tau) + \partial_t\varphi_1(\tau)'Y + \left[\kappa(\theta - Y)\right]'\varphi_1(\tau)$$
$$+\frac{1}{2}tr\left[V'\Sigma'\varphi_1(\tau)\varphi_1(\tau)'\Sigma V\right]$$
$$+\lambda(Y)\left[E_t e^{\varphi_1(\tau)'J} - 1\right] - r(t,Y)f = 0$$

must hold for all values of the state variables, which results in ODEs (3.6)-(3.8).

3.1. One-Factor Jump-Diffusion Vasicek Model

The valuation of fixed income securities requires transition from actual to EMM measure. In general, this task can be accomplished by specifying a stochastic discount factor (SDF) for the economy. In continuous time, the SDF, M_t, places a restriction on a price D_t of a zerocoupon bond according to the following Euler condition (see [47]):

$$E_t\frac{\mathrm{d}D}{D} = -E_t\frac{\mathrm{d}M}{M} - E_t\frac{\mathrm{d}M}{M}\frac{\mathrm{d}D}{D}$$

In a model adapted from [22], the SDF satisfies a jump-diffusion SDE:

$$\frac{\mathrm{d}M}{M} = r\mathrm{d}t - \theta_W\mathrm{d}W - \theta_J\left(\mathrm{d}N - \lambda\mathrm{d}t\right) \tag{3.9}$$

where θ_W is the price of diffusion risk, θ_J is the price of jump risk, $\mathrm{d}N - \lambda\mathrm{d}t$ is a demeaned (compensated) Poisson process with intensity λ jumps per year. They also assume that the sort rate follows O-U process with jumps:

$$r = k\left(\theta - r\right)\mathrm{d}t + \sigma\mathrm{d}W + J\mathrm{d}N.$$

The zero-coupon bond price, being a function of time, *t*, and the short rate, *r*, satisfies the following Ito SDE:

$$\begin{aligned}\mathrm{d}D &= \left(-D_\tau + k\left(\theta - r\right)D_r + \frac{1}{2}\sigma^2 D_{rr}\right)\mathrm{d}t \\ &+ \sigma D_r\mathrm{d}W + \left[D\left(r+J,t\right) - D\left(r,t\right)\right]\mathrm{d}N\end{aligned}$$

Define $q \equiv E_t\left[D\left(r+J,t\right) - D\left(r,t\right)\right]$ and substitute the above differential into the Euler condition to obtain the following PDE for the zero-coupon bond price:

$$\begin{aligned}&E_t\mathrm{d}D \\ &= rD\mathrm{d}t + E_t\left(\theta_W\sigma D_r + \theta_J\lambda\left[D\left(r+J,t\right) - D\left(r,t\right)\right]\right)\mathrm{d}t \\ &\quad - D_\tau + k\left(\theta - r\right)D_r + \frac{1}{2}\sigma^2 D_{rr} + \lambda q \\ &= rD + \theta_W\sigma D_r + \theta_J\lambda q\end{aligned} \tag{3.10}$$

$$-D_\tau + k\Bigg(\underbrace{\theta - \frac{\theta_W \sigma}{k}}_{\theta^Q} - r\Bigg)D_r + \frac{1}{2}\sigma^2 D_{rr}$$
$$+\underbrace{\lambda(1-\theta_J)}_{\lambda^Q} q - rD = 0 \tag{3.11}$$

The resulting PDE describes the evolution of the bond price under EMM. Under this risk-neutral measure the short rate process satisfies

$$\mathrm{d}r = k\left(\theta^Q - r\right)\mathrm{d}t + \sigma \mathrm{d}W^Q + J\mathrm{d}N^Q$$

$$\theta^Q = \theta - \frac{\theta_W \sigma}{k}$$

$$\mathrm{d}W^Q = \mathrm{d}W + \theta_W \mathrm{d}t$$

The transformed counting process has new risk-adjusted intensity $\lambda^Q = \lambda(1-\theta_J)$. The PDE (3.11) belongs to a class that generates exponential-affine bond prices. Thus, we look for a solution to the bond price in the following form:

$$D(r,\tau) = e^{A(\tau)+B(\tau)r}$$

Inserting this expression into the PDE, we have

$$-A' - B'r + k\left(\theta^Q - r\right)B + \frac{1}{2}\sigma^2 B^2$$
$$+\lambda^Q\left(\phi_J(B) - 1\right) - r = 0.$$

where $\phi_J(B) = Ee^{BJ}$ is the MGF of the jump size. The above expression is an identity that must hold for any values of time and short rate. This requirement leads to separate ODEs for coefficients A and B:

$$A' = k\theta^Q B + \frac{1}{2}\sigma^2 B^2 + \lambda^Q\left(\phi_J(B) - 1\right), \tag{3.12}$$

$$B' = -kB - 1, \tag{3.13}$$

$$A(\tau = 0) = B(\tau = 0) = 0. \tag{3.14}$$

The terminal condition (3.14) follows from the fact that the bond has face value of $1 at maturity. Subject to (3.14), solutions to (3.12)-(3.13) are given by

$$B(\tau) = \frac{e^{-k\tau} - 1}{k},$$

$$A(\tau) = \int_0^{\tau} k\left[k\theta^Q B(s) + \frac{1}{2}\sigma^2 B(s)^2 + \lambda^Q \left(\phi_J(B(s)) - 1 \right) \right] \mathrm{d}s.$$

3.2. One-Factor Cox-Ingesoll-Ross Model with Jumps

[4] develops a general equilibrium model for a standard economy with a representative agent with log-utility preferences, and shows that the short rate is a linear function of the single state variable that drives the economy. The short rate therefore inherits the same dynamic properties as the state variable and follows the square root SDE[7] (in this case, we also augment it with a jump-diffusion component; see also [15]):

$$\mathrm{d}r = k^Q\left(\theta^Q - r\right)\mathrm{d}t + \sigma\sqrt{r}\,\mathrm{d}W^Q + J\mathrm{d}N^Q \tag{3.15}$$

According to the first fundamental theorem of finance, discounted prices are martingales. Thus, a zero-coupon bond price satisfies the following PDE:

$$-D_\tau + A^Q D - rD = 0,$$

where $\boldsymbol{A^Q}$ is the infinitesimal generator of the short rate (3.15):

$$A^Q D = k^Q\left(\theta^Q - r\right)D_r + \frac{\sigma^2 r}{2}D_{rr} + \lambda^Q E_t^Q\left[D(r+J,t) - D(r,t)\right].$$

The relation between risk neutral and actual measure parameters is based on the CIR pricing kernel:

$$\frac{\mathrm{d}M}{M} = -r\mathrm{d}t - \theta_W\sqrt{r}\,\mathrm{d}W - \theta_J\left(\mathrm{d}N - \lambda\mathrm{d}t\right),$$

$$\mathrm{d}W^Q = \mathrm{d}W + \theta_W\sqrt{r}\,\mathrm{d}t,$$

where $\lambda^Q = \lambda\left(1 - \theta_J\right)$, $k^Q = k + \theta_W\sigma$ and

$\theta^Q = \dfrac{k\theta}{k+\theta_W\sigma}$. Substituting a solution of the form $e^{A(\tau)-B(\tau)r}$ into the above PDE, just like in the one-factor Vasicek model, we obtain two separate ODEs for the functions $A(\tau)$ and: $A(\tau)$

$$r\left(k^Q B + B_\tau + \frac{\sigma^2}{2}B^2 - 1\right)$$

$$+\left\{-A_\tau - k^Q\theta^Q B + \lambda^Q\left(\phi_J\left(B\right)-1\right)\right\} = 0,$$

$$A(\tau=0) = B(\tau=0) = 0.$$

Integrating the two ODEs, we have the solution for the CIR zero-coupon bond price with jumps in the short rate:

$$D(r,\tau) = e^{A(\tau)-B(\tau)r}$$

where

$$A(\tau) = \int_0^\tau \mathrm{d}s\left(-k^Q\theta^Q B(s) + \lambda^Q\left(\phi_J(B(s))-1\right)\right),$$

$$B(\tau) = \frac{2\left(e^{\xi\tau}-1\right)}{\left(e^{\xi\tau}-1\right)\left(\xi+k^Q\right)+2\xi},$$

and

$$\xi = \sqrt{\left(k^Q\right)^2 + 2\sigma^2}.$$

3.3. A Two-Factor Vasicek Model with Jumps in the Long-Run Mean

In this two-factor model, the two state variables are the short rate, r, and the central tendency (long-run mean) of the short rate, μ. We assume as before that both risks are priced as follows:

$$\frac{\mathrm{d}M}{M} = -r\mathrm{d}t - \theta_W \mathrm{d}W - \theta_J\left(\mathrm{d}N - \lambda\mathrm{d}t\right)$$

where parameters have the same definitions as in (3.19). Under the EMM, the short rate follows a standard Vasicek model, and the long run-mean of the short rate is given by a pure jump process:

$$dr = k\left(\mu - r - \frac{\theta_W \sigma}{k}\right)dt + \sigma dW^Q$$

and

$$d\mu = JdN^Q.$$

The corresponding PDE for a zero-coupon bond price:

$$-D_\tau + A^Q D - rD = 0, \tag{3.16}$$

$$A^Q D = k\left(\mu - r - \frac{\theta_W \sigma}{k}\right)D_r + \frac{\sigma^2}{2}D_{rr}$$

$$+\lambda^Q E_t^Q\left[D(r+J,t) - D(r,t)\right],$$

$$\lambda^Q = \lambda(1-\theta_J).$$

Since we are dealing with an exponential affine model once again, we can write the solution as

$$D(r,\mu,\tau) = e^{A(\tau)-B(\tau)r-C(\tau)\mu}.$$

Substituting this solution into (3.16), the PDE breaks down into three separate ODEs for functions $A(\tau)$, $B(\tau)$, and $C(\tau)$:

$$0 = r(kB - B_\tau - 1) + \mu(C_\tau - kB)$$

$$+\left(-A_\tau + \theta_W \sigma B + \frac{\sigma^2}{2}B^2 + \lambda^Q\left[\phi_J(-C) - 1\right]\right)$$

$$A(0) = B(0) = C(0) = 0$$

The functions $B(\tau)$ and $C(\tau)$ are given by:

$B(\tau) = \frac{1-e^{-k\tau}}{k}$ and $C(\tau) = \tau - B(\tau)$. Moreover, the function $A(\tau)$ can be obtained numerically by using

$$A(\tau) = \int_0^\tau \left(\theta_W \sigma B(s) + \frac{\sigma^2}{2} B(s)^2 + \lambda^Q \left[\phi_J \left(-C(s) \right) - 1 \right] \right) ds. \tag{3.17}$$

Note: When the jump size is normally distributed with mean μ_J, variance σ_J^2, and MGF

$$Ee^{-JC(\tau)} = \exp\left(-\mu_J C(\tau) + \frac{1}{2} \sigma_J^2 C(\tau)^2 \right),$$

by using the following first-order Taylor series approximation as in [18][8],

$$Ee^{-JC(\tau)} - 1 \simeq -\mu_J C(\tau) + \frac{1}{2} \sigma_J^2 C(\tau)^2$$

the function (3.17) turns out to be:

$$\begin{aligned} A(\tau) = & -\frac{4k\lambda^Q \mu_J - \lambda^Q \sigma_J^2 + 4k\sigma\theta_W + 3\sigma^2}{4k^3} \\ & + \tau \frac{2k\left(\lambda^Q \mu_J + \sigma\theta_W\right) + \sigma^2 + \lambda^Q \sigma_J^2}{k} \\ & - \frac{\lambda^Q \left(k\mu_J + \sigma_J^2\right)}{2k} \tau^2 + \frac{1}{6} \lambda^Q \sigma_J^2 \tau^3 \\ & + e^{-k\tau} \frac{k\lambda^Q \mu_J + k\sigma\theta_W - k\tau\lambda^Q \sigma_J^2 + \sigma^2}{k^3} \\ & + e^{-2k\tau} \frac{\sigma^2 + \lambda^Q \sigma_J^2}{4k^3}. \end{aligned}$$

Under the normality assumption of the jump sizes, the closed-form approximate solution for bond prices becomes very useful for calibration, estimation and testing.

4. THE QUADRATIC MODEL

The common problem with most interest rate models is that interest rates can become negative (except in CIRtype models). Even though negative interest rates are not impossible, they are rare. This fact has led to the need to develop models of interest rates that guarantee positive rates. The models in this category are numerous.[9] However, most of these early models suffer from various problems. e.g., a rational log-normal model of

[48] suffers from calibration problems in that bond prices and short rates are bounded both from below and above. As a result this model is arbitrage free only for a finite period of time. This restriction proves too binding in empirical work. Other ways of guaranteeing positive rates is to transform the state variable. Popular class of models that achieve that is log-r models. The short rate is the exponent of a state variable Y_t, $r_t = e^{Y_t}$, which follows an extended Vasicek process. Historically, due to availability of easy lattice implementation tools, the two most widely used models in this class are [51,52]. However, this transformation presents theoretical problems. The problem with these models is that the expected future value of the risk-free bank account $V_t = E\exp\left(\int_0^t r_s \mathrm{d}s\right) \to \infty$ even for arbitrarily small $t > 0$ (see [53]).

4.1. Quadratic-Gaussian Interest Rate Models

Recently, more attention has been paid to a different specification, quadratic-Gaussian, that guarantees positive rates, does not suffer from these problems, and still tractable enough to generate near closed form bond and in some cases even bond option prices. The simplest form of this model is the so-called square Gaussian model in which the short rate is simply the sum of squares of the state variable.[10] A more general model is in the Quadratic-Gaussian (QG) class with the short rate a general quadratic form of the state variables, which follow a process with affine drift and constant volatility. In an n-factor Quadratic-Gaussian model the short rate is

$$r_t = a + b'Y_t + \frac{1}{2}Y_t'cY_t, \tag{4.1}$$

where $a \in \mathbb{R}, b \in \mathbb{R}^{n\times 1}$, and $c \in \mathbb{R}^{n\times n}$. The state variables $\boldsymbol{Y}$ follow an n-dimensional Gaussian process under a risk neutral measure:

$$\mathrm{d}Y_t = \kappa(\theta - Y_t)\mathrm{d}t + \Sigma \mathrm{d}W_t, \tag{4.2}$$

Where $\kappa \in \mathbb{R}^{n\times n}, \theta \in \mathbb{R}^{n\times 1}, \Sigma \in \mathbb{R}^{n\times n}$ constant matrix, and $W_t \in \mathbb{R}^{n\times 1}$ is a vector of standard Wiener processes. We can always rewrite the short rate in (4.1) in the following form after shifting the state variable $\boldsymbol{Y_t}$ by a constant vector:

$$r_t = a + \frac{1}{2}Y_t'cY_t,$$

The state SDE (4.2) does not change from this transformation except for obvious redefinition of parameters. Quadratic-Gaussian models have been studied by [28, 31-35,38,54,55][11] Similarly to affine models, we look for the zero coupon bond price solution to (2.10) in the following form:

$$D_t(t,Y_t) = \exp\left(\varphi_0(\tau) + \varphi_1(\tau)' Y_t + \frac{1}{2} Y_t'\Omega(\tau) Y_t\right) \tag{4.3}$$

where Ω is $n{\times}n$ matrix. The bond price satisfies PDE (2.4) with a=1 and u=0. Using (4.3) in (2.4), we obtain the following PDE:

$$-\partial_\tau \varphi_0 - \partial_\tau \varphi_1(\tau)' Y_t - \frac{1}{2} Y_t' \partial_\tau \Omega(\tau) Y_t$$

$$+\left[\kappa(\theta - Y_t)\right]'\left(\varphi_1(\tau) + \frac{1}{2}(\Omega + \Omega')Y_t\right)$$

$$+\frac{1}{2} tr\left[\Sigma'\left(\varphi_1\varphi_1' + \frac{1}{2}(\Omega+\Omega') + \frac{1}{2}(\Omega+\Omega')Y\varphi_1'\right.\right.$$

$$\left.\left.+\frac{1}{2}\varphi_1 Y'(\Omega+\Omega') + \frac{1}{4}(\Omega+\Omega')YY'(\Omega+\Omega')\right)\Sigma\right]$$

$$-a - \frac{1}{2} Y'cY = 0.$$

Since this equation has to hold for any values of the state variables, it separates into three independent Riccati ODEs for φ_0, φ_1, and Ω:

$$-\partial_\tau \varphi_0 + (\kappa\theta)' \varphi_1 + \frac{1}{4} tr\left[\Sigma'(\Omega+\Omega')\Sigma\right] + \frac{1}{2}\varphi_1'\Sigma\Sigma'\varphi_1 - a = 0,$$

$$-\partial_\tau \varphi_1(\tau) - \kappa'\varphi_1 + \frac{1}{2}(\Omega+\Omega')\kappa\theta + \frac{1}{2}(\Omega+\Omega')\Sigma\Sigma'\varphi_1 = 0,$$

$$-\partial_\tau \Omega - \kappa'(\Omega+\Omega') + \frac{1}{4}(\Omega+\Omega')\Sigma\Sigma'(\Omega+\Omega'') - c = 0.$$

Moreover, if the matrix Ω is symmetric, the system reduces to

$$-\partial_\tau \varphi_0 + (\kappa\theta)' \varphi_1 + \frac{1}{2} tr[\Sigma'\Omega\Sigma]$$

$$+\frac{1}{2}\varphi_1'\Sigma\Sigma'\varphi_1 - a = 0,$$

$$-\partial_\tau \varphi_1(\tau) - \kappa'\varphi_1 + \Omega\kappa\theta + \Omega\Sigma\Sigma'\varphi_1 - b = 0,$$

$$-\partial_\tau \Omega - 2\kappa'\Omega + \Omega\Sigma\Sigma'\Omega - c = 0.$$

Example 1. For a one-factor square-Gaussian model of interest rates considered in [6] and [5]:

$$r_t = Y_t^2,$$

$$dY_t = (\theta(t) - aY_t)dt + \sigma dW_t. \tag{4.4}$$

For this model, the bond prices, as well as bond option prices, have closed form expressions given by the following:

1) The zero coupon bond price is

$$D_t(t,T,Y_t)$$
$$= \exp(\varphi_0(t,T) - \varphi_1(t,T)X_t - \Omega(t,T)X_t^2),$$

where

$$X_t = Y_t - \alpha_t,$$

$$\alpha_t = e^{-at}\left(\sqrt{r_0} + \int_0^t e^{as}\theta(s)ds\right);$$

$$\varphi_0(t,T) = \int_t^T \left(\frac{\sigma^2}{2}\varphi_1^2(s,T) - \sigma^2\Omega(s,T) - \alpha_s^2\right)ds,$$

$$\varphi_1(t,T)=\int_t^T \frac{e^{2\gamma\left(T-\frac{s}{2}\right)}(a+\gamma)+\gamma-a}{e^{2\gamma\left(T-\frac{t}{2}\right)}(a+\gamma)+\gamma-a}\alpha_s \mathrm{d}s,$$

$$\Omega(t,T)=\frac{e^{2\gamma(T-t)}-1}{e^{2\gamma(T-t)}(a+\gamma)+\gamma-a}$$

and

$$\gamma^2=a^2+2\sigma^2.$$

2) The European call option price with maturity T_1 and strike K written on a zero-coupon bond with maturity T_2 is given by

$$c(T_1,T_2)=D_t(t,T_2,Y_t)(N(d_1)-N(d_2))$$
$$-KD_t(t,T_1,Y_t)(N(d_3)-N(d_4))$$

with parameters given by the following expressions:

$$d_1=\frac{h\tau-v}{\sigma\sqrt{\tau C(t,T_1)}},$$

$$d_2=\frac{l\tau-v}{\sigma\sqrt{\tau C(t,T_1)}},$$

$$d_3=\frac{h-M(t,T_1)}{\sigma\sqrt{\tau C(t,T_1)}},$$

$$d_4=\frac{l-M(t,T_1)}{\sigma\sqrt{\tau C(t,T_1)}},$$

$$v = M(t,T_1) - \varphi_1(T_1,T_2)\sigma^2 C(t,T_1),$$

$$\tau = 1 + 2\sigma^2 C(t,T_1) C(T_1,T_2),$$

$$h = \frac{\sqrt{d} - \varphi_1(T_1,T_2)}{2C(T_1,T_2)},$$

$$l = \frac{-\sqrt{d} - \varphi_1(T_1,T_2)}{2C(T_1,T_2)},$$

$$l = \frac{-\sqrt{d} - \varphi_1(T_1,T_2)}{2C(T_1,T_2)},$$

$$d = \varphi_1^2(T_1,T_2) + 4C(T_1,T_2)(\varphi_0(T_1,T_2) - \ln K),$$

$$M(t,T) = N(t,T)X - \int_t^T \sigma^2 N(s,T)\varphi_1(s,T)\mathrm{d}s, N(t,T) = \frac{2\gamma e^{\gamma(T-t)}}{e^{2\gamma(T-t)}(a+\gamma)+\gamma-a}.$$

4.2. Linear-Quadratic Interest Rate Model

The difficulty of fitting term structures of interest rates with diffusion models has led to the study of the impact of jumps in the interest rates on spot rate properties. [25, 26,37,57,58] point out that including jumps in a model helps better explain properties of various interest rates. When extending a model in any way, we are also concerned about whether the model will still have a closed form solution for bond prices, bond option (cap, floor, swaption) prices, and whether the extension helps improve the fit of volatility term structures to those implied by swaptions and caps when the simpler model has failed to do so. In the case of including jumps into diffusion models, [59] use three years of interest rate cap price data and show that within a three factor SVJ DTSM significant negative jumps in interest rates are required to fit implied volatility smile.

[36] argue that most of the previous research is too restrictive in the way it treats jump intensity. Usual assumptions range from constant intensity to a rather restrictive functional form of underlying state variables that may include economic variables as in [37]. [36] introduce a new class of interest

rate models, linear-quadratic term structure model (LQTSM). The new feature is that jump intensity is now a separate state variable that follows it own SDE. In their empirical implementation, they use essentially a 3-factor model with the state vector that includes the short rate, the random volatility, and the stochastic intensity. The rate on 3-month T-bills serves as a proxy for the short rate. To estimate parameter values [36] relies on generalized method of moments (GMM) approach of [60]. At each iteration of the GMM optimization routine, however, they have to compute the remaining two latent state variables, volatility and intensity, for every date of their dataset. They do it by resorting to the so-called implied-state GMM (IS-GMM) procedure of [61]. For given parameter values at each iteration they use high frequency futures market data on the changes in T-Bill yields to compute conditional variance and kurtosis. Stochastic volatility and intensity are then identified from these two higher moments. They find that introducing random intensity substantially improves both the model fit (through better fit of short term kurtosis) as well as the dynamics of the interest rate volatility term structure. In a VAR model run on daily data they also establish that jumps are not only associated with macroeconomic announcements or FOMC meeting dates but also with unanticipated news.

Below, we show how LQTSM class is designed to have a closed-form solution, and we give a simple example of a special case closely related to the SVJT model that [36] studies. We start with the extended transform (2.6). LQTSM class is designed (or defined) to have the following exponential LQ form for extended transform[12]:

$$f(t,Y)=E_t\left[\exp\left(-\alpha\int_t^T r(s,Y_s)ds+u'Y_T\right)\right]$$

$$=e^{A(\tau)+B(\tau)'Y_t+Y_t'C(\tau)Y_t} \tag{4.5}$$

$$\mathrm{d}Y_t=(\mu-\lambda\mu_J)\mathrm{d}t+\Sigma\mathrm{d}W+J\mathrm{d}N \tag{4.6}$$

$$r(t,Y)=a_r+b_r'Y+Y'c_rY \tag{4.7}$$

where $Y\in\mathbb{R}^{n\times 1}$ is a vector of state variables, $\mu\in\mathbb{R}^{n\times 1}$, $\lambda\in\mathbb{R}$is jump intensity, jump size vector $J\in\mathbb{R}^{n\times 1}$ with time-homogeneous MGF $\phi_J(u)=Ee^{u'J}$, the mean of the jump size vector $\mu_J=E(J)\in\mathbb{R}^{n\times 1}$, $\Sigma\in\mathbb{R}^{n\times n}$, $W\in\mathbb{R}^{n\times 1}$ is a vector of standard Wiener processes, Poisson process N is independent of jump size and of the diffusion vector W, a_r, $A\in\mathbb{R}$, b_r, $B\in\mathbb{R}^{n\times 1}$ and c_r, $C\in\mathbb{R}^{n\times n}$ [36] derives risk-neutral corrections to model parameters based on power-utility pricing kernel by following the approach of [4,38]. We skip this stage in the derivation and assume that all processes are already expressed under an EMM. Extended transform (4.5) must satisfy the following PDE:

$$-\partial_{\tau} f + Af - \alpha r(t,Y) f = 0 \tag{4.8}$$

$$f(T,Y) = e^{u'Y} \tag{4.9}$$

$$Af = (\mu - \lambda\mu_J)' \partial_Y f + \frac{1}{2} tr[\Sigma' \partial_{YY'} f \Sigma] + \lambda(Y)[E_t f(t,Y^+) - f(t,Y)] \tag{4.10}$$

where Y^+ is the value of the state vector immediately after a jump. Similar to affine JD models, the expectation in the integro differential equation (4.8) should break up into the product of the MGFs for continuous and discontinuous parts of the state vector process as these parts are assumed independent in this specification. The only complication is the quadratic form in the definition of extended transform for LQTSM class in (4.5). [36] works around this complication by assuming that in the state vector only a subvector of, say, the first n_1 components is discontinuous, i.e., follows a JD process. The remaining $n_2=N-n_1$ components are diffusions:

$$J = \begin{pmatrix} J_{n_1\times 1} \\ 0_{n_2\times 1} \end{pmatrix}$$

$$Y = \begin{pmatrix} Y_{1_{n_1\times 1}} \\ Y_{2_{n_2\times 1}} \end{pmatrix}$$

The extended transform (4.5) must be exponential quadratic only in Y_2:

$$f(t,Y) = e^{A(\tau)+B(\tau)' Y_t + Y_2' C(\tau) Y_2} \tag{4.11}$$

In order to obtain separate ODEs for three functions $A(\tau), B(\tau)$, and $C(\tau)$, the next obvious requirement is that once we substitute (4.11) into (4.8) there must be no terms other than affine in Y (i.e., $A(Y)$) and quadratic in Y_2 (i.e., $Q(Y_2)$). Then, the functional forms of the short rate, r, intensity, λ, and drift and diffusion coefficients of the transition equation (4.6) are dictated by the above requirement. For example, the second term in PDE

(4.8) contains the gradient of the transform and the jump part is proportional to λ:

$$\partial_Y f = \begin{pmatrix} f_{Y_1} \\ f_{Y_2} \end{pmatrix} \to \begin{pmatrix} {}_{n_1\times 1} Const \\ {}_{n_2\times 1} A(Y_2) \end{pmatrix}$$

These functional forms of the gradient and the jump term jointly imply that the first n_1 components of vector μ in (4.6) and intensity λ must be $A(Y_1)$ and $Q(Y_2)$ and the last n_2 components of vector μ must be $A(Y_1)$:

$$(\mu - \lambda\mu_J)' \partial_Y f = \begin{pmatrix} {}_{n_1\times 1}\mu_1 - \lambda\mu_{J1} \\ {}_{n_2\times 1}\mu_2 \end{pmatrix}' \begin{pmatrix} f_{Y_1} \\ f_{Y_2} \end{pmatrix},$$

$$\lambda = a_\lambda + b_\lambda' Y + Y_2' c_\lambda Y_2,$$

$$(\mu_1)_k = (a_\mu + b_\mu' Y + Y' c_\mu Y_2)_k,$$

$$(\mu_2)_k = (d_\mu + e_\mu Y_2)_k,$$

$$k = \overline{1, n_1}, k = \overline{1, n_2}.$$

Just like λ, the short rate (last term in the PDE) must be $A(Y_1)$ and $Q(Y_2)$:

$$r(t, Y) = a_r + b_r' Y + Y_2' c_r Y_2,$$

with obvious definitions for a_r, b_r, c_r. Further, the trace part of the PDE has the following form:

$$tr[\Sigma' \partial_{YY'} f \Sigma] = tr[\Sigma\Sigma' \partial_{YY'} f] = tr[\Omega \partial_{YY'} f]$$

$$= tr\left[\begin{pmatrix} \Omega_{11} & \Omega_{12} \\ \Omega_{21} & \Omega_{22} \end{pmatrix} \partial_{YY'} f\right]$$

$$= tr\left(\Omega_{11} \underbrace{f_{Y_1 Y_1'}}_{const^* f} + 2\Omega_{12} \underbrace{f_{Y_1 Y_2'}}_{A(Y_2)^* f} + \Omega_{22} \underbrace{f_{Y_2 Y_2'}}_{Q(Y_2)^* f} \right)$$

where ΩΣΣ⍰ is a symmetric matrix. The above expression implies that

$$\Omega_{11} \sim A(Y_1)$$

and

$$Q(Y_2), \Omega_{12} \sim A(Y_2), \Omega_{22} \sim const.$$

Substituting (4.11) into PDE (4.8) we have the following identity that must hold for all ***Y***:

$$-\partial_\tau A - \partial_\tau B(\tau)' Y - Y_2' \partial_\tau C(\tau) Y_2 + \begin{pmatrix} a_\mu - a_\lambda \mu_{J1} + (b_\mu - b_\lambda \mu_{J1})' Y + Y_2'(c_\mu - c_\lambda \mu_{J1}) Y_2 \\ d_\mu + e_\mu' Y_2 \end{pmatrix}' \begin{pmatrix} B_1 \\ B_2 + (C + C') Y_2 \end{pmatrix}$$
$$+ \frac{1}{2} tr\left(\Omega_{11} B_1 B_1' + 2\Omega_{12} B_1 \left[B_2 + (C + C') Y_2 \right]' + \Omega_{22} \left[(B_2 + (C + C') Y_2)(B_2 + (C + C') Y_2)' + C + C' \right] \right)$$
$$+ (a_\lambda + b_\lambda' Y + Y_2' c_\lambda Y_2)\left[\phi_J(B_1) - 1 \right] - \alpha (a_r + b_r' Y + Y_2' c_r Y_2) = 0$$

where

$$B = \begin{pmatrix} (B_1)_{n_1 \times 1} \\ (B_2)_{n_2 \times 1} \end{pmatrix}$$

Terms independent of ***Y***, affine in $\boldsymbol{Y_1}$, affine in $\boldsymbol{Y_2}$, and quadratic in $\boldsymbol{Y_2}$ must each be equal to zero in this identity. This restriction results in a system of four ODEs for functions $A(\tau)$, $B_1(\tau)$, $B_2(\tau)$, and $C(\tau)$. For details, the reader can consult [36]. The initial conditions are

$$A(0) = 0, B_1(0) =_{n_1 \times 1} u_1,$$

$$B_2(0) =_{n_2 \times 1} u_2, C(0) = 0_{n_2 \times n_2}.$$

The zero coupon bond price has the same form as in (4.5) with $u = (u_1, u_2)' = 0_N$.

Example (LQTSM of [36]). Assume that the state vector consists of short rate, volatility, and random jump intensity. These three state variables are assumed to solve the following transition equations:

$$\mathrm{d}Y = \mathrm{d}\begin{pmatrix} (Y_1)_{1 \times 1} \\ (Y_2)_{2 \times 1} \end{pmatrix} = \mathrm{d}\begin{pmatrix} r_t \\ v_t \\ \lambda_t \end{pmatrix}$$
$$= \begin{pmatrix} \mu_r - \kappa_{rr} r - \mu_{rJ} Y_2 + Y_2' \Psi_1 Y_2 \\ \mu_{v\lambda} - \Psi_2 Y_2 \end{pmatrix} \mathrm{d}t$$
$$+ \Sigma \begin{pmatrix} \mathrm{d}W_r \\ \mathrm{d}W_v \\ \mathrm{d}W_\lambda \end{pmatrix} + J \begin{pmatrix} \mathrm{d}N \\ 0 \\ 0 \end{pmatrix}, \tag{4.12}$$

with the following definitions and notation:

$$\mu_{rJ} = \begin{pmatrix} 0 \\ \mu_J \end{pmatrix}, \mu_{v\lambda} = \begin{pmatrix} \mu_v \\ \mu_\lambda \end{pmatrix},$$

$$\Psi_1 = \begin{pmatrix} \varsigma_{vv} & 0 \\ 0 & 0 \end{pmatrix}, \Psi_2 = \begin{pmatrix} \kappa_{vv} & 0 \\ 0 & \kappa_{\lambda\lambda} \end{pmatrix},$$

$$\Sigma = \begin{pmatrix} \nu & 0 & 0 \\ 0 & \sigma_v & 0 \\ 0 & 0 & \sigma_\lambda \end{pmatrix},$$

$$\operatorname{var}\begin{pmatrix} dW_r \\ dW_v \\ dW_\lambda \end{pmatrix} = \begin{pmatrix} 1 & \rho_{rv} & \rho_{r\lambda} \\ \rho_{rv} & 1 & \rho_{\lambda v} \\ \rho_{r\lambda} & \rho_{\lambda v} & 1 \end{pmatrix} dt,$$

$$\Omega = \frac{1}{dt}\left[\Sigma * \operatorname{var}\begin{pmatrix} dW_r \\ dW_v \\ dW_\lambda \end{pmatrix}\Sigma\right]$$

$$= \begin{pmatrix} \nu^2 & \rho_{rv}\nu\sigma_v & \rho_{r\lambda}\nu\sigma_\lambda \\ \rho_{rv}\nu\sigma_v & \sigma_v^2 & \rho_{\lambda v}\sigma_v\sigma_\lambda \\ \rho_{r\lambda}\nu\sigma_\lambda & \rho_{\lambda v}\sigma_v\sigma_\lambda & \sigma_\lambda^2 \end{pmatrix}$$

$$= \begin{pmatrix} \Omega_{11} & \Omega_{12} \\ \Omega_{21} & \Omega_{22} \end{pmatrix}.$$

We look for a zero coupon bond price in the following form:

$$D(t,\tau,Y) = \exp\left(A(\tau) + B_1(\tau) r_t + B_2(\tau)' Y_2 + Y_2' C(\tau) Y_2\right). \tag{4.13}$$

As in [36] we assume that $C(\tau)$ is symmetric. In the Appendix we derive the following ODEs for functions, $A(\tau)$, $B_1(\tau)$, and $C(\tau)$

$$-\partial_\tau A + \mu_r B_1 + \mu'_{v\lambda} B_2 + B_1 \Omega_{12} B_2 + \frac{1}{2} tr\left[\Omega_{22}\left(B_2 B'_2 + 2C\right)\right] = 0, \tag{4.14}$$

$$-\partial_\tau B_1 - B_1 \kappa_{rr} - 1 = 0, \tag{4.15}$$

$$-\partial_\tau B_2 - B_1 \mu_{r,J} - \Psi'_2 B_2 + 2C\mu_{v\lambda} + 2B_1 C\Omega'_{12} + 2C\Omega_{22} B_2 + \begin{pmatrix} 0 \\ \phi_J(B_1) - 1 \end{pmatrix} = 0, \tag{4.16}$$

$$-\partial_\tau C + B_1 \Psi'_1 - 2\Psi'_2 C + 2C\Omega_{22} C + \begin{pmatrix} \frac{B_1^2}{2} & 0 \\ 0 & 0 \end{pmatrix} - 0, \tag{4.17}$$

subject to the following initial conditions:

$$A(0) - 0, B_1(0) = 0, \tag{4.18}$$

$$B_2(0) = 0_{2\times 1}, C(0) = 0_{2\times 2}. \tag{4.19}$$

Here, $\phi_J(B_1) = Ee^{B_1 J}$ is the moment generating function of the jump size.

In order to conclude the discussion of LQTSMs, we derive the price of a European call option from a generalized transform using a technique similar to the one used by [39,62,63]. We are looking for the price of a European call option on a zero coupon bond. The call has $\tau = T - t$ years to maturity and a strike price, $\boldsymbol{K}$. The zero-coupon bond has maturity date, $\boldsymbol{T_D}$, the face value of $\$1$, and the current price $\boldsymbol{D_t}$. We provide the derivation of the option price for the 3-factor LQTSM from [36] given in the above example. We assume that the state dynamics in (4.12) are given under an EMM. The value of the European call option is given by the following risk neutral expectation with respect to (4.12):

$$c_t = E_t\left[\exp\left(-\int_t^T r_s \mathrm{d}s\right)\left(D_T - K\right)\theta\left(\ln D_T - \ln K\right)\right], \tag{4.20}$$

where ▯ is the Heaviside function defined as

$$\theta\left(\ln D_T - \ln K\right) = \begin{cases} 1, \text{ if } D_T \geq K \\ 0, \text{ if } D_T < K \end{cases}.$$

In order to handle the truncation under the expectation in (4.20), we use a well known complex integral representation of the Heaviside function

$$\theta\left(\ln D_T - \ln K\right) = \int_{-\infty}^{+\infty} \frac{d\xi}{2\pi} \frac{i}{\xi + i\eta} e^{-i\xi\left(\ln D_T - \ln K\right)}$$

Using this representation we write the call price as (4.21)

where

$$\phi_k\left(\xi, \tau\right) = e^{i\xi \ln \ln K} E_t\left[\exp\left(-\int_t^T r_s \mathrm{d}s + g_k \ln D_T\right)\right], \tag{4.22}$$

And $g_k = 1 - k - i\xi$, k=0, 1 To complete the derivation, we use a well-known result from the theory of generalized functions:

$$\frac{1}{\xi + i\eta} = \frac{P}{\xi} - i\pi\delta\left(\xi\right) \tag{4.23}$$

where $\frac{P}{\xi}$ is the principal value of the integral over $\frac{1}{\xi}$:

$$\frac{P}{\xi} = \lim_{\eta \to 0}\left(\int_{-\infty}^{-\eta} + \int_{\eta}^{+\infty}\right) \mathrm{d}\xi \frac{\varphi(\xi)}{\xi}.$$

Each term in (4.23) is understood as a linear functional acting on a finite infinitely differentiable function ?:

$$\lim_{\eta \to 0} \int_{-\infty}^{+\infty} \mathrm{d}\xi \frac{\varphi(\xi)}{\xi + i\eta}$$

$$= -i\pi\varphi(0) + \lim_{\eta \to 0}\left(\int_{-\infty}^{-\eta} + \int_{\eta}^{+\infty}\right) \mathrm{d}\xi \frac{\varphi(\xi)}{\xi}$$

$$= -i\pi\varphi(0) + \int_{-\infty}^{+\infty} \mathrm{d}\xi \frac{\varphi(\xi) - \varphi(0)}{\xi},$$

where we use the fact that the principal value of the integral over ??? vanishes as this function is odd. Subtracting $\frac{\varphi(0)}{\xi}$ from the integrand makes it regular, does not require principal value specification, and is suitable for numerical integration.

The expression for the call price can now be written as follows:

$$c_t = \Pi_0(\tau) - K\Pi_1(\tau)$$

where

$$\Pi_k(\tau) = \frac{\phi_k(0,\tau)}{2} - \int_{-\infty}^{+\infty} \frac{d\xi}{2\pi} \frac{\phi_k(\xi,\tau) - \phi_k(0,\tau)}{i\xi},$$

$$k = 0,1.$$

Function ϕ_k (□□□ in (4.22) can be viewed as a generalized transform for the zero coupon bond yield. Since we already have the result for the bond price for the LQTSM class, we look for the expectation in (4.22) in the exponential quadratic form: (4.24)

Here, □□□=T_D□T is maturity at the time the option matures. Recall that □□=T□T is the time remaining to the maturity of the option as of today's date ***t***. For a given ***k***, the four functions in (4.24), $\Lambda_{m,k}$(□) □(m=0,3), satisfy the same Riccati ODEs (11)-(16) as $A(\tau)$, $B_1(\tau)$, $B_2(\tau)$, and $C(\tau)$, respectively, and the following initial conditions:

$$\Lambda_{0,k}(0) = g_k A(\tau'), \quad \Lambda_{1,k}(0) = g_k B_1(\tau'),$$
$$\Lambda_{2,k}(0) = g_k B_2(\tau'), \quad \Lambda_{3,k}(0) = g_k C(\tau'),$$

$$\begin{aligned} c_t &= E_t\left[\exp\left(-\int_t^T r_s ds\right) D_T \theta(\ln D_T - \ln K)\right] - KE_t\left[\exp\left(-\int_t^T r_s ds\right)\theta(\ln D_T - \ln K)\right] \\ &= \int_{-\infty}^{+\infty} \frac{d\xi}{2\pi}\frac{i}{\xi + i\eta}\left\{E_t\left[\exp\left(-\int_t^T r_s ds + \ln D_t - i\xi(\ln D_T - \ln K)\right)\right] - KE_t\left[\exp\left(-\int_t^T r_s ds - i\xi(\ln D_T - \ln K)\right)\right]\right\} \\ &= \int_{-\infty}^{+\infty} \frac{d\xi}{2\pi}\frac{i}{\xi + i\eta}\left[\phi_0(\xi,\tau) - K\phi_1(\xi,\tau)\right]. \end{aligned}$$

(4.21)

$$\begin{aligned} \phi_k(\xi,\tau) &= e^{i\xi \ln K} \times E_t\left[\exp\left(-\int_t^T r_s ds + g_k\left[A(\tau') + B_1(\tau') r_T + B_2(\tau')' Y_2(T) + Y_2'(T) C(\tau') Y_2(T)\right]\right)\right] \\ &= \exp\left(i\xi \ln K + \Lambda_{0,k}(\tau) + \Lambda_{1,k}(\tau) r_t + \Lambda_{2,k}(\tau)' Y_2 + Y_2' \Lambda_{3,k}(\tau) Y_2\right). \end{aligned}$$

(4.22)

where $g_k = 1 - k - i\xi, k = 0,1.$

Alex Paseka, Theodoro Koulis,, Aerambamoorthy Thavaneswaran, (finite difference techniques) and direct methods (infinitesimal perturbation analysis and likelihood ratio).

REFERENCES

1. O. Vasicek, "An Equilibrium Characterization of the Term Structure," Journal of Financial Economics, Vol. 5, No. 2, 1977, pp. 177-188. doi:10.1016/0304-405X(77)90016-2
2. J. C. Hull and A. D. White, "Pricing Interest-Rate-Derivative Securities," Review of Financial Studies, Vol. 3, No. 4, 1990, pp. 573-592. doi:10.1093/rfs/3.4.573
3. J. C. Hull and A. D. White, "One-Factor Interest-Rate Models and the Valuation of Interest-Rate Derivative Securities," Journal of Financial and Quantitative Analysis, Vol. 28, No. 2, 1993, pp. 235-254. doi:10.2307/2331288
4. J. C. Cox, J. E. Ingersoll and S. A. Ross, "A Theory of the Term Structure of Interest Rates," Econometrica, Vol. 53, No. 2, 1985, pp. 385-407. doi:10.2307/1911242
5. F. Jamshidian, "A Simple Class of Square Root Interest Rate Models," Applied Mathematical Finance, Vol. 2, No. 1, 1995, pp. 61-72. doi:10. 1080/ 13504869500000004
6. Pellser, "Eccient Method for Valuing and Managing Interest Rate and Other Derivative Securities," Ph.D. Thesis, Erasmus University, Rotterdam, 1996.
7. F. A. Longsta, "The Valuation of Options on Yields," Journal of Financial Economics, Vol. 26, No. 1, 1990, pp. 97-121. doi:10.1016/0304-405X(90)90014-Q
8. F. A. Longsta and E. S. Schwartz, "Interest Rate Volatility and the Term Structure: A Two-Factor General Equilibrium Model," The Journal of Finance, Vol. 47, No. 4, 1992, pp. 1259-1282. doi:10.2307/2328939
9. R. R. Chen and L. O. Scott, "Pricing Interest Rate Options in a Two-Factor Cox-Ingersoll-Ross Model of the Term Structure," Review of Financial Studies, Vol. 5, No. 4, 1992, pp. 613-636. doi:10.1093/rfs/5.4.613
10. P. Balduzzi, et al., "Stochastic Mean Models of the TermStructure of Interest Rates," In: B. Tuckman and N. Jegadeesh, Eds., Advanced Tools for the Fixed-Income Professional, Wiley, Chichester, 2000.
11. P. Balduzzi, et al., "A Simple Approach to Three-Factor Affine Term Structure Models," The Journal of Fixed Income, Vol. 6, No. 3, 1996, pp. 43-53.doi:10.3905/jfi.1996.408181
12. L. Chen, "Interest Rate Dynamics, Derivatives Pricing, and Risk Management," Lecture Notes in Economics and Mathematical Systems, Vol. 435, 1996. doi: 10.1007/978-3-642-46825-4
13. J. James and N. Webber, "Interest Rate Modelling," John Wiley & Sons, New York, 2000.
14. Y. Maghsoodi, "Solution of the Extended CIR Term Structure and Bond Option Valuation," Mathematical Finance, Vol. 6, No. 1, 1996, pp. 89-109. doi:10. 1111/j.1467-9965.1996.tb00113.x
15. C. M. Ahn and H. E. Thompson, "Jump-Diffusion Processes and the Term Structure of Interest Rates," The Journal of Finance, Vol. 43, No. 1, 1988, pp. 155-174.doi:10.2307/2328329
16. S. H. Babbs and N. J. Webber, "Term Structure Modelling Under Alternative Official Regimes," In: M. Dempster and S. Pliska, Eds., Mathematics of Derivative Securities, Cambridge University Press, Cambridge, 1998, pp. 394-422.
17. D. Backus, S. Foresi and L. Wu, "Macroeconomic Foundations of Higher Moments in Bond Yields," New York University, New York, 1997.

18. J. Baz and S. R. Das, "Analytical Approximations of the Term Structure for Jump-Diffusion Processes: A Numerical Analysis," The Journal of Fixed Income, Vol. 6, No. 1, 1996, pp. 78-86. doi:10.3905/jfi.1996.408164
19. G. Chacko, "A Stochastic Mean/Volatility Model of Term Structure Dynamics in a Jump-Diffusion Economy," Harvard University, Cambridge, 1996.
20. S. R. Das, "A Direct Discrete-Time Approach to Poisson-Gaussian Bond Option Pricing in the Heath-JarrowMorton Model," Journal of Economic Dynamics and Control, Vol. 23, No. 3, 1998, pp. 333-369. doi:10.1016/S0165-1889(98)00031-1
21. S. Das, "Discrete-Time Bond and Option Pricing for JumpDiffusion Processes," Review of Derivatives Research, Vol. 1, No. 3, 1997, pp. 211-243. doi: 10.1007/BF01531143
22. S. R. Das and S. Foresi, "Exact Solutions for Bond and Option Prices with Systematic Jump Risk," Review of Derivatives Research, Vol. 1, No. 1, 1996, pp. 7-24.doi:10.1007/BF01536393
23. S. Heston, "A Model of Discontinuous Interest Rate Behavior, Yield Curves, and Volatility," Review of Derivatives Research, Vol. 10, No. 3, 2007, pp. 205-225. doi:10.1007/s11147-008-9020-3
24. C. Zhou, "The Term Structure of Credit Spreads with Jump Risk," Journal of Banking & Finance, Vol. 25, No. 11, 2001, pp. 2015-2040. doi:10.1016/S0378-4266(00)00168-0
25. S. R. Das, "The Surprise Element: Jumps in Interest Rates," Journal of Econometrics, Vol. 106, No. 1, 2002, pp. 27- 65. doi:10.1016/S0304-4076(01)00085-9
26. M. Johannes, "The Statistical and Economic Role of Jumps in Continuous-Time Interest Rate Models," The Journal of Finance, Vol. 59, No. 1, 2004, pp. 227-260.doi:10.1111/j.1540-6321.2004.00632.x
27. G. Chacko and S. Das, "Pricing Interest Rate Derivatives: A General Approach," The Review of Financial Studies, Vol. 15, No. 1, 2002, pp. 195-241.doi: 10.1093/rfs/15.1.195
28. D. R. Beaglehole and M. S. Tenney, "General Solutions of Some Interest Rate-Contingent Claim Pricing Equations," The Journal of Fixed Income, Vol. 1, No. 2, 1991, pp. 69-83. doi:10.3905/jfi.1991.408014
29. F. Jamshidian, "Bond, Futures and Option Valuation in the Quadratic Interest Rate Model," Applied Mathematical Finance, Vol. 3, No. 2, 1996, pp. 93-115.doi:10.1080/13504869600000005
30. N. El Karoui, R. Myneni and R. Viswanathan, "Arbitrage Pricing and Hedging of Interest Rate Claims with State Variables: Theory," University of Paris, Paris, 1992.
31. F. A. Longsta., "A Nonlinear General Equilibrium Model of the Term Structure of Interest Rates," Journal of Financial Economics, Vol. 23, No. 2, 1989, pp. 195-224.doi:10.1016/0304-405X(89)90056-1
32. G. M. Constantinides, "A Theory of the Nominal Term Structure of Interest Rates," The Review of Financial Studies, Vol. 5, No. 4, 1992, pp. 531-552. doi:10.1093/rfs/5.4.531
33. B. Lu, "An Empirical Analysis of the Constantinides Model of the Term Structure," University of Michigan, Ann Arbor, 2000.
34. D. H. Ahn, R. F. Dittmar and A. R. Gallant, "Quadratic Term Structure Models: Theory and Evidence," The Review of Financial Studies, Vol. 15, No. 1, 2002, pp. 243-288. doi:10.1093/rfs/15.1.243
35. M. Leippold and L. Wu, "Asset Pricing under the Quadratic Class," Journal of Financial and Quantitative Analysis, Vol. 37, No. 2, 2002, pp. 271-295. doi:10.2307/3595006
36. G. Jiang and S. Yan, "Linear-Quadratic Term Structure Models—Toward the Understanding of Jumps in Interest Rates," Journal of Banking & Finance, Vol. 33, No. 3, 2009, pp. 473-485. doi:10.1016/j.jbankfin.2008.08.018 [Citation Time(s):12]
37. M. Piazzesi, "Bond Yields and the Federal Reserve," Journal of Political Economy, Vol. 113, No. 2, 2005, pp. 311- 344. doi:10.1086/427466

38. P. Cheng and O. Scaillet, "Linear-Quadratic Jump-Diffusion Modeling," Mathematical Finance, Vol. 17, No. 4, 2007, pp. 575-598. doi:10.1111/j.1467-9965.2007.00316.x
39. D. Duffe, J. Pan and K. Singleton, "Transform Analysis and Asset Pricing for Affine Jump-Diffusions," Econometrica, Vol. 68, No. 6, 2000, pp. 1343-1376. doi:10.1111/1468-0262.00164
40. S. Heston, "A Closed-Form Solution for Options with Stochastic Volatility with Applications to Bond and Currency Options," Review of Financial Studies, Vol. 6, No. 2, 1993, pp. 327-343, 1993. doi:10.1093/rfs/6.2.327
41. P. Collin-Dufresne and R. S. Goldstein, "Pricing Swaptions within an Affine Framework," Journal of Derivatives, Vol. 10, No. 1, 2002, pp. 9-26. doi:10.3905/jod.2002.319187
42. D. Lando, "Credit Risk Modeling: Theory and Applications," Princeton University Press, Princeton, 2004.
43. R. C. Merton, "On the Pricing of Corporate Debt: The Risk Structure of Interest Rates," Journal of Finance, Vol. 29, No. 2, 1974, pp. 449-470. doi:10.2307/2978814
44. T. S. Y. Ho and S. B. Lee, "Term Structure Movements and Pricing Interest Rate Contingent Claims," The Journal of Finance, Vol. 41, No. 5, 1986, pp. 1011-1029.doi:10.2307/2328161
45. Y. Ait-Sahalia, "Testing Continuous-Time Models of the Spot Interest Rate," Review of Financial Studies, Vol. 9, No. 2, 1996, pp. 385-426. doi:10.1093/rfs/9.2.385
46. R. Gallant and G. Tauchen, "Reprojecting Partially Observed Systems with Application to Interest Rate Diffusions," Journal of the American Statistical Association, Vol. 93, No. 441, 1998, pp. 10-24.
47. J. Cochrane, "Asset Pricing," Princeton University Press, Princeton, 2005.
48. B. Flesaker and L. Hughston, "Positive Interest," Risk, Vol. 9, No. 1, 1996, pp. 46-49.
49. M. Muslela and M. Rutkowski, "Martingale Methods in Financial Modelling," Springer-Verlag, Berlin, 2004.
50. M. Rutkowski, "A Note on the Flesaker-Hughston Model of the Term Structure of Interest Rates," Applied Mathematical Finance, Vol. 4, No. 3, 1997, pp. 151-163.doi:10.1080/135048697334782
51. F. Black, E. Derman and W. Toy, "A One-Factor Model of Interest Rates and Its Application to Treasury Bond Options," Financial Analysts Journal, Vol. 46, No. 1, 1990, pp. 33-39. doi:10.2469/faj.v46.n1.33
52. F. Black and P Karasinski, "Bond and Option Pricing When Short Rates Are Lognormal," Financial Analysts Journal, Vol. 47, No. 4, 1991, pp. 52-59. doi:10.2469/faj.v47.n4.52
53. M. Hogan and K. Weintraub, "The Log-Normal Interest Rate Model and Eurodollar Futures," Citibank, New York. 1993.
54. M. Leippold and W. Liuren, "Design and Estimation of Quadratic Term Structure Models," European Finance Review, Vol. 7, No. 1, 2003, pp. 47-73. doi:10.1023/A:1022502724886
55. C. Gourieroux and R. Sufana, "Wishart Quadratic Term Structure Models," University of Toronto, Toronto, 2003.
56. K. Singleton, "Empirical Dynamic Asset Pricing: Model Specification and Econometric Assessment," Princeton University Press, Princeton, 2006.
57. T. Andersen, L. Benzoni and J. Lund, "Stochastic Volatility, Mean Drift, and Jumps in the Short-Term Interest Rate," Northwestern University, Chicago, 2004.
58. H. Farnsworth and R. Bass, "The Term Structure with Semi-Credible Targeting," Journal of Finance, Vol. 58, No. 2, 2003, pp. 839-866. doi:10.1111/1540-6261.00548
59. R Jarrow, H. Li and F Zhao, "Interest Rate Caps 'Smile' Too! But Can the LIBOR Market Models Capture the Smile?" Journal of Finance, Vol. 62, No. 1, 2007, pp. 345-382. doi:10.1111/j.1540-6261.2007.01209.x
60. L. P. Hansen, "Large Sample Properties of Generalized Method of Moments Estimators," Econometrica, Vol. 50, No. 4, 1982, pp. 1029-1054. doi:10.2307/1912775

61. J. Pan, "The Jump-Risk Premia Implicit in Options: Evidence from an Integrated Time-Series Study," Journal of Financial Economics, Vol. 63, No. 1, 2002, pp. 3-50.doi:10.1016/S0304-405X(01)00088-5
62. Lewis, "Option Valuation under Stochastic Volatility: With Mathematica Code," Finance Press, 2000.
63. P. Carr and D. B. Madan, "Option Valuation Using the Fast Fourier transform," Journal of Computational Finance, Vol. 2, No. 4, 1999, pp. 61-73.
64. L. Cao and Z. F. Guo, "A Comparison of Gradient Estimation Techniques for European Call Options," Accounting & Taxation, Vol. 4, No. 1, 2012, pp. 75-82.
65. L. Cao and Z. F. Guo, "Delta Hedging with Deltas from a Geometric Brownian Motion Process," Proceedings of International Conference on Applied Financial Economic, Samos Island, March 2011.

APPENDIX

Derivation of Bond Price for Cox-Ingersoll-Ross Model

Proof. The bond price process is

$$B(t,T)=\mathrm{E}\left[\exp\left\{-\int_t^T r(u)\,\mathrm{d}u\right\}\middle|\mathcal{F}(t)\right].$$

Because

$$\exp\left\{-\int_0^t r(u)\,\mathrm{d}u\right\}B(t,T)=\mathrm{E}\left[\exp\left\{-\int_0^T r(u)\,\mathrm{d}u\right\}\middle|\mathcal{F}(t)\right]$$

the tower property implies that this is a martingale. The Markov property implies that $B(t,T)=B(r(t),t,T)$.

Because $\exp\left\{-\int_0^t r(u)\,\mathrm{d}u\right\}B(r(t),t,T)$ is a martingaleits differential has no drift term. We compute

$$\mathrm{d}\left(\exp\left\{-\int_0^t r(u)\,\mathrm{d}u\right\}B(r(t),t,T)\right)$$

$$=\exp\left\{-\int_0^t r(u)\,\mathrm{d}u\right\}\left(-r(t)B(r(t),t,T)\mathrm{d}t+B_r(r(t),t,T)\mathrm{d}r(t)+\frac{1}{2}B_{rr}(r(t),t,T)\mathrm{d}r(t)+B_t(r(t),t,T)\mathrm{d}t\right)$$

$$=\exp\left\{-\int_0^t r(u)\,\mathrm{d}u\right\}\left(-rB\mathrm{d}t+B_r(\alpha-\beta r)\mathrm{d}t+B_r\sigma\sqrt{r}\mathrm{d}W+\frac{1}{2}B_{rr}\sigma^2 r\mathrm{d}t+B_t\mathrm{d}t\right)$$

Setting the drift term to zero, we obtain the partial differential equation

$$-rB(r,t,T)+B_t(r,t,T)$$
$$+(\alpha-\beta r)B_r(r,t,T)+\frac{1}{2}\sigma^2 rB_{rr}(r,t,T)=0$$

The terminal condition is $B(r,T,T)=1, r\geq 0$. We look for a solution of the form $B(r,t,T)=e^{-rC(t,T)-A(t,T)}$, where $C(T,T)=0, A(T,T)=0$. Then we have $B_t=(-rC_t-A_t)B$, $B_r=-CB$, $B_{rr}=C^2B$, so that the partial differential equation becomes

$$0=rB\left(-1-C_t+\beta C+\frac{1}{2}\sigma^2C^2\right)-B(A_t+\alpha C).$$

The ordinary differential equations become

$$-1-C_t(t,T)+\beta C(t,T)+\frac{1}{2}\sigma^2C^2(t,T)=0,$$

With $C(T,T)=0$, and

$$A(t,T)=\alpha\int_t^T C(u,T)\mathrm{d}u,$$

where $A(T,T)=0$ and $A_t(t,T)=-\alpha C(t,T)$. This is equivalent to solving

$$\frac{\mathrm{d}C(s,T)}{\left(C(s,T)+\frac{\beta-\sqrt{\beta^2+2\sigma^2}}{\sigma^2}\right)\left(C(s,T)+\frac{\beta+\sqrt{\beta^2+2\sigma^2}}{\sigma^2}\right)}$$
$$=\frac{\sigma^2}{2}\mathrm{d}s.$$

Integrating both sides of the above equation with respect to *s*, we obtain

$$\int_t^T\frac{\mathrm{d}C(s,T)}{C(s,T)+\frac{\beta-\sqrt{\beta^2+2\sigma^2}}{\sigma^2}}-\int_t^T\frac{\mathrm{d}C(s,T)}{C(s,T)+\frac{\beta+\sqrt{\beta^2+2\sigma^2}}{\sigma^2}}$$
$$=\sqrt{\beta^2+2\sigma^2}(T-t).$$

Letting $\delta=\sqrt{\beta^2+2\sigma^2}$, we have

$$\frac{\sigma^2 C(t,T)+\beta+\delta}{\sigma^2 C(t,T)+\beta-\delta}=e^{\delta(T-t)}\left(\frac{\beta+\delta}{\beta-\delta}\right)$$

Therefore, $C(t,T)$ is given by

$$(t,T)=\frac{\left(e^{\delta(T-t)}-1\right)(\beta+\delta)}{\sigma^2\left(1-e^{\delta(T-t)}\left(\frac{\beta+\delta}{\beta-\delta}\right)\right)}$$

$$=\frac{2\left(e^{\delta(T-t)}-1\right)}{(\beta+\delta)e^{\delta(T-t)}-(\beta-\delta)} \text{ since } \sigma^2=\frac{\delta^2-\beta^2}{2}$$

$$=\frac{2\left(e^{2\gamma(T-t)}-1\right)}{2\gamma\left(e^{2\gamma(T-t)}+1\right)+\beta\left(e^{2\gamma(T-t))}-1\right)} \text{ by letting } \delta=2\gamma$$

$$=\frac{\sinh(\gamma(T-t))}{\gamma\cosh(\gamma(T-t))+\frac{1}{2}\beta\sinh(\gamma(T-t))}.$$

Derivation of ODEs (4.14)-(4.17).

To arrive at the result (4.14)-(4.17), we need some intermediate expressions for partial derivatives of the zero coupon bond price given by (4.13) with respect to the state vector:

$$\partial_{Y_2} D=(B_2+2CY_2)D$$

$$\partial_{Y_2Y_2'} D=\left\{(B_2+2CY_2)(B_2+2CY_2)'+2C\right\}D$$

$$\partial_{Y_2r} D=B_1(B_2+2CY_2)D$$

$$\partial_{rr} D=B_1^2 D$$

$$\Rightarrow \Omega_{11}\partial_{rr} D=v^2B_1^2 D=Y_2'\begin{pmatrix} B_1^2 & 0 \\ 0 & 0 \end{pmatrix}Y_2 D$$

Insert zero price expression (4.13) in to PDE (4.8) with a=1 and $\boldsymbol{u}=0_{3\times1}$:

$$-\partial_\tau A-\partial_\tau B_1 r-\partial_\tau B_2' Y_2-Y_2'\partial_\tau C Y_2+\begin{pmatrix}\mu_r-\kappa_{rr} r-\mu_{rJ}' Y_2+Y_2'\Psi_1 Y_2\\ \mu_{v\lambda}-\Psi'_2 Y_2\end{pmatrix}'\begin{pmatrix}B_1\\ B_2+2CY_2\end{pmatrix}$$

$$+\frac{1}{2}tr\left\{Y_2'\begin{pmatrix}B_1^2 & 0\\ 0 & 0\end{pmatrix}Y_2+2\Omega_{12}B_1\left(B_2+2CY_2\right)+\Omega_{22}\left[\left(B_2+2CY_2\right)\left(B_2+2CY_2\right)'+2C\right]\right\}+\begin{pmatrix}0\\ \phi_J\left(B_1\right)-1\end{pmatrix}'Y_2-r=0.$$

The above equation must hold for arbitrary values of the state vector. This condition gives rise to a system of four Riccati ODEs for functions $A(?)$, $B_1(?)$, $B_2(?)$ and $C(t)$:

$$-\partial_\tau A+\mu_r B_1+\mu_{v\lambda}' B_2+B_1\Omega_{12}B_2$$
$$+\frac{1}{2}tr\left[\Omega_{22}\left(B_2B_2'+2C\right)\right]=0$$

$$-\partial_\tau B_1-B_1\kappa_{rr}-1=0$$

$$-\partial_\tau B_2-B_1\mu_{rJ}-\Psi_2' B_2+2C\mu_{v\lambda}+2B_1C\Omega_{12}'$$
$$+2C\Omega_{22}B_2+\begin{pmatrix}0\\ \phi_J\left(B_1\right)-1\end{pmatrix}=0$$

$$-\partial_\tau C+B_1\Psi_1'-2\Psi_2' C+2C\Omega_{22}C+\begin{pmatrix}\frac{B_1^2}{2} & 0\\ 0 & 0\end{pmatrix}=0$$

The system is subject to initial conditions (4.18)-(4.19).

NOTES

[1]Examples include the original one-factor model [1], extended Vasicek ([2,3]), one-factor Cox-Ingersoll-Ross models ([2,4-7]), two-factor CoxIngersoll-Ross models ([2,8,9]), three-factor models ([10-12]). For a more detailed list of models we refer the reader to [13].

[2]The number of models in this class is vast. Examples include [15-23], [19-21,24-26] provide evidence that jumps are essential in modeling interest rate distribution. [27] provide a general approach to interest rate derivative pricing for exponential affine jump-diffusion models of interest rates.

[3][27,39-42] used the extended transform to price derivatives in closedform using inverse Fourier transform.

[4]This expression holds even if the terminal time, T, is random. In this case, the formula is known as Dynkin's formula.

[5]In (2.9), the expectation is usually computed under the actual measure. In practice, the actual moments of the state variables are used instead of the risk neutral ones.

[6]Jump-diffusion models of interest rates are numerous. e.g., [15] generalise the equilibrium model of CIR by including jumps in the dynamics of underlying state variables. [21,22], and [18] augment the Vasicek model with jumps in the short rate. [39] consider a general class of affine jump-diffusion models.

[7]We assume here that the process is already under EMM.
[8][18] argue that approximation is reasonable because jump sizes are typically small so that JC is expected to be small.
[9][48] describes interest rate models based on a price kernel that guarantees positive rates. [49,50] provide a more detailed and general description of the model. For more detailed discussion of these and other models see [13].
[10]The models in this class have been studied by [28-30], among others.
[11]See [56] for a general specification and a discussion of canonical form of a QG model.
[12][36] assume that the jump size distribution is Bernoulli taking a positive value $\mu_+ > 0$ with probability p and a negative value $\mu_- < 0$ with probability $1 - p$. We do not make any specific assumption on the jump size distribution and keep the discussion general.

CHAPTER 3

Management Earnings Forecasts and IPO Performance: Evidence of a Regime Change

Sabri Boubaker[1,2] Dimitrios Gounopoulos[3] Antonios Kallias[1] Konstantinos Kallias[4]

[1] *Champagne School of Management, Groupe ESC Troyes, Troyes, France*

[2] *IRG, Universite´ Paris Est, Cre´teil, France*

[3] *Newcastle University Business School, Newcastle University, Newcastle Upon Tyne, UK*

[4] *Portsmouth Business School, University of Portsmouth, Portsmouth, UK*

ABSTRACT

Companies undertaking initial public offerings (IPOs) in Greece were obliged to include next-year profit forecast in their prospectuses, until the regulation changed in 2001 to voluntary forecasting. Drawing evidence from IPOs issued in the period 1993–2015, this is the first study to investigate the effect of disclosure regime on management earnings forecasts and IPO long-term performance. The findings show mainly positive forecast errors (forecasts are lower than actual earnings) and higher long-term returns during the mandatory period, suggesting that the mandatory disclosure requirement causes issuers to systematically bias profit forecasts downwards as they opt for the safety of accounting conservatism. The mandatory disclosure requirement artificially improves IPO share performance. Overall, our results show that mandatory disclosure of earnings forecasts can impede capital market efficiency once it goes beyond historical financial information to involve compulsory projections of future performance.

Keywords: Management earnings forecasts Mandatory disclosure Voluntary disclosure Forecast accuracy IPOs Greece

1 INTRODUCTION

An initial public offering (IPO) brings a firm's shares to the public for the first time and it anticipates mass participation during the subscription period. Public interest is highly dependent on evaluation of available information from the newly issued prospectus, since this information is often the first window on the firm's past and its projected future performance. Few investors have the luxury of time to read the entirety of this official document, whereas others prefer to focus on key points disclosed in the prospectus. One figure that attracts special attention is the management earnings forecast.[1] Investors assess the accuracy of earnings forecasts, as they can serve as a performance indicator for the newly listed firm.

Accuracy in future earnings forecasts is difficult to achieve, as there are many unpredictable events that can take place between the forecast day and the day of the official announcement of actual earnings. Such events can include market instability, unforeseen political events, exchange rate fluctuations, liquidity problems, and misunderstanding of market trends. Despite efforts by senior management of many newly listed IPOs to provide accurate earnings forecasts, few firms succeed. Many firms end up with a large difference between forecast earnings and actual results. This forecast error is dualistic in nature. It can be optimistic if the earnings forecast in the prospectus is higher than the actual earnings presented in the first company annual report, or pessimistic if the earnings forecast appears to be less than actual earnings. The nature of the difference between actual and forecast earnings may lead to entirely different implications in the long term.

Motivated by the importance of earnings forecasts and their tendency to signal future performance, the aim of this study is to assess the usefulness of management earnings forecast disclosure to investors. Previous literature concentrates on the impact of various accounting decisions on long-term returns (Mitchell and Stafford 2000; Yi 2001; Liao et al. 2015). For example, the literature on public equity offerings focuses on the effect of discretionary accruals on the long-term performance of these offerings (Teoh et al. 1998a). In a related study, Chou et al. (2010) show how earnings management practices affect private placements of equity. Profit forecasts may include unwarranted optimism when managers exercise their discretion in an abusive way (Jaggi et al. 2006). Additional research draws a link between earnings forecast direction (i.e., positive or negative compared to actual earnings) and a number of post-IPO dimensions, such as firm organization and changes in competitive strategy of the firm (e.g., Bhabra and Pettway 2003).

This paper builds on the management earnings forecast literature in that it analyzes the effect that an exogenous change in IPO disclosure environment can have on the reliability of available information and investor behavior. Motivated by heavy market reliance on initial forecasts during IPOs (Dechow et al. 2000), we investigate the use of this mechanism, which is supposed to cause less friction in an IPO and to facilitate efficient allocation of capital. Basic questions that this paper addresses include: Does a flexible disclosure status brought about by regulatory change (mandatory to voluntary) contribute to increased forecast accuracy? Can earnings forecasts provided in IPO prospectuses constitute an important determinant of future market performance? Do investors penalize inaccurate forecasts by IPO firms with strong negative returns?

This paper differs from the existing literature on the effect of earnings forecasts on IPO performance in several ways. First, it focuses on management errors by contrasting forecast earnings in the prospectus with actual earnings as reported in the first annual report. The difference between the two figures constitutes forecast error; this is the central focus of this study. Previous studies do not explore this error, but compare earnings published before going public with earnings presented in annual reports up to three years after going public. However, the magnitude and sign of management forecast errors constitute valuable information. If the initial forecast is optimistic, realized earnings will come as a disappointment to investors and the market will correct earlier overvaluations (hence, lower long-run performance). Inversely, pessimistic forecasts should pave the way for higher future returns.

Second, this study analyzes management earnings forecasts under two different regulatory regimes (mandatory and voluntary) within the same market (Greece). This gives us the opportunity to conduct an experiment with direct comparison between two samples, one in which management is required to provide earnings forecasts in the prospectus (mandatory disclosure), and a second in which management has the choice of providing earnings forecasts (voluntary disclosure). The results shed light on which disclosure mechanism better serves the interests of investors, and will assist regulators in formulating decisions that are aligned with well-functioning capital markets.

The empirical evidence indicates the mandatory disclosure of earnings forecasts is associated with the production of inaccurate management earnings forecast. Moreover, the three-year returns (starting with the end of the first day of trading) become significantly more negative in the voluntary period compared to the mandatory disclosure period. By and large, IPOs that do not disclose earnings forecasts experience considerably inferior long-term returns.[2] Moreover, IPO firms with high positive forecast errors (i.e., actual profits exceeding forecast profits) are associated with lower returns. In sum, the evidence shows considerable accounting conservatism during the mandatory period as a form of insurance on the issuer side.

Overall, the findings have important policy implications, since this is, to the best of our knowledge, the first study to analyze the effect of the transition to a voluntary disclosure regime on earnings forecast accuracy and long-term performance. This regime change provides an opportunity for policymakers to observe the reaction of the market and for researchers to explore whether firms signal their own quality through earnings forecasts. In addition, the findings have major implications on how financial markets create managerial incentives to reduce forecast errors and to avoid costs associated with potential legal actions by shareholders when reported earnings diverge considerably from earnings forecasts (e.g., Skinner 1994; Frankel et al. 1995; Teoh et al. 1998b; Frankel et al. 2002; Baginski et al. 2002; Karamanou and Vafeas 2005; Jaggi et al. 2006; Shen et al. 2014).

The remainder of this paper proceeds as follows. Section 2 presents a literature review and develops the research hypotheses. Section 3 presents an overview of the institutional background. Section 4 describes the data and presents the methodology. Section 5 presents the descriptive statistics and interprets the empirical results. We test the robustness of our results in Sect. 6. Finally, Sect. 7 sets forth our conclusions.

2 LITERATURE REVIEW

2.1 Theoretical framework

Numerous studies examine post-issue market performance of IPOs (e.g., Ritter 1991; Chen et al. 2000; Dimovski and Brooks 2004). Gajewski and Gresse (2006), in their survey of European IPO markets, report significant underperformance of newly listed firms at the three-year horizon in all countries, except Greece and Portugal. Thomadakis et al. (2012) confirm earlier results by reporting that the Greek case differs from international evidence. The authors reveal that long-term outperformance continues for a long period after listing. The unique case of Greek IPOs represents an interesting setting to explore post-issue performance in more depth and in light of management earnings forecasts.

Theories such as window-dressing, market timing,[3] market optimism,[4] price support,[5] and investor sentiment explain long-term underperformance of firms going public. Some of these theories, like window-dressing and investor sentiment, may explain both IPO long-run underperformance and management earnings forecasts. According to Teoh et al. (1998a), the window-dressing theory supports the view that managers are ready to manipulate reported earnings around IPOs to give the stock market a false signal on the future profitability of the candidate firm, thus reducing the cost of capital. These authors observe that the more earnings management there is around IPOs, the more stock price performance decreases after three years.[6] Investor sentiment theories focus on the impact of "irrational" or "sentiment" investors on stock prices. In the case of IPOs, this potential impact appears strong because IPO firms are more likely to suffer from information opacity, and hence are more difficult to value (Ljungqvist 2007). Ljungqvist et al. (2006) model an IPO firm's response to the presence of sentiment investors. These authors assume that some sentiment investors are optimistic on the future prospects of the IPO firm. The objective of the issuer is to capture a bigger portion of the "surplus" attributable to the presence of sentiment investors, resulting in offer price maximization relative to the intrinsic value of the stock. Nevertheless, regulatory restrictions provide a low level of flexibility to issuers.

2.2 The literature on management earnings forecasts and long-run performance

Management earnings forecasts have received much research attention, because under specific circumstances, this information can signal the direction of long-term performance of IPOs. There are a number of studies that shed light on inefficiencies in the IPO market. For instance, Hughes (1986) and Guo (2005) highlight the importance of the information asymmetry between investors and the issuer of an IPO.[7] Li and McConomy (2004) show that retained ownership and management earnings forecasts are credible, value-relevant signals. However, in contrast to the Greek IPO market, forecasts in Canada never became compulsory. Jelic et al. (2001) are the first to explore the link between management forecast errors and long-term performance of IPOs. Their results suggest positive and statistically significant long-term returns up to three years after listing, which contradicts the consensus in the IPO literature for significant, negative long-term performance. Further, their evidence shows a negative association between upward bias in management earnings forecasts and IPO performance in the first 12 months following the IPO. Hensler et al. (1997) and Jain and Kini (2000) explain the usefulness of IPO prospectus data to potential investors. Their results indicate that prospectus information remains useful in the aftermarket over the short window of one year, although the value of this information declines rapidly with time. While the prospectus clues that were considered by these researchers comprise a variety of firm- and IPO-specific features, management projections and forward-looking information are discarded.

From a different point of view, Bhabra and Pettway (2003) argue that the reason for the lack of underperformance in the first year post-IPO is the price support provided by underwriters in the immediate aftermarket, leaving no room for behavioral interpretations. Jaggi et al. (2006) show that mandatory disclosure of earnings forecasts in Taiwanese IPOs results in more optimistic forecasts than pessimistic forecasts, especially for firms expecting better performance in the forecast year compared to the previous year. Overall, the findings from Malaysia (Jelic et al. 1998) and Singapore (Firth et al. 1995) suggest that managers tend to be pessimistic forecasters, whereas findings from Canada (Pedwell et al. 1994), New Zealand (Mak 1989; Firth and Smith 1992), and Australia (Lee et al. 1993; Firth et al. 2012) suggest that managers tend to overestimate future earnings in their forecasts. Evidence for Hong Kong is inconclusive. Selva et al. (1994) report optimistic forecasts, while Chan et al. (1996) report that management forecasts are conservative and pessimistic.

More recently, Gong et al. (2009) find a positive association between management earnings forecasts and long-term performance for US IPO firms, which is stronger for firms operating in a more uncertain business environment. These authors identify the length of the firm's operating cycle, the volatility of turnover growth, and operating cash flow as the causes of uncertainty. These findings motivate us to study Greek IPO firms, which operate in an uncertain business environment, and to consider the transition from a mandatory to a voluntary disclosure regime. The regulatory change that took place early in the 2000s provides the unique setting we need to examine the link between management earnings forecasts and long-term performance of IPO firms.

2.3 Hypothesis development

Greece was the first country in the world to change its mandatory IPO forecasting regime to a voluntary one.[8] These two disclosure regimes embody entirely different philosophies and follow distinct principles, which affect the behavior of management during the IPO decision-making process. The overall Greek framework offers a unique opportunity to explore earnings forecasts prior to and after the regulatory change.

Prior literature documents a mandatory disclosure environment for earnings forecasts in Singapore, Malaysia, and New Zealand, and a voluntary one in other Commonwealth countries (Firth and Smith 1992; Firth et al. 1995; Hartnett and Romcke 2000; Chen et al. 2001; Li and McConomy 2004; Jelic 2011). The results on earnings forecasts reveal low levels of error in countries with voluntary disclosure (Australia: 34.49 %, Hong Kong: 12.79 %, UK: 11 %) compared with countries with mandatory earnings forecasts (Malaysia: 54.1 % and New Zealand: 111 %). This initial observation provides a first indication that moving from a mandatory disclosure environment to a voluntary one improves the accuracy of earnings forecasts.

Hutton et al. (2003) document that good-news forecasts are informative only when accompanied with verifiable forward-looking information that improves their credibility, whereas bad-news earnings forecasts are always informative. Huang et al. (2014) examine the efficacy of a forecast regulation in China and report that voluntary forecasts are timelier and more precise than mandatory forecasts, suggesting that voluntary forecasts are of higher quality than mandatory forecasts. In the same vein, Horton et al. (2013) report that the quality of the information environment, proxied by forecast accuracy, significantly increased following mandatory adoption of International Financial Reporting Standards.

In addition to this important international evidence in support of a voluntary regime, the environment of acute uncertainty characterizing an IPO incentivizes managers to act in ways that insulate them from any negative implications (e.g., reputational damage or litigation costs) in the post-IPO period. Consequently, if all issuers are required to disclose a forecast, then the average forecast is likely to be pessimistic, reflecting the general conservatism in estimates. On the basis of these arguments, we propose the following hypothesis:

H_1 **Regulatory change from mandatory to voluntary earnings forecasts increases the accuracy of IPO earnings forecasts.**

Degeorge and Zeckhauser (1993) argue that companies prefer to go public after unusually high earnings performance. They contend that superior earnings reduce the risk faced by investors and secure wide participation (see, also, Brav and Gompers 1997). Clearly, however, not all issuers are capable of deferring an IPO until their accounting bottom line is deemed satisfactory. For example, smaller firms or those belonging to technology industries condition their profitability on the proceeds to be raised so that positive earnings follow a successful listing and not the other way around. Forecasts of next-year profits could be important in these situations. Degeorge and Derrien (2001a) examine earnings forecasts published in IPO prospectuses and report that forecast error is the main driver of IPO stock price performance, as it embodies investor expectations at the time of the IPO. Gounopoulos et al. (2015) document that IPO firms with accurate forecasts experience lower levels of underpricing than those that are unable to provide accurate earnings forecasts. However, forecasts can hardly represent a global response to the problem of ex-ante uncertainty, given that the forecast horizon is confined to the next year's profits. In fact, in the eyes of investors, IPO firms are likely to remain question marks until a sequence of future reporting periods. Consequently, resolution of such uncertainty should more apparently be reflected in long-term rather than first-day or other short-term returns.

Our focus during the post-issue period is also on firms that include a forecast in their prospectus and fail to meet earnings expectations. Such firms, by intentionally inflating their earnings forecasts, may obtain a higher share price on the IPO date. If there were no subsequent costs to managers of such firms, more firms would prefer to issue optimistic earnings forecasts. However, taking into account the risk of subsequent legal actions, the incentives for accounting conservatism through production of pessimistic forecasts are substantial. In addition, IPO firms that do not meet earnings expectations pay a price in the form of significantly lower post-issue performance than forecasters that meet their earnings forecasts (Jog and McConomy 2003). This line of reasoning leads to the following testable hypothesis:

H_2 **IPO firms that provide accurate earnings forecasts exhibit significantly higher post-issue long-term performance.**

3 INSTITUTIONAL BACKGROUND

3.1 Regulatory switch

By 2000, Greece appeared to be on track to qualify for the first eurozone enlargement and join the pioneer 11 Member States, which had already replaced their national currencies with the euro. Key to Greece's candidacy has been the work of the Hellenic Capital Market Commission to identify possibilities for convergence with the institutional and regulatory framework of Greece's eurozone partners. In Greece's market for new equities, a great deal of friction resulted from a regulation forcing issuers to forecast profitability and disclose estimates in the IPO prospectus. This requirement caused much discontent among prospective IPO issuers, due to its significant compliance costs as well as a dearth of operational experience, which often resulted in highly precarious projections.

The subsequent switch to a voluntary disclosure environment came as a long-anticipated remedy to the above concerns. Given the limited depth of the Athens Stock Exchange (ASE) in terms of member companies and trading volumes, this regulatory flexibility was implemented to foster participation by eliminating some of the listing barriers. It was also intended to convey in an unambiguous manner the political will to implement structural reforms in the interest of market efficiency and harmonization.

In line with the economic and political objectives, the transition from mandatory to voluntary disclosure of management earnings forecasts marked a stepping stone toward the adoption of the more comprehensive International Accounting Standards (IAS). IAS assesses regulations on the basis of three criteria: first, that it fulfills the essential requirements imposed by the directives of the Council of the European Union (i.e., implementation of the regulation is conducive to a fair and accurate representation of a firm's financial standing and performance); second, that, in the spirit of the 17 July 2000 Council decisions, it advances the European public interest; and third, that it adheres to quality standards that considers financial statement information both relevant and functional to end users. Maintenance of the status quo, that is, systematic production of biased earnings forecasts under a mandatory regime, would plausibly alienate the Greek capital market from IAS philosophy and principles. This was a discrepancy that Greek authorities were quick to understand.

3.2 Development of the IPO market in Greece

The IPO market in Greece experienced three historical periods of major development. The first is the period 1925–1926 with 9 IPOs, the second includes 1972–1974 with 32 IPOs, and the third is the period 1993–2001 with 243 IPOs. Each of these periods has unique characteristics and is associated with growth in the Greek economy. Specifically, Thomadakis et al. (2016) show that from 1924 to 1940, the ASE experienced a boom and an unprecedented record of new listings. In total, 71 companies were admitted to trading in that period,[9] as a result of the increasing public spending (including foreign aid) on refugee assistance from Asia Minor, which created high domestic demand for food, clothing, and housing. The pace of growth picked up; 1924–1928 was a distinct period of rapid economic development in Modern Greek history. Inflation was also high, but was, on the whole, much lower than in the preceding decade, averaging an annual rate of 13 % in the period 1924–1927.

The second IPO wave was a direct result of the development of the Greek economy during the 1972–1974 period. This was an era of GDP growth at an average annual rate of 6.2 % in an environment of monetary stability, while maintaining improved national production, employment, and exports, whose composition shifted to manufacturing, brought major increases in financial saving, and low inflation. This period of prosperity was interrupted by the first energy crisis and the simultaneous political (and geopolitical) crisis of Cyprus in 1974.

The third period is the most prominent and played a critical role in the rapid growth of the IPO market in Greece. In particular, 1997 was a very significant year, a landmark in the history of the Greek economy and in the history of the ASE. In the 1997–1999 period, the ASE witnessed its greatest phase of growth. Overall, the period 1993–2001 was characterized by readjustment of macroeconomic indicators, with the main goal being maintenance of the inflation rate lower than 3 % and reduction of the fiscal deficit. By the end of 2000, the Greek economy had transformed into a "modern" economy, with an updated structure and strong dynamism. Healthy conditions were present in the economy in the 1997–1999 period, as economic growth, monetary stability, investment in infrastructure, growth in exports, and reform of the business sector motivated many companies to seek higher growth rates through IPOs. As a result, a record 31.68 billion euros was raised through new public offerings during this period.

4 DATA AND METHODOLOGY

4.1 Sample

The study sample covers the period January 1993 to December 2015 and consists of 303 Greek IPOs. Companies that operate as closed-end funds are excluded from the sample, because they are subject to different regulations and present unique characteristics. A strong effort was made to collect data for all remaining firms listed on the ASE. Toward that end, a variety of sources were used to gather the needed information, including Compust at, Data stream, and Thomson Financial Securities Data Corporation (SDC). To be included in the sample, a firm must have a prospectus with required financial statements of at least one year prior to going public, data on earnings forecasts, and stock prices for at least 36 months (or 756 trading days) following the IPO.

Forecast earnings for one year after going public are manually retrieved from the IPO prospectuses and crosschecked with the official statistical "Bulletin of the ASE." Actual earnings for years $t + 1$, $t + 2$, and $t + 3$ are collected from the respective annual reports and crosschecked with data provided by the Capital Market Commission.[10] Market returns and stock prices are collected from Data stream, while ASEGI (ASE General Index) is retrieved from the ASE database. Necessary adjustments are made for dividends and stock splits. Financial data are derived from published financial statements at the end of fiscal year $t - 1$. Other variables used in the analysis are hand-collected from IPO prospectuses.

4.2 Methodology

The forecast error measure is the difference between actual and predicted earnings for a given period. It can be calculated with the error sign (forecast error) or without (absolute forecast error). Jaggi (1997), Cheng and Firth (2000), and Chen et al. (2001), among others, consider that the signed *FE* captures the bias in the forecast, as it indicates whether the manager has been optimistic or pessimistic in forecasting. Prior literature addresses this issue and partially explains why earnings forecasts are generally over-optimistic. A positive value for the signed forecast error implies that, on average, IPO firms have a pessimistic bias, while a negative value for the signed forecast error represents an optimistic bias.

Management forecast error for an IPO firm is calculated as

$$FE_{it} = (AP_{it} - FP_{it}) / |FP_{it}|, \tag{1}$$

where FE_{it} is forecast error for company i at date t, AP_{it} stands for actual profit, and FP_{it} is forecast profit.

Average forecast error using the signed *FE* measure does not give credible information on the average size of the error, since positive and negative errors cancel each other out. The absolute forecast error (*AFE*) is, therefore, a more appropriate measure of the accuracy of forecasts. *AFE* represents the magnitude of the error, while average *FE* (inclusive of sign) measures the bias in forecasts. *AFE* provides an indication of how close forecasts are to actual profits in absolute terms. *AFE* is measured as follows:

$$AFE_{it} = |(AP_{it} - FP_{it})| / |FP_{it}|, \tag{2}$$

where *AFE* is absolute forecast error.

We consider the cumulative abnormal returns (*CAR*) from the end of the first day of trading until the earlier of the stock's delisting date or its third anniversary as a measure of the long-run performance. The measure follows the method of Ritter (1991). We use the general market index to adjust stock returns on a monthly basis. Monthly market-adjusted returns are calculated as monthly raw returns on a stock over the monthly market return for the corresponding period. The market-adjusted return for stock *i* in event month *t* is given as

$$Adjusted\ Return_{it}(ar_{it}) = Monthly\ raw\ return_{it}(r_{it}) - Monthly\ market\ return_{it}(r_{mt}) \quad ,(3)$$

The average adjusted return on a portfolio of *n* stocks for event month *t* is the equally-weighted arithmetic average of the market-adjusted returns:

$$AR_t = \frac{1}{n}\sum_{i=1}^{n} ar_{it} \quad (4)$$

The cumulative market-adjusted aftermarket return (*CAR*) from event month *q* to event month *s* is the sum of the average market-adjusted returns over this period:

$$CAR_{q,s} = \sum_{t=q}^{s} AR_t \quad (5)$$

Abnormal returns depend on the benchmark used. To calculate the abnormal return we use the Fama and French (1993) model. Fama and French (1993)[11] show that when the standard three-factor model (without the momentum factor) is estimated in randomly chosen sample firms with small size and low book-to-market ratio, the null hypothesis of zero abnormal performance is over-rejected. The majority of our sample consists of small and growth firms, thus this potential problem can be particularly severe. Mitchell and Stafford (2000) also raise the possibility that the intercept under the null hypothesis may be biased under the standard calendar-time approach. We examine the performance of these monthly portfolios by calculating the subsequent excess return and running the following calendar-time regression. Under this model, abnormal returns are calculated using the Fama and French (1993) three-factor model as follows:

$$(R_{pt} - R_{ft}) = a + \beta(R_{mt} - R_{ft}) + \gamma SMB_t + \delta HML_t + \varepsilon_{pt} \quad (6)$$

where R_{pt} is the calendar-time portfolio return, R_{ft} is the one-month Treasury-bill yield during month *t*, R_{mt} is the value-weighted market index return, SMB_t is the difference in returns of value-weighted portfolios of small firms and large firms during month t, HML_t is the return differential of value-weighted portfolios of high and low book-to-market firms in month t, β, γ, and δ are regression parameters specific to the portfolio, and ε_{pt} is the error term. We construct the *SMB* and *HML* factors as in Fama and French (1993).

4.3 Control variables

There is some evidence on factors affecting the quality of earnings forecasts; however there is thus far no study that associates this observed accuracy with the long-term performance of IPO firms. Addressing this gap, we identify a number of IPO- and firm-specific variables that can plausibly offer incremental explanatory power to the reliability of managerial predictions.

Size: Size, commonly proxied by market capitalization, has been extensively used as a determinant of the quality of management earnings forecasts and firm long-run performance.[12]In terms of earnings forecast, it has been argued that large companies are able to use the best expertise and modern, sophisticated forecasting techniques to attain more accuracy (e.g., Eddy and Seifert 1992; Firth and Smith 1992; Mak 1994). Some authors also argue that large companies may even have greater control over their market settings and they might have gained comparative economies of scale, making them less susceptible to economic fluctuations (e.g., Firth et al. 1995). Taken together, these arguments converge in the direction of enhanced accuracy for larger corporate issuers.

Age: The operating history of a firm prior to going public plays a significant role in its long-run stock price performance and in the accuracy of its forecasts. Firm age has been suggested as a proxy for the risk level of the IPO firm (Ritter 1984; Carter et al. 1998; Goergen et al. 2007). Firms with a long history are shown to provide more detailed management earnings forecast disclosure and more accurate forecasts about their future performance, as they are likely to have better control over their operations and a better feeling for the market environment (e.g., Firth and Smith 1992; Cazavan-Jeny and Jeanjean 2007). This is expected to affect their future performance and to offer better returns to their long-term investors. *Age* is measured as the number of years starting from the year of establishment until the year of going public.

Timelag: Time lag is the number of days from prospectus registration to listing date. In many countries, time lag is short; however, it generally ranges from 5 to 70 days. During this period, changes in market conditions may affect the price performance of IPO firms, as well as aftermarket returns. Many studies (e.g., Lee et al. 1993; Firth et al. 1995; Jelic et al. 1998; Hartnett and Romcke 2000; Lonkani and Firth 2005) support the notion that the length of the forecasting period influences the degree of forecast accuracy. Forecasts with a shorter time horizon are expected to be more accurate and to have better long-term performance, while those with longer time horizons are associated with greater uncertainty and less satisfactory returns in the long term (Jaggi 1997).

Privatization: Megginson et al. (2000) examine long-run aftermarket performance for a 36-country sample of 264 privatization IPOs for the period 1981–1997 and show a positive performance irrespective of the considered period (one, three, and five years). Brav and Gompers (1997) show empirically decreasing long-term underperformance of privatized IPO firms. It is expected that these firms are publicly recognizable and that any government, as the principal shareholder, would like to see a successful issue and, at the same time, secure positive feedback from the markets and the media. Because of the expected high public participation, management makes an effort to provide an accurate earnings forecast, which enhances the reputation of the firm and helps secure good long-term returns. We include in the model a dummy variable, *Privatization*, which is set to 1 if the firm is a privatized entity, and zero otherwise.

Oversubscription: The degree of forecast accuracy is likely to decrease with the level of oversubscription. It is expected that uninformed investors create high demand for highly underpriced, low-quality IPOs associated with high forecast error. These IPOs become part of the speculative attitude by well-informed investors in the aftermarket and experience amplified flipping activity. This behavior is well explained by the speculative bubble hypothesis, which is an alternative approach toward explaining the post-listing return behavior of IPO firms. When investors do not behave rationally, they may over- or under-react to information about IPO prospects and they may temporarily over- or undervalue the price of the initial offerings. As a result, investor demand will be high in the short-run, leading to poor long-term performance.

Underwriter reputation[13]: Most theoretical studies regarding long-term performance of IPOs and the role of underwriters' during the going-public period concentrate on asymmetric information and its effect on stock price performance. Baron (1982) and Dimovski et al. (2011) argue that asymmetric information exists between the well-informed underwriters and the less-informed issuers; therefore, underwriters are able to price new issues below the market equilibrium to reduce the probability that they will absorb losses due to unsold shares. Chen and Firth (1999) report a positive relation between underwriter reputation and accuracy of earnings forecasts. Ghikas et al. (2008) find that underwriters align their interests with those of the pre-IPO stockholders, affecting the quality of the earnings forecast by not incorporating adverse information provided by auditing of qualified reports. Overall, the aim of reputable underwriters is to maintain their reputation by associating themselves with more accurate disclosure information and better long-term returns to their loyal customers and retail investors.

Market conditions (*Hot/Cold*): This study uses a dummy variable to account for market conditions. This dummy variable takes the value of 1 for IPO firms that are listed during a hot period, and zero for those listed during a cold period. The "hot period" and "cold period" classifications are based on the intensity of IPO listing activity. The rationale underlying this control variable is that if an earnings forecast takes place during a hot period (i.e., periods with large numbers of listings on the ASE) and the actual earnings appear in the cold period, then the results can be lower than expected (optimistic forecast), leading to a high forecast error. Conversely, if an earnings forecast occurs during a cold period and the actual earnings are announced during a hot period, then the actual earnings can be much higher than expected (pessimistic forecast), leading to a high forecast error. Thomadakis et al. (2012) show a strong relation between cold IPO periods and long-term performance. These authors point out that some IPOs are issued in hot markets, when long-term expectations are low and the general level of the stock market is decreasing, while other issues take place during cold periods, when long-term return projections are promising.

Given ownership: The monitoring of managerial actions, such as management earnings forecasts, depends to a large extent on firm ownership structure. The literature indicates that a high (low) percentage of retained (given) ownership signals high firm quality (e.g., Li and McConomy 2004).[14] Managements of such IPO firms believe in the strength and potential of their companies and have strong motivation to provide accurate earnings forecasts. Accuracy is appreciated by long-term investors and encourages them to keep their position in the firm, supporting its investment plans. This creates better conditions for good long-term returns. The evidence from Hansen and Torregrosa (1992) indicates a direct relationship between the proportion of shares retained by owners and long-term performance of IPO firms. Jain and Kini (1994) argue that long-run underperformance can be partly explained by low managerial incentives following an IPO.

4.4 Description of the model

AFE and FE vary across companies. We construct cross-sectional models to explain these variations and regress AFE and FE on macroeconomic and firm-specific factors as follows:

$$AFE \text{ or } FE = a_0 + a_1 Size + a_2 Age + a_3 TimeLag + a_4 Privatization + a_5 Oversubscription + a_6 Underwriter + a_7 HotCold + a_8 Ownership + a_9 Industry + a_{10} Mandatory + \varepsilon \quad (7)$$

Negative signs are expected on size (*Size*), age (*Age*), privatized firms (*Privatization*), level of oversubscription (*Oversubscription*), and underwriter quality (*Underwriter*); plausibly, these factors as proxies of an issuer's quality and perceived prestige result in the production of more accurate forecasts. Positive signs with AFE are expected for those factors that are likely to instill greater uncertainty in IPO investors, such as length of the listing period (*TimeLag*), market conditions (*HotCold*), given ownership (*Ownership*), and industrial classification (*Industry*).

Further, we test for long-term performance using the following model:

Long –term performance=

$$a_0 + a_1 Size + a_2 Age + a_3 TimeLag + a_4 Privatization + a_5 Oversubscription + a_6 Underwriter + a_7 HotCold + a_8 Ownership + a_9 Industry + a_{10} Mandatory + \varepsilon \quad (8)$$

5 EMPIRICAL RESULTS ON PROFIT FORECASTS

5.1 Descriptive statistics

Panel A of Table 1 provides an annualized listing of the 303 Greek IPOs that have been listed on the ASE since 1993. It also classifies earnings forecasts into mandatory and voluntary disclosure. Among the 83 IPOs listed during the voluntary disclosure period, 22 did not include earnings forecasts in their prospectuses. Overall, the sample of IPOs listed under the voluntary disclosure environment is relatively small, as the Greek debt crisis of 2009 to 2015 has affected the number of new listings. It is noteworthy that there have been no IPOs in Greece in the last five years of the study period. The last time the ASE experienced three consecutive years without an IPO was in the 1980s.

Panel B of Table 1 reports average values for market capitalization, firm age, oversubscription, and given ownership for each year in the sample period. Further, it partitions the sample based on whether earnings forecasts are optimistic or pessimistic. The optimistic (pessimistic) group includes firms with high (low) forecasts compared to the actual earnings realized after the IPO. The optimistic sample typically comprises larger firms (mean market capitalization is € 62.06 million versus € 26.05 million for the pessimistic sample). The optimistic (pessimistic) group is also associated with a more (less) appeal to IPO investors, evidenced by an

Table 1 Sample description

Year	IPO firms	Forecasted profits		
	Full sample	Mandatory	Voluntary	
			Provide a forecast	No forecast
Panel A: Breakdown of Greek IPOs depending on the disclosure regime				
1993	10	10		
1994	45	45		
1995	19	19		
1996	20	20		
1997	14	14		
1998	23	23		
1999	38	38		
2000	51	51		
2001	23		19	4
2002	21		15	6
2003	16		11	5
2004	11		8	3
2005	6		2	4
2006	2		2	
2007	3		3	
2008	0		0	
2009	1		1	
2010–2015	0		0	
Total	303	220	61	22

Year	Market Capitalisation (million Euros)		*Age (years)*		*Oversubscription (times)*		*Ownership (%)*	
	Optimistic	Pessimistic	Optimistic	Pessimistic	Optimistic	Pessimistic	Optimistic	Pessimistic
Panel B: Descriptive statistics depending on earnings forecasts (optimistic versus pessimistic)								
1993	22.32	9.80	19.5	12.00	13.80	17.50	23.27	22.30
1994	5.98	6.63	19.36	16.21	95.41	123.27	21.64	22.09
1995	3.34	4.02	15.84	19.00	26.79	6.56	20.21	29.96
1996	36.80	3.92	17.25	19.00	38.69	32.59	14.88	20.78
1997	3.67	2.33	13.12	12.80	39.19	36.70	16.97	19.62
1998	63.03	19.38	19.53	22.50	201.55	122.85	19.88	16.11
1999	28.29	35.15	17.33	25.81	279.81	215.46	16.33	21.12
2000	72.54	43.79	16.66	25.32	169.29	87.69	16.73	16.77
2001	123.20	63.03	26.13	21.13	2.66	7.23	31.13	20.45
2002	4.73	6.81	11.83	12.88	40.21	22.83	20.83	19.33
2003	7.57	3.39	12.60	11.66	3.30	12.07	19.60	20.83
2004	11.82	9.81	23.00	16.00	6.86	6.86	20.31	20.31
2005	–	7.37	–	4.00	–	3.67	–	15.33
2006	238.34	–	20.66	–	7.47	–	27.77	–
2007	1085.70	330.63	45.66	13.00	4.53	6.13	33.16	22.92
2008–15	74.13	–	10.00	–	2.13	–	29.13	–
Average	**62.06**	**26.05**	**18.20**	**19.55**	**110.50**	**79.80**	**19.74**	**20.67**
Δ mean (*p* value)	0.01***		0.15		0.03**		0.23	

This table reports descriptive statistics for sample IPOs. Panel A breaks down Greek IPOs depending on the disclosure regime. Panel B divides the sample based on earnings forecasts (optimistic versus pessimistic). Optimistic group includes firms with high forecasts compared to actual earnings. Pessimistic group includes firms with low forecasts compared to actual earnings. 'Market capitalization' is the offer price times the number of shares outstanding at the first day of public trading. '*Age*' is the age of the firm from incorporation to IPO date. '*Oversubscription*' is a demand multiple estimated as demand for IPO shares over shares offered. '*Ownership*' is the proportion of given ownership by the pre-IPO shareholders

The bold value shows average optimistic and pessimistic values of market capitalization, age, oversubscription and ownership

The symbols *, **, and *** denote statistical significance at the 10, 5, and 1 % levels, respectively average subscription multiple of 110.50 (79.80). In contrast, firm age and the given ownership fail to generate statistically significant differences between the two groups. Table 2 provides the distribution of IPOs per industry with a further breakdown by disclosure period (mandatory versus voluntary). The main industries represented in the sample are finance, insurance and real estate (25.41 %) and mining and construction (19.80 %). The chemical and health industries each account only for 2.64 % of sample IPOs.

Table 2 Industry classification of Greek IPOs

Industry classification	Full sample (N = 303)		Mandatory period (N = 220)		Voluntary period (N = 83)	
	N	%	N	%	N	%
Chemicals	8	2.64	6	2.72	2	2.40
Commercial	15	4.95	10	4.54	5	6.02
Food	30	9.90	22	10.00	8	9.63
Health	8	2.64	6	2.72	2	2.40
Mining and construction industries	60	19.80	50	22.72	10	12.04
Manufacturing	23	7.59	17	7.72	6	7.22
Wholesale and retail trade	15	4.95	10	4.54	5	6.02
Finance, insurance and real estate	77	25.41	54	24.54	23	27.71
Service industries	13	4.29	6	2.72	7	8.43
Technology	36	11.88	23	10.45	13	15.66
Tourism and Travel	18	5.94	16	7.27	2	2.40
Total	303	100	220	100	83	100

This table shows the distribution of the 303 Greek IPOs across industries based on the ASE industry classification for the period 1993–2015. There are 220 IPOs during the mandatory disclosure period (1993–2000) and 83 IPOs during the voluntary disclosure period (2001–2015)

The summary statistics for absolute forecast errors, forecast errors, and adjusted returns for one, two, and three years after going public are shown in Table 3. Means, medians, and standard deviations are broken down by mandatory and voluntary periods. The mean (median) forecast error for the total sample is 3.66 % (−0.38 %). The positive sign reveals that, on average, forecasts are less than the actual profits, leading to pessimistic forecasts. When breaking down forecast error by mandatory and voluntary disclosure regime, the results reveal a positive mean of 8.65 % for firms listed during the mandatory period and a negative mean of −9.58 % for firms listed during the voluntary period. This tells us that firms are more conservative when compelled to provide forecasts in their prospectuses, and therefore, actual profits are typically higher than forecast. Once the disclosure environment turns voluntary, Greek IPO firms behave differently and the forecast error sign changes, indicating more optimistic forecasts than the actual earnings. This lends support to the argument that earnings forecasts during mandatory (voluntary) periods are generally pessimistic (optimistic).

Table 3 Summary statistics of forecast errors and return measures

	AFE	*FE*	*ER1Y1D*	*ER2Y1D*	*ER3Y1D*	*ER1Y1M*	*ER2Y1M*	*ER3Y1M*
Panel A: All sample								
Mean	39.72	3.66	3.94	−14.42	4.45	−15.09	−28.90	−16.26
t-value	(5.68)***	(−1.46)	(0.78)	(−2.38)**	(0.45)	(−4.17)***	(−6.12)***	(−1.76)*
Median	31.03	−0.38	−19.69	−26.43	−23.66	−23.57	−30.40	−23.87
z-value	(8.66)***	(8.05)***	(3.35)***	(2.95)***	(3.65)***	(2.46)***	(2.10)***	(3.04)***
Standard-deviation	34.52	52.81	84.23	101.13	165.68	60.25	78.01	146.92
Maximum	234.82	234.82	556.95	559.61	746.86	329.41	306.83	728.83
Minimum	0.38	−132.32	−152.52	−366.11	−454.20	−155.47	−310.84	−504.85
Number	303	303	303	303	303	303	303	303
Panel B: Mandatory disclosure regime (1993–2000)								
Mean	40.32	8.65	7.25	−7.86	16.34	−16.41	−27.64	−8.88
t-value	(18.01)***	(2.27)	(0.91)	(−1.29)	(1.29)	(−3.94)***	(−4.92)***	(−0.82)
Median	31.65	2.96	−19.85	−25.58	2.16	−25.22	−30.40	−23.53
z-value	(1.70)***	(1.06)	(3.23)***	(3.08)***	(3.23)***	(2.38)***	(2.22)***	(2.50)***
Standard-deviation	34.38	52.35	89.80	109.59	176.37	62.71	83.81	157.14
Maximum	234.82	234.82	556.95	559.61	746.86	329.41	306.83	728.83
Minimum	0.38	−84.82	−152.52	−366.11	−454.20	−155.47	−310.84	−504.85
Number	220	220	220	220	220	220	220	220
Panel C: Voluntary disclosure regime (2001–2015)								
Mean	36.83	−9.58	−2.63	−32.26	−34.68	−10.09	−31.79	−32.53
t-value	(9.29)***	(−1.69)*	(−0.23)	(−4.08)***	(−2.36)***	(−1.40)	(−4.52)***	(−2.36)*
Median	26.40	−9.11	−18.59	−38.01	−52.26	−19.69	−26.17	−27.47
z-value	(1.44)*	(0.99)	(1.35)*	(0.54)	(1.48)**	(0.91)	(0.90)	(1.76)***
Standard-deviation	32.96	48.72	60.98	60.65	113.81	50.66	53.30	102.50
Maximum	163.30	163.30	223.23	115.84	567.76	159.59	59.80	593.05

	AFE	*FE*	*ER1Y1D*	*ER2Y1D*	*ER3Y1D*	*ER1Y1M*	*ER2Y1M*	*ER3Y1M*
Minimum	0.57	−90.30	−96.96	−149.65	−207.83	−90.79	−143.25	−191.77
Number	61	61	61	61	61	61	61	61
Panel D: Test of difference between Mandatory and Voluntary means and medians								
Mean	(−0.13)	(1.81)*	(0.22)	(2.69)***	(3.89)***	(0.13)	(1.92)*	(3.22)***
Median	(−0.29)	(1.89)*	(0.84)	(2.50)**	(3.62)***	(0.49)	(1.46)	(3.19)***

FE: forecast error; FE = (Actual Profit (AP) − Forecast Profit (FP))/|Forecast Profit (FP)|. AFE: absolute forecast error; AFE = |FE|. ERNY1D: Adjusted returns from first day price to N years after going public (N = 1, 2, 3). ERNY1M: Adjusted returns from first month to N years after going public (N = 1, 2, 3). The symbols *, **, and *** denote statistical significance at the 10, 5, and 1 % levels, respectively

Long-term performance of Greek IPOs varies depending on the disclosure environment. IPOs in the mandatory regime period with pessimistic forecasts present positive long-term returns, with an average of 16.34 % after three years (*ER3Y1D*). When the disclosure regime changes to voluntary and the forecasts turn optimistic, the average three-year long-term return becomes negative, −34.68 %. This result indicates that investors reward a pessimistic earnings forecast approach by management, as they prefer to see better actual earnings than forecast one year after the IPO.

Long-term returns from the end of the first month of trading are negative for all IPOs. More specifically, firms going public under a mandatory disclosure environment experience negative returns one (*ER1Y1M*), two (*ER2Y1M*), and three (*ER3Y1M*) years after going public of −16.41, −27.64, and −8.88 %, respectively. Those firms that voluntarily provide earnings forecasts show negative returns of −10.09, −31.79, and −32.53 % over the same periods. IPOs in the mandatory regime period provide better long-term returns compared to those in the voluntary regime period two and three years after going public. The difference in long-term returns is statistically significant at conventional levels.

In summary, Table 3 indicates that long-term returns during the mandatory regime outperform those during the voluntary regime two (three) years after the IPO. The rationale underlying this result is that actual earnings tend to be better than the forecasts in the mandatory regime (hence good news, better returns), but lower than the forecasts in the voluntary regime (hence bad news, worse returns).

Table 4 (Panel A) shows the frequency distribution of forecast errors. Forecast errors are, on average, mainly positive (negative) during the mandatory (voluntary) period, meaning that forecasts for most firms are lower (higher) than actual results. It appears that most Greek firms providing mandatory (voluntary) forecast earnings have been underestimating (overestimating) their forecast profits and are generally more (less) conservative during the mandatory (voluntary) disclosure period. The mean AFE for Greek companies is 40.32, 36.83, and 39.72 % for mandatory, voluntary, and combined samples, respectively. The majority of IPOs in the mandatory and voluntary samples experience an AFE of less than 25 % (92 IPOs listed under the mandatory regime and 30 IPOs listed under the voluntary regime). Since the regulation changed, allowing IPO firms to provide an earnings forecast only when they feel confident to do so, 30 out of 61 IPOs (49.18 %) announced AFE lower than 25 %, which constitutes preliminary evidence in support of improved forecast accuracy.

Table 4 Forecast errors and measures of return by listing year

Listing year	*AFE* (%)	<25 %	25–50 %	>50 %	*FE* (%)	<−20 %	−20 to 20 %	>20 %
Panel A: Summary of FE and AFE by year of listing and by disclosure regime (mandatory and voluntary)								
Mandatory (1993–2000)								
1993	21.56	9	1	0	4.17	4	4	2
1994	37.22	18	14	13	−0.02	16	13	16
1995	48.86	7	5	7	33.36	4	5	10
1996	47.17	10	4	6	11.28	7	8	5
1997	32.58	6	4	4	3.77	5	4	5
1998	35.75	11	4	8	15.98	4	10	9
1999	50.70	14	7	17	10.09	14	9	15
2000	38.18	21	15	15	4.90	18	18	15
1993–2000	40.32	92	68	70	8.65	72	71	77
Voluntary (2001–2015)								
2001	37.37	8	6	5	−26.61	9	8	2
2002	50.60	6	3	6	−4.05	7	6	2
2003	39.62	5	3	3	−4.32	5	4	2
2004	25.81	6	0	2	−7.10	2	4	2
2005	46.63	1	0	1	−31.41	1	1	0
2006	8.62	2	0	0	8.62	0	1	1
2007	19.81	2	0	1	−0.55	1	1	1
2008	–	–	–	–	–	–	–	–
2009	28.31	0	1	0	28.31	0	1	0
2010–2015	–	–	–	–	–	–	–	–
2001–2015	36.83	30	13	18	−9.58	25	26	10
Total	39.72	122	71	88	3.98	97	97	87

Listing year	*ER1Y1D*	N	*ER2Y1D*	N	*ER3Y1D*	N	*ER1Y1M*	N	*ER2Y1M*	N	*ER3Y1M*	N
Panel B: Long term performance by year of listing and by disclosure regime (mandatory and voluntary)												
Mandatory (1993–2000)												
1993	−13.45	10	2.32	10	90.01	10	−19.59	10	−13.88	10	53.05	10
1994	−1.04	45	29.61	45	137.50	45	−8.08	45	20.12	45	85.56	45
1995	−0.75	19	−54.07	19	−49.97	19	4.87	19	−51.21	19	−62.60	19
1996	−42.00	20	−71.89	20	−136.35	20	−57.17	20	−134.48	20	−159.26	20
1997	51.01	14	−32.22	14	−62.25	14	9.26	14	−104.10	14	−123.98	14
1998	69.45	23	7.80	23	11.43	23	−45.57	23	−35.09	23	−34.23	23
1999	22.98	38	4.19	38	10.09	38	−13.25	38	−14.50	38	−5.10	38
2000	−9.63	51	−10.59	51	4.34	51	−11.10	51	−11.34	51	8.81	51
1993–2000	7.25		−7.86		16.34		−16.41		−27.64		−8.88	
Voluntary (2001–2015)												
2001	15.47	19	29.91	19	19.66	19	11.39	19	23.93	19	19.47	19
2002	1.36	15	−46.42	15	−92.04	15	−17.60	15	−28.59	15	−57.32	15
2003	−9.84	11	−78.93	11	−58.30	11	−18.76	11	−87.14	11	−67.21	11
2004	−49.44	8	−85.50	8	−63.30	8	−49.98	8	−89.89	8	−71.93	8
2005	6.33	2	−23.41	2	3.88	2	9.43	2	−16.45	2	26.42	2
2006	−11.76	2	−38.35	2	−20.39	2	−22.23	2	−41.91	2	−22.82	2
2007	4.44	3	−27.15	3	30.34	3	13.45	3	−29.46	3	−34.03	3
2009	−2.54	1	−23.71	1	–	0	7.83	1	−12.02	1	–	0
2001–2015	−2.63		−32.26		−34.68		−10.09		−31.79		−32.53	

Panel B of Table 4 analyzes long-term performance based on year of issuance and disclosure of earnings forecasts (mandatory versus voluntary). There is clear evidence that IPO firms that mandatorily disclose earnings forecasts experience much better long-term returns than others. More specifically, the IPOs that mandatorily provide earnings forecasts have an average return three years after the IPO of −8.88 % compared to −32.53 % for firms that voluntarily provide earnings forecasts. Similar results are found using alternative long-term performance measures over two and three years following the IPO (*ER2Y1D*, *ER3Y1D*, and *ER2Y1M*). Individual-year cases are sometimes surprising. For instance, newly listed firms in the hot period of 1999 enjoyed good long-term returns of 22.98, 4.19, and 10.09 %, one, two, and three years after going public, respectively. Further, IPO firms with voluntary earnings disclosure do not seem to be good long-term investments, because of the severe negative long-term returns. For instance, IPO firms listed in 2002 present a positive one-year return, 1.36 %, but extremely large second- and third-year negative returns of −46.42 and −92.04 %, respectively.

Table 5 presents the frequency distribution of the forecast errors in relation to long-term performance. IPOs with low forecast errors ($0 < FE \leq 0.1$) during the mandatory period are rewarded with positive long-term returns (12.86, 38.93, and 30.15 %, for the first, second, and third year after listing, respectively), while newly listed firms with forecast error above 80 % ($FE > 0.8$) have strongly negative returns (−13.54, −35.38, and −69.81 %, for the first, second, and third year after listing, respectively). Similarly, IPOs with AFE below 10 % ($0 < AFE \leq 0.1$) offer positive long-term returns of 42.33, 36.38, and 25.22 % depending on the horizon, while those with inaccurate forecasts (AFE ≥ 1.00) have negative returns of −38.53, −15.92, and −42.55 % depending on the horizon.

These results do not qualitatively change when we examine the voluntary sample. Only IPOs with low forecast error ($0 < FE \leq 0.1$) offer positive three-year long-term returns (16.14 %). New issues with AFE lower than 10 % offer marginally positive returns to their investors (0.67 %). As expected, when management is unable to offer accurate absolute forecasts (AFE ≥ 0.8), the returns are, on average, strongly negative for two and three years after the IPO, suggesting that investors punish IPOs that do not provide accurate earnings forecasts in their prospectuses during the voluntary period. The 22 IPOs that chose to avoid announcing management earnings forecasts experienced severe negative long-term returns of −20.38, −40.20, and −50.35 % one, two, and three years after going public, respectively.

Table 5 IPO long-term returns depending on forecast accuracy

Distribution of *FE*	Number of IPOs	Year 1 (%)	Year 2 (%)	Year 3 (%)	Distribution of *AFE*	Number of IPOs	Year 1 (%)	Year 2 (%)	Year 3 (%)
Panel A: Mandatory (1993–2000)									
$FE \le -0.6$	18	6.86	−5.61	22.12	$0 < AFE \le 0.1$	28	42.33	36.38	25.22
$-0.6 < FE \le 0.4$	21	−7.25	−6.96	2.59	$0.1 < AFE \le 0.2$	41	−1.64	1.39	58.79
$-0.4 < FE \le -0.2$	32	−11.32	−15.09	65.96	$0.2 < AFE \le 0.3$	32	18.06	23.54	49.10
$-0.2 < FE \le -0.1$	18	−15.56	−14.98	−33.60	$0.3 < AFE \le 0.4$	20	−6.98	−29.94	10.02
$-0.1 < FE \le 0$	16	66.33	15.63	10.09	$0.4 < AFE \le 0.5$	20	−10.53	−11.05	19.03
$0 < FE \le 0.1$	19	12.86	38.93	30.15	$0.5 < AFE \le 0.6$	19	−10.09	3.42	−4.68
$0.1 < FE \le 0.2$	21	−23.76	−13.60	82.13	$0.6 < AFE \le 0.7$	17	61.14	−25.41	6.53
$0.2 < FE \le 0.4$	20	−20.12	−44.84	−11.34	$0.7 < AFE \le 0.8$	10	20.98	−57.76	−84.03
$0.4 < FE \le 0.6$	18	−15.93	6.56	18.62	$0.8 < AFE \le 0.9$	12	−22.64	−59.83	−15.97
$0.6 < FE \le 0.8$	17	35.72	−51.34	−33.00	$0.9 < AFE \le 1.00$	10	−17.39	−21.31	−75.70
$FE \ge 0.8$	20	−13.54	−35.38	−69.81	$AFE \ge 1.00$	11	−38.53	−15.92	−42.55
						220	**7.25**	**−7.86**	**16.34**
Panel B: Voluntary (2001–2015)									
$FE \le -0.6$	8	−27.05	−36.00	−51.15	$0 < AFE \le 0.1$	10	16.30	−0.06	0.67
$-0.6 < FE \le 0.4$	11	−8.08	−70.48	−94.89	$0.1 < AFE \le 0.2$	12	5.69	−27.14	−38.49
$-0.4 < FE \le -0.2$	5	−9.04	−27.45	−41.82	$0.2 < AFE \le 0.3$	4	−38.10	−75.55	−67.86
$-0.2 < FE \le -0.1$	5	−26.29	−42.13	−26.46	$0.3 < AFE \le 0.4$	4	6.87	−1.99	−28.35
$-0.1 < FE \le 0$	6	44.15	15.88	−10.76	$0.4 < AFE \le 0.5$	8	16.22	−56.07	−60.18
$0 < FE \le 0.1$	6	−7.50	−10.60	16.14	$0.5 < AFE \le 0.6$	6	−28.26	−69.16	−90.57
$0.1 < FE \le 0.2$	11	18.12	−13.95	−44.20	$0.6 < AFE \le 0.8$	7	−28.29	4.32	−48.67
$FE \ge 0.2$	9	7.27	−53.51	−29.82	$AFE \ge 0.8$	10	1.37	−36.46	−46.89
						61	**−2.63**	**−32.26**	**−34.68**
Panel C: No forecast									
						22	**−20.38**	**−40.20**	**−50.35**

The bold value in Panel A shows the following: 220 is the total number of IPOs (in the Mandatory (1993–2000) period) and 7.25, −7.86 and 16.34 are the IPO returns for Year 1, Year 2 and Year 3, respectively

The bold value in Panel B shows the following: 61 is the total number of IPOs (in the Voluntary (2001–2015) period) and −2.63, −32.26 and −34.68 are the IPO returns for Year 1, Year 2 and Year 3, respectively

The bold value in Panel C shows the following: 22 is the total number of IPOs (in the non-forecast sample) and −20.38, −40.20 and −50.35 are the IPO returns for Year 1, Year 2 and Year 3, respectively

In summary, the findings shown in Table 5 indicate that post-IPO long-run performance (in particular the last column of Table 5) is highest for IPOs during the mandatory period (16.34 %), followed by IPOs during the voluntary period with voluntary management forecasts (−34.68 %), and is lowest for IPOs during the voluntary period without management forecasts (−50.35 %). Such differences in long-run performance may be at least partially attributable to market sentiment rather than a poorer quality of IPOs during the voluntary period, or other confounding factors. Consequently, it is a mistaken strategy to ignore the prevailing disclosure regime when analyzing IPO performance.

Table 6 reports pairwise correlations for all explanatory variables. The correlation coefficients are low, suggesting that multicollinearity is unlikely to be a problem.

Table 6 Correlation matrix

	Size	*Age*	*TimeLag*	*Privatization*	*Over subscription*	*Underwriter*	*HotCold*	*Ownership*	*Industry*	*Mandatory*
Size	1.000									
Age	0.294***	1.000								
TimeLag	−0.237***	−0.073	1.000							
Privatization	0.396***	0.202***	−0.089*	1.000						
Oversubscription	−0.133**	0.058*	0.023	−0.101***	1.000					
Underwriter	0.229***	0.128**	−0.025	−0.043	−0.042	1.000				
HotCold	0.073	−0.027**	0.036	0.044	0.091	−0.061	1.000			
Ownership	0.180***	−0.038**	0.104*	0.092	−0.212***	0.053	0.039	1.000		
Industry	0.036	0.026	0.001	0.046	−0.012	0.105*	−0.034	0.045	1.000	
Mandatory	−0.027	0.125**	0.263***	−0.092*	0.354***	0.109*	−0.052	−0.029	−0.058	1.000

The table presents Pearson pairwise correlations of the independent variables. The sample consists of Greek initial public offerings listed over the period January 1, 1993 to December 31, 2015. All variables are defined in "Appendix". The symbols *, **, and *** denote statistical significance at the 10, 5, and 1 % levels, respectively

5.2 Regression analysis

Table 7 reports results from regressing absolute forecast error and forecast error on disclosure regime (mandatory versus voluntary), along with a number of control variables suggested by the literature. These regressions are in the spirit of prior literature and are estimated using ordinary least squares with robust standard errors to control for heteroskedasticity. The adjusted R-squared are similar to existing studies.

Table 7 The effect of the disclosure regime on absolute forecast error and forecast error

Variables	Full sample		Size and industry matched		Industry and age matched	
	(1) *AFE*	(2) *FE*	(3) *AFE*	(4) *FE*	(5) *AFE*	(6) *FE*
Constant	-11.540	4.132	-5.797	-127.855	31.635	-76.511
	(0.653)	(0.899)	(0.922)	(0.187)	(0.355)	(0.294)
Size	2.765*	-0.022	3.596	4.496	0.460	2.450
	(0.084)	(0.991)	(0.268)	(0.389)	(0.788)	(0.504)
Age	−0.135	−0.187	−0.370*	−0.244	−0.120	−0.173
	(0.438)	(0.184)	(0.074)	(0.475)	(0.410)	(0.539)
TimeLag	0.357**	−0.184	0.019	0.453	0.391*	0.284
	(0.045)	(0.412)	(0.950)	(0.414)	(0.088)	(0.557)

Privatization	0.664	−5.090	9.091	−3.719	2.389	−0.201
	(0.937)	(0.676)	(0.491)	(0.872)	(0.836)	(0.992)
Oversubcription	0.0167	0.0200	0.0222	0.0181	0.0222	0.0227
	(0.467)	(0.387)	(0.330)	(0.438)	(0.306)	(0.330)
Underwriter	−4.136	−4.541	−1.293	13.976	7.632	14.199
	(0.298)	(0.365)	(0.824)	(0.194)	(0.150)	(0.193)
HotCold	−13.950***	1.096	0.695	−1.198	−2.655	−4.560
	(0.005	(0.837)	(0.928)	(0.932)	(0.585)	(0.705)
Ownership	−0.350*	−0.413	−0.546	0.938	−0.638***	0.358
	(0.071)	(0.121)	(0.143)	(0.144)	(0.002)	(0.469)
Industry	4.698	8.562	−4.944	−0.699	1.716	4.813
	(0.342)	(0.207)	(0.425)	(0.953)	(0.756)	(0.696)
Mandatory	10.460**	17.890***	15.746**	34.878***	13.408**	26.123**
	(0.018)	(0.009)	(0.019)	(0.004)	(0.016)	(0.025)
Adjusted R^2	0.073	0.513	0.131	0.098	0.160	0.064
Number of IPOs	281	281	122	122	122	122

This table presents the regression results of the effect of the disclosure regime on the absolute forecast error and forecast error for the 281 Greek IPOs made between January 1993 and December 2015 (220 IPOs listed during the mandatory disclosure regime: January 1993–December 2000; 61 IPOs listed during the voluntary disclosure regime: January 2001–December 2015). Columns 1 and 2 use the full sample; Columns 3 and 4 (Columns 5 and 6) use a sample that matches the 61 IPOs of the voluntary period with 61 from the mandatory period using firm size and industry (firm age and industry). The dependent variables are *AFE* and *FE*. Independent variables are: *Size* proxied by the natural logarithm of the market capitalization. Market capitalization is measured as the total number of outstanding shares after the IPO multiplied by price per share. *Age* is the age of the firm from incorporation to IPO date. *TimeLag* is the number of days between the date of issuance of the prospectus and the first day of the stocks' listing. *Privatization* is a dummy variable that equals 1 for privatized IPOs, and zero otherwise. *Oversubscription* is a demand multiple estimated as demand for IPO shares over shares offered. *Underwriter* is a dummy variable that equals 1 for reputable underwriters, and zero otherwise. Reputable underwriters are classified based on three criteria: (i) The number of IPOs they advised as lead underwriters, (ii) the fee requested for the offered services, and (iii) the capital raised for the public offerings. *HotCold* is a dummy variable that equals 1 for IPOs listed during a hot period, and zero for IPOs that were listed during a cold period. The classification "Hot period" and "Cold period" is based on the intensity of IPO listing activity and market returns. *Ownership* is the proportion of given ownership by the pre-IPO shareholders. *Industry* is a dummy variable that equals 1 for industrial IPO firms, and zero otherwise. *Mandatory* is a dummy variable that equals 1 for IPOs listed during 1993–2000, and zero otherwise. The *t*-statistics are corrected for heteroskedasticity using the Newey-West method. The symbols *, **, and *** denote statistical significance at the 10, 5, and 1 % levels, respectively

Columns 1 and 2 use the full sample of 281 IPO firms that provide a forecast during the mandatory (220 IPO firms) and the voluntary (61 IPO firms) regime periods. They shed light on the effect of disclosure regime on AFE and FE. Consistent with H_1, mandatory disclosure of earnings forecasts is associated with production of inaccurate management earnings forecasts. The coefficient on the *Mandatory* dummy is positive and statistically significant (at the 5 % level) when using AFE as the dependent variable. Our conclusions are reinforced when FE is the dependent variable. The coefficient on *Mandatory* is positive and statistically significant at the 1 % level, providing evidence that management earnings forecasts of IPOs issued during the mandatory period are lower than actual earnings, and suggesting that managers under-forecast earnings when they are mandated to provide earnings forecasts. This result is in line with the argument that managers provide less optimistic forecasts during the mandatory period as a form of insurance against possible criticism and legal action against the company in the post-IPO period.

As a robustness check, we use a matched-sample approach to ensure that our results are not driven by firm characteristics that may affect the choice to go public during the mandatory regime period. More specifically, for each of the 61 IPOs in the voluntary regime period, we assign the closest-neighbor from the 220 IPOs available in the mandatory regime period. In columns 3 and 4, firm size and industry are used as matching criteria, whereas in columns 5 and 6, matching is performed using firm age and industry at the time of IPO. The empirical results show that the coefficient on *Mandatory* remains positive and statistically significant irrespective of the matching criteria. Mandatory disclosure of earnings forecasts in IPO prospectuses is associated with greater forecast error (*AFE* and *FE*). Managers seem to underestimate profits more when the disclosure regime is mandatory than when it is voluntary.

Turning to the other determinants of forecast accuracy, we report, for the full sample, a positive effect of firm size on *AFE*. Contrary to our expectation, *AFE*s increase with firm size. This is not the first study to document a similar result. Lonkani and Firth (2005) generate similar evidence for their sample of Thai IPOs. The absolute forecast errors seem to increase with the period that separates publication of the prospectus and the first listing day (*TimeLag*). The longer this time interval is, the greater the likelihood of erroneous estimates due to possible changes in the underlying assumptions. The coefficient on the *HotCold* dummy is negative and statistically significant at the 1 % level. This result stands in sharp contrast to the notion that management forecast accuracy becomes a lesser concern in hot markets. The percentage of given ownership inversely relates to AFE, suggesting that managers of firms with large retained ownership are more careful in their forecasts. It is noteworthy that firm age and underwriter reputation do not seem to affect the quality of management earnings forecasts.

Table 8 shows the determinants of long-term performance by disclosure period (mandatory and voluntary). This table explores the explanatory factors (other than forecast accuracy) of IPO long-term returns under the two disclosure regimes. To capture possible time trends, we use first-, second-, and third-year returns as dependent variables. Further, to reduce the effect of any aberration in the early trading period (such as investment banker price support), we measure returns beginning from the first day of trading (*ER1Y1D*, *ER2Y1D*, and *ER3Y1D*) and from the first month of trading (*ER1Y1M*, *ER2Y1M*, and *ER3Y1M*). The results obtained from the mandatory sample show that large firms realize negative returns up to one year post-IPO (−0.118) and a positive return over a two-year period (0.145). During voluntary regime forecasts, the results are consistently positive and statistically significant irrespective of the measure of long-term returns. Firm age is negatively associated with long-term returns under the voluntary period. The coefficient sign is consistent with the intuition that longer operational experience reduces risk, and this, in turn, limits realized returns.

The privatization dummy exhibits a positive sign, indicating the presence of higher risk underpinning the long-run viability of firms that have existed only under state control and, following an IPO, are expected to compete in a market-driven business environment. In addition, high oversubscription during the book building period is a signal for good long-term returns. The mandatory period (exclusively) illustrates this. The negative association with first-year returns becomes a positive and statistically significant relation (at least at the 5 % level) for the returns of subsequent years. The finding that reputable underwriters barely give rise to a discernible pattern of performance is surprising, albeit the same under both regimes. The limited Greek IPO market and the even more limited number of local underwrites may explain this result. As expected, IPOs listed during a cold period are associated with better returns over a three-year period, highlighting the significance of market conditions during the public offering period. In contrast to the mandatory sample, a high percentage of given ownership is rewarded with good investor returns over the voluntary period. Finally, we document no statistically significant evidence for *TimeLag* and *Industry*.

Table 8 The determinants of the IPO long-term performance by disclosure regime

Variables	*ER1Y1D* 1993–2000 Mandatory period	*ER1Y1M* 1993–2000	*ER2Y1M* 1993–2000	*ER3Y1M* 1993–2000	*ER1Y1D* 2001–2015 Voluntary period	*ER2Y1D* 2001–2015	*ER3Y1D* 2001–2015	*ER1Y1M* 2001–2015	*ER2Y1M* 2001–2015	*ER3Y1M* 2001–2015
Constant	2.153**	0.666	−3.215***	−2.635	−3.862***	−2.872**	−6.850***	−4.029***	−2.749***	−4.335**
	(0.026)	(0.442)	(0.002)	(0.133)	(0.003)	(0.011)	(0.002)	(0.002)	(0.007)	(0.022)
Size	−0.118**	−0.032	0.145***	0.123	0.210***	0.157***	0.340***	0.221***	0.138***	0.200**
	(0.022)	(0.442)	(0.008)	(0.157)	(0.004)	(0.005)	(0.002)	(0.002)	(0.007)	(0.034)
Age	−0.0019	−0.005*	0.002	−0.001	−0.011***	−0.011***	−0.016**	−0.010***	−0.007**	−0.011*
	(0.693)	(0.067)	(0.586)	(0.918)	(0.003)	(0.001)	(0.026)	(0.002)	(0.022)	(0.079)
TimeLag	0.004	0.003	0.006	0.003	0.005	−0.009	0.003	0.003	0.001	0.005
	(0.334)	(0.420)	(0.128)	(0.746)	(0.497)	(0.275)	(0.821)	(0.582)	(0.876)	(0.675)
Privatization	0.440*	0.393*	−0.017	0.250	0.168	0.354**	0.162	0.071	0.356**	0.049
	(0.055)	(0.095)	(0.956)	(0.757)	(0.368)	(0.044)	(0.690)	(0.703)	(0.021)	(0.855)
Oversubscription	−0.0002	−0.0004**	0.001**	0.001***	0.0014	0.001	−0.004	−0.001	0.001	−0.002
	(0.469)	(0.014)	(0.011)	(0.009)	(0.716)	(0.917)	(0.317)	(0.628)	(0.855)	(0.776)
Underwriter	−0.032	−0.163*	0.079	0.228	−0.069	0.023	0.119	0.016	0.081	0.072
	(0.832)	(0.098)	(0.569)	(0.322)	(0.656)	(0.885)	(0.583)	(0.860)	(0.514)	(0.750)
HotCold	0.580***	−0.019	−0.295**	−0.530***	−0.115	−0.141	0.008	−0.003	−0.005	−0.019
	(0.004)	(0.837)	(0.037)	(0.002)	(0.420)	(0.229)	(0.976)	(0.979)	(0.967)	(0.937)
Ownership	−0.015**	−0.009**	0.002	0.011	0.011*	0.009*	0.025	0.006	0.001	0.021
	(0.043)	(0.025)	(0.674)	(0.410)	(0.083)	(0.087)	(0.142)	(0.277)	(0.804)	(0.233)
Industry	0.048	0.155	0.279*	−0.158	−0.194	−0.184	−0.049	−0.070	−0.125	0.0323
	(0.755)	(0.147)	(0.099)	(0.557)	(0.197)	(0.189)	(0.858)	(0.553)	(0.360)	(0.907)
Adjusted R^2	0.091	0.047	0.051	0.052	0.089	0.086	0.107	0.209	0.060	0.045
Number of IPOs	220	220	220	220	61	61	61	61	61	61

This table presents the regression results of the determinants of long-term performance for the 281 Greek IPOs made between January 1993 and December 2015 (220 IPOs listed during the mandatory disclosure regime: January 1993–December 2000; 61 IPOs listed during the voluntary disclosure regime: January 2001–December 2015). Long-term performance is calculated using the Fama and French (FF) three-factor model. The dependent variables are: *ERNY1D*—Adjusted returns from first day price to N years after going public (N = 1, 2, 3). *ERNY1M*—Adjusted returns from first month price to N years after going public (N = 1, 2, 3). Independent variables are: *Size* proxied by the natural logarithm of the market capitalization. Market capitalization is measured as the total number of outstanding shares after the IPO multiplied by price per share. *Age* is the age of the firm from incorporation to IPO date. *TimeLag* is the number of days between the date of issuance of the prospectus and the first day of the stocks' listing. *Privatization* is a dummy variable that equals 1 for privatized IPOs, and zero otherwise. *Oversubscription* is a demand multiple estimated as demand for IPO shares over shares offered. *Underwriter* is a dummy variable that equals 1 for reputable underwriters, and zero otherwise. Reputable underwriters are classified based on three criteria: (i) The number of IPOs they advised as lead underwriters, (ii) the fee requested for the offered services, and (iii) the capital raised for the public offerings. *HotCold* is a dummy variable that equals 1 for IPOs listed during a hot period, and zero for IPOs that were listed during a cold period. The classification "Hot period" and "Cold period" is based on the intensity of IPO listing activity and market returns. *Ownership* is the proportion of given ownership by the pre-IPO shareholders. *Industry* is a dummy variable that equals 1 for industrial IPO firms, and zero otherwise. The *t*-statistics are corrected for heteroskedasticity using the Newey-West method. The symbols *, **, and *** denote statistical significance at the 10, 5, and 1 % levels, respectively

Table 9 explores how the prevailing disclosure regime (captured by the *Mandatory* dummy variable) affects the long-term performance of IPOs using the cumulative abnormal returns of the Fama and French (1993) three-factor model as dependent variables. The results confirm that IPOs in the mandatory period realize superior returns rather than in the voluntary period (mainly columns 3 and 6). The fact that statistical significance is obtained only for coefficients in the regressions with two-year, and especially three-year returns shows that time accentuates the positive relation. This finding is consistent with the rationale that management earnings forecasts can be a strong tool in the hands of investors and can help them make future investment decisions. In particular, with a greater number of subsequent accounting periods to report earnings that consistently and sizably outperform the expectations (initial forecast in the IPO prospectus), the euphoric sentiment of investors reflects the upward trend of long-term returns. In parallel, the extent of conservatism that, on average, surrounds profit forecasts is fully revealed. This finding is mainly significant two, and in particular, three years after IPO.

Table 9 The effect of earnings forecasts accuracy on the IPO long-term performance

Variables	*ER1Y1D* 1993–2015	*ER2Y1D* 1993–2015	*ER3Y1D* 1993–2015	*ER1Y1M* 1993–2015	*ER2Y1M* 1993–2015	*ER3Y1M* 1993–2015	*ER3Y1D* 1993–2015	*ER3Y1M* 1993–2015	*ER3Y1D* 1993–2015	*ER3Y1M* 1993–2015
Constant	1.087	−1.276	−2.832*	−0.070	−2.636***	−2.748**	−4.363***	−3.977***	−3.765**	−3.167**
	(0.146)	(0.261)	(0.011)	(0.912)	(0.000)	[illegible]	[illegible]	[illegible]	(0.011)	(0.019)
Size	−0.063	0.042	0.094	0.006	0.123***	0.115*	0.169**	0.156**	0.149*	0.134**
	(0.118)	(0.479)	(0.217)	(0.852)	(0.001)	(0.069)	(0.025)	(0.021)	(0.050)	(0.049)
Age	−0.003	0.003	0.002	−0.004**	−0.0003	−0.002	−0.007	−0.006	−0.006	−0.004
	(0.365)	(0.493)	(0.965)	(0.034)	(0.880)	(0.599)	(0.257)	(0.318)	(0.384)	(0.442)
TimeLag	0.004	0.004	0.006	0.002	0.003	0.002	0.011	0.005	0.009	0.004
	(0.226)	(0.453)	(0.461)	(0.547)	(0.389)	(0.723)	(0.233)	(0.497)	(0.291)	(0.598)
Privatization	0.526***	0.212	0.311	0.431**	0.116	0.266	0.434	0.370	0.320	0.278
	(0.002)	(0.352)	(0.509)	(0.018)	(0.579)	(0.640)	(0.464)	(0.533)	(0.586)	(0.641)
Oversubscription	−0.000	0.001**	0.001*	−0.0004**	0.001***	0.001***	0.001**	0.002***	0.001*	0.002***
	(0.970)	(0.021)	(0.065)	(0.011)	(0.006)	(0.001)	(0.044)	(0.004)	(0.051)	(0.007)
Underwriter	−0.003	0.128	0.316	−0.086	0.072	0.115	0.476**	0.245	0.468**	0.223
	(0.975)	(0.320)	(0.146)	(0.222)	(0.463)	(0.519)	(0.036)	(0.220)	(0.044)	(0.277)
HotCold	0.462***	−0.051	−0.283*	0.012	−0.225**	−0.462***	−0.372*	−0.476***	−0.368*	−0.482***
	(0.000)	(0.748)	(0.067)	(0.876)	(0.042)	(0.004)	(0.089)	(0.003)	(0.089)	(0.003)
Ownership	−0.008	−0.003	0.016	−0.005	0.001	0.016	0.018	0.018	0.0149	0.016
	(0.113)	(0.586)	(0.190)	(0.169)	(0.786)	(0.135)	(0.219)	(0.150)	(0.262)	(0.152)
Industry	−0.001	0.028	0.082	0.096	0.064	−0.105	0.202	−0.021	0.170	−0.069
	(0.993)	(0.819)	(0.753)	(0.255)	(0.591)	(0.648)	(0.465)	(0.932)	(0.533)	(0.776)
Mandatory	0.280	0.355*	0.588**	−0.050	0.028	0.334*	1.190***	0.892***	0.621**	0.267
	(0.770)	(0.086)	(0.011)	(0.442)	(0.754)	(0.070)	(0.001)	(0.008)	(0.019)	(0.283)
AFE							0.003	0.006		
							(0.672)	(0.234)		
FE									0.005	0.003
									(0.250)	(0.476)
*AFE*Mandatory*							−0.014*	−0.015**		
							(0.057)	(0.048)		
*FE*Mandatory*									−0.010**	−0.005
									(0.044)	(0.315)
Adjusted R^2	0.088	0.081	0.055	0.068	0.075	0.052	0.061	0.064	0.063	0.058
Number of IPOs	281	281	281	281	281	281	281	281	281	281

This table presents the regression results of the effect of AFE and FE on the long-term performance for the 281 Greek IPOs made between January 1993 and December 2015 (220 IPOs listed during the mandatory disclosure regime: January 1993–December 2000; 61 IPOs listed during the voluntary disclosure regime: January 2001–December 2015). Long term performance is calculated using the Fama and French (FF) three-factor model. The dependent variables are: *ERNY1D*—Adjusted returns from first day price to N years after going public (N = 1, 2, 3). *ERNY1M*—Adjusted returns from first month price to N years after going public (N = 1, 2, 3). Independent variables are: *Size* proxied by the natural logarithm of the market capitalization. Market capitalization is measured as the total number of outstanding shares after the IPO multiplied by price per share. *Age* is the age of the firm from incorporation to IPO date. *TimeLag* is the number of days between the date of issuance of the prospectus and the first day of the stocks' listing. *Privatization* is a dummy variable that equals 1 for privatized IPOs, and zero otherwise. *Oversubscription* is a demand multiple estimated as demand for IPO shares over shares offered. *Underwriter* is a dummy variable that equals 1 for reputable underwriters, and zero otherwise. Reputable underwriters are classified based on three criteria: (i) The number of IPOs they advised as lead underwriters, (ii) the fee requested for the offered services, and (iii) the capital raised for the public offerings. *HotCold* is a dummy variable that equals 1 for IPOs listed during a hot period, and zero for IPOs that were listed during a cold period. The classification "Hot period" and "Cold period" is based on the intensity of IPO listing activity and market returns. *Ownership* is the proportion of given ownership by the pre-IPO shareholders. *Industry* is a dummy variable that equals 1 for industrial IPO firms, and zero otherwise. *Mandatory* is a dummy variable that equals 1 for IPOs listed during 1993–2000, and zero otherwise. FE: forecast error; FE = (Actual Profit (AP) − Forecast Profit (FP))/Forecast Profit (FP)|, AFE: absolute forecast error; AFE = |FE|. The *t*-statistics are corrected for heteroskedasticity using the Newey-West method. The symbols *, **, and *** denote statistical significance at the 10, 5, and 1 % levels, respectively

Focusing on marginal effects constitutes an alternative way to shed light on the association with long-term performance. To this end, Columns 7 and 8 augment the baseline specification with AFE and its interaction with the *Mandatory* dummy. The resulting coefficients on the new terms convey important intuition on market investors' perceptions of forecast accuracy. In particular, focusing on the *ER3Y1D* regression, we find a negative sign for the interaction term and a slightly positive coefficient for *AFE*. Taken together, the effect of *AFE* on *ER3Y1D*, given a mandatory period, is negative (0.003 – 0.014 = –0.011). That is, forecast accuracy is rewarded by means of higher long-term returns, which is consistent with H_2. This suggests that issuers should always disseminate reliable information regardless of the prevailing regulatory regime. In addition, conservative forecasts should not be used as a tool for attaining higher market valuations, since the market is shown to be able to identify and penalize speculative behavior. Consistent with this relation, Columns 9 and 10 report qualitatively similar results when we substitute *AFE* for *FE*.

As for control variables, the results reveal a positive relation between firm size and long-term performance. The coefficient for firm age is negative only when returns are calculated from the end of the first month of trading and only for one year. This indicates that investors in companies with short operational history before listing enjoy better one-year returns. The privatization variable has the expected sign; the positive coefficient indicates that privatized firms experience superior one-year returns. Thus, investors should take advantage of and participate in privatized company issues. The coefficient on market condition (*HotCold*) shows that firms going public during hot periods offer better returns up to the one-year period, whereas those going public during cold periods provide better returns to investors after two and three listing years.

6 ADDITIONAL ROBUSTNESS CHECKS

6.1 A propensity score matching approach

Although our findings lend support to more pronounced IPO underperformance under voluntary disclosure, one cannot rule out the possibility that the effect is partially distorted by endogeneity. In particular, it is possible that the mandatory and voluntary periods feature distinct firm- or market-level characteristics that affect the aftermarket performance. We address this concern using a propensity score matching (PSM) technique. Through inclusion of possible confounding characteristics (*Size, Age, Privatization, Oversubscription, Underwriter, HotCold, Ownership*, and *TimeLag*), PSM matches observations from two samples in a simultaneous procedure. Therefore, framing the mandatory (voluntary) regime as the treatment (control) group, the observed difference in means captures the effect that is exclusively due to treatment. We apply four distinct PSM approaches: (i) nearest neighbor, (ii) radius, (iii) Kernel and (iv) stratification for returns calculated from the first day as well as the end of the first month. Table 10 summarizes the results.

Table 10 Propensity score matching

IPO returns (%)	Mandatory group (1993–2000)	Voluntary group (2001–2015)	Difference	*t*-statistic
	M	V	M–V	H_o: M = V
Panel A: Nearest-neighbor matching				
ER1Y1D	7.25	-20.56	27.81	2.02***
ER2Y1D	-7.86	-49.36	41.50	3.38***
ER3Y1D	16.34	-66.42	82.76	3.94***
ER1Y1M	-16.41	-21.07	4.66	0.60
ER2Y1M	-27.64	-37.04	9.40	0.89
ER3Y1M	-8.88	-57.62	48.74	4.83***
Panel B: Radius (with 0.1 caliper) matching				
ER1Y1D	7.25	-8.95	16.20	1.74*
ER2Y1D	-7.86	-39.75	31.89	3.04***
ER3Y1D	16.34	-45.03	61.37	3.46***
ER1Y1M	-16.41	-13.27	-3.14	-0.54
ER2Y1M	-27.64	-34.62	6.98	0.85
ER3Y1M	-8.88	-39.91	31.03	2.70***
Panel C: Kernel matching				
ER1Y1D	7.25	-15.54	22.79	3.08***
ER2Y1D	-7.86	-46.51	38.65	7.47***
ER3Y1D	16.34	-53.02	69.36	2.94***
ER1Y1M	-16.41	-17.81	1.40	0.17
ER2Y1M	-27.64	-36.74	9.10	1.01
ER3Y1M	-8.88	-44.09	35.21	2.39***
Panel D: Stratification matching				
ER1Y1D	7.25	-15.16	22.41	3.07***

ER2Y1D	-7.86	-45.74	37.88	4.03***
ER3Y1D	16.34	-55.22	71.56	4.23***
ER1Y1M	-16.41	-18.83	2.42	0.64
ER2Y1M	-27.64	-34.63	6.99	0.62
ER3Y1M	-8.88	-44.34	35.46	2.74***

This table reports the results from the propensity score matching test where IPOs from the voluntary disclosure period of 2001–2015 are matched to those from the mandatory disclosure period of 1993–2000. The matching is made using the methods of one-to-one nearest-neighbor (Panel A), radius with 0.1 caliper (Panel B), Kernel (Panel C), and stratification (Panel D). Each panel presents mean IPO returns and difference in mean IPO returns for the groups under comparison. The statistical significance is given by the *t*-statistics. The symbols *, **, and *** denote statistical significance at the 10, 5, and 1 % levels, respectively

The nearest-neighbor matching method (Panel A) assigns each voluntary disclosure IPO firms a mandatory counterpart based on propensity score proximity. The results on the matched sample show that long-term returns considerably deteriorate in the voluntary period compared to the mandatory period. Moreover, the long-term underperformance during the voluntary period is lower the longer the interval over which returns are measured. The long-term performance difference is largest for a three-year period and has the highest statistical significance (*ER3Y1D*: 82.76, *t*-statistic 3.94; *ER3Y1M*: 48.74, *t*-statistic 4.83). The radius (Panel B), Kernel matching method (Panel C), and stratification matching method (Panel D) provide results consistent with the nearest-neighbor matching method. The statistical significance appears troublesome only for *ER1Y1M* and *ER2Y1M*, while *ER1Y1D* and *ER2Y1D* are statistically significant at conventional levels. The evidence on three-year returns is invariably robust.[15]

6.2 Management earnings forecasts during the voluntary period

The main conclusion of this study is that mandatory disclosure of earnings forecasts in IPO prospectuses is associated with better long-term returns. We attribute this association to the fear of managers failing to meet their forecast expectations. However, our reporting conservatism argument may be questionable if aftermarket performance depends on the management decision of whether or not to issue earnings forecasts for IPO firms listed during the voluntary period. To test the robustness of our results to this source of bias, we regress long-term returns on the *Voluntary Forecast* dummy, which takes 1 for IPO firms providing an earnings forecast during 2001–2015 and zero otherwise, along with control variables. The empirical results in Table 11 show that the coefficient on *Voluntary Forecast* dummy is not statistically significant irrespective of the measure of long-term performance. This result is consistent with the argument that the decision to issue voluntary earnings forecasts in IPO prospectuses is not related to long-term performance, suggesting that this issue is not serious in our case.[16]

Table 11 Effect of the decision to issue management earnings forecasts on IPO long-run performance

Variables	*ER1Y1D*	*ER2Y1D*	*ER3Y1D*
	2001–2015	2001–2015	2001–2015
Constant	−195.500*	−202.401**	−623.802***
	(0.066)	(0.048)	(0.003)
Size	13.350*	14.110**	39.890***
	(0.055)	(0.032)	(0.002)
Age	−0.948**	−0.513	−1.323
	(0.046)	(0.223)	(0.258)
TimeLag	0.206	−0.887	0.115
	(0.755)	(0.235)	(0.946)
Privatization	10.810	21.350	−18.080
	(0.619)	(0.253)	(0.702)
Oversubscription	−0.241	0.0413	−0.382
	(0.313)	(0.909)	(0.363)
Underwriter	−33.190**	−38.140***	−36.250
	(0.023)	(0.006)	(0.144)
HoldCold	−15.450	−2.958	−20.970
	(0.240)	(0.872)	(0.356)
Ownership	−0.234	−0.506	−0.880
	(0.699)	(0.436)	(0.482)

Industry	−3.466	0.870	19.300
	(0.818)	(0.951)	(0.641)
Voluntary Forecast	13.030	−1.439	3.475
	(0.285)	(0.923)	(0.869)
Adjusted R^2	0.073	0.101	0.124
Number of IPOs	83	83	83

The table presents the regression results of the effect of the earnings forecast disclosure decision on IPO long-term performance for a sample of 83 Greek firms that went public during the voluntary forecast disclosure period (2001–2015). Long term performance is calculated using the Fama and French (FF) three-factor model. The dependent variables are: *ERNY1D*—Adjusted returns from first day price to N years after going public (N = 1, 2, 3). Independent variables are: *Size* proxied by the natural logarithm of the market capitalization. Market capitalization is measured as the total number of outstanding shares after the IPO multiplied by price per share. *Age* is the age of the firm from incorporation to IPO date. *TimeLag* is the number of days between the date of issuance of the prospectus and the first day of the stocks' listing. *Privatization* is a dummy variable that equals 1 for privatized IPOs, and zero otherwise. *Oversubscription* is a demand multiple estimated as demand for IPO shares over shares offered. *Underwriter* is a dummy variable that equals 1 for reputable underwriters, and zero otherwise. Reputable underwriters are classified based on three criteria: (i) The number of IPOs they advised as lead underwriters, (ii) the fee requested for the offered services, and (iii) the capital raised for the public offerings. *HotCold* is a dummy variable that equals 1 for IPOs listed during a hot period, and zero for IPOs that were listed during a cold period. The classification "Hot period" and "Cold period" is based on the intensity of IPO listing activity and market returns. *Ownership* is the proportion of given ownership by the pre-IPO shareholders. *Industry* is a dummy variable that equals 1 for industrial IPO firms, and zero otherwise. *Voluntary Forecast* is a dummy variable that equals 1 for IPOs providing earnings forecasts during 2001–2015, 0 otherwise. The *t*-statistics are corrected for heteroskedasticity using the Newey-West method. The symbols *, **, and *** denote statistical significance at the 10, 5, and 1 % levels, respectively

7 CONCLUSION

Greek IPOs were required to disclose next-year profit forecast in their prospectuses until the regulation changed in 2001 to make it voluntary. This study uses all Greek IPOs issued during 1993–2015 to investigate the effect of disclosure regime on management earnings forecasts and IPO long-term performance. The empirical results indicate that IPO aftermarket performance (excluding the first day of trading) significantly deteriorates in the voluntary period, suggesting that investors during the voluntary regime punish firms that provide inaccurate earnings forecasts in their prospectuses. In addition, the post-IPO long-run performance is highest for IPOs during the mandatory period, followed by IPOs with voluntary management forecasts and those without management forecasts. These differences in long-run performance do not seem to be due to poorer quality of IPOs in the voluntary period, but to decreased management earnings forecast accuracy during during the mandatory disclosure regime. The evidence is consistent with the systematic production of pessimistic forecasts during the mandatory period, which is mainly driven by accounting conservatism and lawsuit avoidance.

IPO firms that provide inaccurate forecasts in the mandatory regime exhibit poorer long-term performance. Therefore, setting initial expectations at a low level so as to exceed them effortlessly over the subsequent reporting periods is clearly a shortsighted strategy. Each IPO issuer should make efforts to produce accurate earnings forecasts. Firms with low forecast errors during the mandatory period offer higher long-term returns, suggesting that management efforts to provide accurate forecasts is rewarded by shareholders.

A closer look at AFE shows positive long-term returns for firms with low errors during the mandatory period, while long-term returns turn negative when errors are high. Similar results are found for the voluntary period, where only firms presenting low AFE are associated with positive long-term returns over a three-year period. It appears that investors are one step ahead of regulators and penalize, with strong negative returns, firms that are unable to provide reliable earnings forecasts.

In summary, we find positive forecast errors and higher long-term performance during the mandatory period, suggesting that the mandatory disclosure requirement causes IPO firms to systematically bias profit forecasts downwards. The mandatory disclosure requirement, hence, artificially improves long-term IPO share performance. Moreover, we find that IPO long-term performance decreases with management earnings forecast error during the mandatory period.

FOOTNOTES

1 International evidence covering management earnings forecasts around IPOs includes Lee et al. (1993) and Hartnett and Romcke (2000) for Australia; Li and McConomy (2004) for Canada; Chen and Firth (1999), Firth et al. (2008), Tan et al (2015) for China; Cormier and Martinez (2006) for France; Gounopoulos et al. (2015) for Greece; Jaggi (1997), Cheng and Firth (2000), and Chen et al. (2001) for Hong Kong; Jelic et al. (2001) for Malaysia; Firth and Smith (1992) for New Zealand; Firth et al. (1995) for Singapore; Jaggi (2006) for Taiwan; Firth and Lonkani (2005) for Thailand; Jelic (2011) for the UK; and Drobetz et al. (2015) for the maritime industry.

2 We follow Ritter (1984), who considers cold periods as those with low average initial return.

3 According to the market timing hypothesis, managers choose a window of opportunity to launch an IPO. This window is identified as a function of the firm's performance and market conditions. Generally, managers prefer to take their firms public when they have performed well, and probably the IPO date is conditional on the firm's cycle of activity and operational performance. Further, the window of opportunity for an IPO may be determined by market conditions. In a bullish market, the number of IPOs tends to increase, because the placement of stocks is easier, the risk of failure of an IPO is lower, and securities are priced higher.

4 Market optimism theory states that active "buy and sell trades" in the aftermarket during the first day of trading, so-called "flipping activity" is a good indication of future stock price performance. The theory suggests that there are periods when investors are particularly confident about firms' future projects and profits, and that managers are induced to make offerings in these periods.

5 The price support hypothesis indicates that underwriters stabilize stock prices during a short period of time after the IPO so as to avoid failure of the issue. Prices are artificially supported at a high level in the short-run, but at the end of the stabilization period, performance decreases.

6 This hypothesis assumes that investors are not able to correctly estimate firm value at the time of the offering. After a three-year period, investors would be able to identify the accounting adjustments and would reallocate their portfolios.

7 Hughes (1986) highlights that, to avoid market failure, the issuer should make a disclosure about firm value that can be verified by the investment bank.

8 The regulatory switch from mandatory to voluntary management earnings forecasts was motivated by its expected contribution to the efficient and cost-effective functioning of the capital markets. Protection of investors and maintenance of confidence in the Greek financial market were also important issues. This regulatory change was also intended to reinforce the freedom of movement of capital within the Greek market and to help small family companies to go public. The change involved the forecast for the next year-end.

9 Among the 71 listings, the most prominent were banking firms, with 15 listings, textiles with 12, construction with 8, and chemicals and food, each with 7 listings.

10 *t* corresponds to the year of IPO.

11 Fama and French (1996) find that many market "anomalies" can be explained by taking into account size and book-to-market effects through the use of a three factor benchmark.

12 Priori literature also shows that firm size is an important determinant of IPO underpricing (e.g., Hussein and Zhou 2014; Wu 2014; Deng and Zhou 2016) and long-term performance (e.g., McGuinness 2016).

13 Underwriters in the Greek stock market are either large banks (e.g., the National Bank of Greece, EFG Eurobank, Alpha Bank, and Piraeus Bank) or major securities firms.

14 Li and McConomy (2004) show that earnings forecast and retained ownership decisions are jointly determined by managers after controlling for other factors that affect each decision independently. These authors document a substitution effect between these two decisions.

15 We use this PSM matched sample to rerun our regressions in Table 7. The coefficient on *Mandatory* dummy remains positive and statistically significant, suggesting that management earnings forecasts during IPOs issued during the mandatory period are lower than actual earnings. In other words, managers under-forecast earnings when they are constrained to provide earnings forecasts. The empirical results remain qualitatively similar when we rerun regressions in Table 9 using the PSM matched sample. In particular, the coefficient on *Mandatory* dummy remains positive and statistically significant, suggestion that firms that went public during the mandatory period realize superior returns compared to those floated in the voluntary period. These results are not tabulated, for the sake of brevity, but are available from the authors upon request.

16 The results remain qualitatively similar when we use the fitted values of '*Voluntary Forecast*' from a probit model that determine the decision of managers to disclose earnings forecasts or not during the voluntary period. These results are not reported in the paper, for the sake of brevity, but are available from the authors.

ACKNOWLEDGMENTS

We would like to thank Francois Derrien, Michael Firth, Dimitris Ghikas, Bikki Jaggi, Kevin Keasey, Gopal Krishnan, Jay Ritter, Maureen McNichols, Paul McGuinness, Theodore Sougiannis, Nikos Vafeas, and an anonymous referee for their useful comments and suggestions. The usual disclaimer applies.

Appendix

See Table 12.

Table 12 Variables definition

Variable	Definition
Panel A: Measures of forecast errors and long term performance	
AFE	The Absolute Forecast Error is measured as the absolute difference between the actual earnings (published in the annual report) and the forecast earnings (announced in the prospectus of the newly listed firm) scaled by the absolute value of the forecast earnings. It is based on forecast of the next year earnings
FE	The Forecast Error is measured as the difference between the actual earnings (AE) and the forecast earnings (FE) scaled by the absolute value of the forecast earnings |FE|
ERNY1D	Adjusted returns from the first day to N years after going public (N = 1, 2, 3)
ERNY1M	Adjusted returns from first month to N years after going public (N = 1, 2, 3)
Panel B: IPO characteristics	
Size	The natural logarithm of the market capitalization. Market capitalization is measured as the total number of outstanding shares after the IPO multiplied by price per share
Age	Age of the firm from incorporation to IPO date
TimeLag	Number of days between the date of issuance of the prospectus and the first day of the stocks' listing
Privatization	A dummy variable that equals 1 for privatized IPOs, and zero otherwise
Oversubscription	Oversubscription is a demand multiple estimated as demand for IPO shares over shares offered
Underwriter	A dummy variable that equals 1 for reputable underwriters, and zero otherwise. Reputable underwriters are classified based on three criteria: (i) The number of IPOs they advised as lead underwriters, (ii) the fee requested for the offered services, and (iii) the capital raised for the public offerings
HotCold	A dummy variable that equals 1 for IPOs listed during a hot period, zero for IPOs that were listed during a cold period. The classification "Hot period" and "Cold period" is based on the intensity of IPO listing activity and market returns
Ownership	The proportion of given ownership by the pre-IPO shareholders
Mandatory	A dummy variable that equals 1 for IPOs listed during 1993–2000, and zero otherwise
Voluntary Forecast	A dummy variable that equals 1 for IPOs providing earnings forecasts during 2001–2015, zero otherwise
Industry	A dummy variable that equals 1 for industrial IPO firms, and zero otherwise

We follow Yung et al. (2008), Boehme and Colak (2012), and Thomadakis et al. (2016) to define market conditions (hot versus cold) based on the intensity of IPO listing activity and market returns. Hot periods are 1993–1994, 1997–2000, and 2004–2006. Cold periods are 1995–1996, 2001–2003, and 2007–2015

REFERENCES

1. Baginski S, Hassell J, Kimbrough M (2002) The effect of legal environment on voluntary disclosure: evidence from management earnings forecasts issued in U.S. and Canadian markets. Acc Rev 77:25–50
2. Baron D (1982) A model of the demand for investment banking advising and distribution services for new issues. J Financ 37:955–976
3. Bhabra H, Pettway R (2003) IPO prospectus information and subsequent performance. Financ Rev 38:369–397

4. Boehme R, Colak G (2012) Primary market characteristics and secondary market frictions of stocks. J Financ Mark 15:286–327
5. Brav A, Gompers P (1997) Myth or reality? The long run underperformance of initial public offerings: evidence from venture and non venture capital backed companies. J Financ 32:1791–1821
6. Carter B, Dark F, Singh R (1998) Underwriter reputation, initial returns, and the long run performance of IPO stocks. J Financ 53:285–311
7. Cazavan-Jeny A, Jeanjean T (2007) Levels of voluntary disclosure in IPO prospectuses: an empirical analysis. Rev Acc Financ 6:131–149
8. Chan A, Cora S, Millie T, Wong D, Chan R (1996) Possible factors of the accuracy of prospectus earnings forecasts in Hong Kong. Int J Acc 31:381–398
9. Chen G, Firth M (1999) The accuracy of profit forecasts and their roles and associations with IPO firm valuations. J Int Financ Manag Acc 10:202–226
10. Chen G, Firth M, Kim J-B (2000) The post-issue market performance of initial public offerings in China's new stock markets. Rev Quant Financ Acc 14:319–339
11. Chen G, Firth M, Krishnan G (2001) Earnings forecast errors in IPO prospectuses and their associations with initial stock returns. J Multinatl Financ Manag 11:225–240
12. Cheng T, Firth M (2000) An empirical analysis of the bias and rationality of profit forecasts published in new issue prospectuses. J Bus Financ Acc 27:423–446
13. Chou D-W, Gombola M, Liu F-Y (2010) Earnings management and long-run stock performance following private equity placements. Rev Quant Financ Acc 34:225–245
14. Cormier D, Martinez I (2006) The association between management earnings forecasts, earnings management, and stock market valuation: evidence from French IPOs. Int J Acc 41:209–236
15. Dechow PM, Hutton A, Sloan R (2000) The relation between analysts' forecasts of long-term earnings growth and stock price performance following equity offerings. Contemp Acc Res 17:1–32
16. DeGeorge F, Derrien F (2001a) IPO performance and earnings expectations: some French evidence. Working paper, HEC Paris
17. DeGeorge F, Zeckhauser R (1993) The reverse lbo decision and firm performance: theory and evidence. J Financ 48:1323–1348
18. Deng Q, Zhou Z-G (2016) The pricing of the first day opening price returns for ChiNext IPOs. Rev Quant Financ Acc. doi:10.1007/s11156-015-0500-x
19. Dimovski W, Brooks R (2004) Initial public offerings in australia 1994 to 1999, recent evidence of underpricing and underperformance. Rev Quant Financ Acc 22:179–198
20. Dimovski W, Philavanh S, Brooks R (2011) Underwriter reputation and underpricing: evidence from the Australian IPO market. Rev Quant Financ Acc 37:409–426
21. Drobetz W, Gounopoulos D, Merika A, Merikas A (2015) Determinants of management earnings forecasts: the case of global shipping IPOs. European Financial Management (forthcoming)
22. Eddy A, Seifert B (1992) An examination of hypothesis concerning earning forecasts errors. Quart J Bus Econ 31:22–37
23. Fama E, French K (1993) Common risk factors in the returns on bonds and stocks. J Financ Econ 33:3–56
24. Fama E, French K (1996) Multifactor explanations of asset pricing anomalies. J Financ 51:55–84
25. Firth M, Smith A (1992) The accuracy of profits forecasts in initial public offerings prospectuses. Acc Bus Res 22:239–247
26. Firth M, Kwok B, Liau-Tan C, Yeo G (1995) Accuracy of profit forecasts in IPO prospectus. Acc Bus Rev 2:55–83

27. Firth M, Li Y, Wang SS (2008) Valuing IPOs using price-earnings multiples disclosed by IPO firms in an emerging capital market. Rev Pac Basin Financ Mark Polic 11:429–463
28. Firth M, Gounopoulos D, Pulm J (2012) IFRS adoption and management earnings forecasts of Australian IPOs. In: European accounting association conference
29. Frankel R, McNichols M, Wilson GP (1995) Discretionary disclosure and external financing. Acc Rev 70:135–150
30. Frankel RM, Johnson MF, Nelson KK (2002) The relation between auditor's fees for nonaudit services and earnings management. Acc Rev 77:71–115
31. Gajewski JF, Gresse C (2006) A survey of the European IPO market. ECMI paper
32. Ghikas D, Papadaki A, Siougle G, Sougiannis T (2008) The relevance of quantifiable audit qualifications in the valuation of IPOs. Rev Acc Stud 13:512–550
33. Goergen M, Khurshed A, Mudambi R (2007) The long run performance of IPOs: can it be predicted. Manag Financ 33:401–419
34. Gong G, Li L, Xie H (2009) The association between management earnings forecast errors and accruals. Acc Rev 84:497–530
35. Gounopoulos D, Kraft A, Skinner F (2015) Voluntary versus. Mandatory earnings management in IPOs. Newcastle University Working Paper
36. Guo R (2005) Information collection and IPO underpricing. Rev Quant Financ Acc 25:5–19
37. Hansen R, Torregrosa P (1992) Underwriter compensation and corporate monitoring. J Financ 47:1537–1555
38. Hartnett N, Romcke J (2000) The predictability of management forecast error: a study of Australian IPO disclosure. Multinatl Financ J 4:101–132
39. Hensler D, Ruherford R, Springer T (1997) The survival of initial public offerings in the aftermarket. J Financ Res 20:93–110
40. Horton J, Serafeim G, Serafeim I (2013) Does mandatory IFRS adoption improve the information environment. Contemp Acc Res 30:388–423
41. Hughes P (1986) Signalling by direct disclosure under asymmetric information. J Acc Econ 8:199–242
42. Hussein MM, Zhou ZG (2014) The initial return and its conditional return volatility: evidence from the Chinese IPO market. Rev Pac Basin Financ Mark Polic 17:1–32
43. Hutton A, Miller G, Skinner D (2003) The role of supplementary statements with management earnings forecast. J Acc Res 41:867–890
44. Jaggi B (1997) Accuracy of forecast information disclosed in the IPO prospectuses of Hong Kong companies. Int J Acc 32:301–319
45. Jaggi B, Chin C, Lin W, Lee P (2006) Earnings forecast disclosure regulation and earnings management: evidence from Taiwan IPO firms. Rev Quant Financ Acc 26:275–299
46. Jain B, Kini O (1994) The post-issue operating performance of IPO firms. J Financ 49:1699–1726
47. Jain B, Kini O (2000) Does the presence of venture capitalists improve the survival profile of IPO firms? J Bus Financ Acc 27:1139–1183
48. Jelic R (2011) Management forecasts and IPO performance. University of Sussex Working Paper
49. Jelic R, Saadouni B, Briston R (1998) The accuracy of earnings forecast in IPO prospectuses on the Kuala Lumpur stock exchange 1984–1995. Acc Bus Res 29:57–72
50. Jelic R, Saadouni B, Briston R (2001) Performance of Malaysian IPOs: underwriters reputation and earnings forecast. Pac-Bas Financ J 9:457–486
51. Jog V, McConomy B (2003) Voluntary disclosures of management earnings forecast in ipos and the impact on underpricing and post-issue return performance. J Bus Financ Acc 30:125–167
52. Karamanou I, Vafeas N (2005) The association between corporate boards, audit committees, and management earnings forecasts: an empirical analysis. J Acc Res 43:453–486

53. Lee I, Taylor N, Yee C, Yee M (1993) Prospectus forecast earning: evidence and explanations. Austr Acc Rev 3:21–32
54. Li Y, McConomy B (2004) Simultaneous signaling in IPOs via management earnings forecasts and retained ownership: an empirical analysis of the substitution effect. J Acc Audit Financ 19:1–28
55. Liao T-L, Huang C-J, Liu H-C (2015) Earnings management during lockup in explaining IPO operating underperformance. Rev Pac Basin Financ Mark Polic 18:1–21
56. Ljungqvist A (2007) IPO underpricing. In: Eckbo B (ed) Handbook of corporate finance: empirical corporate finance. Elsevier, Amsterdam
57. Ljungqvist A, Nanda V, Singh R (2006) Hot markets, investor sentiment, and IPO pricing. J Bus 79:1667–1702
58. Lonkani R, Firth M (2005) The accuracy of IPO earnings forecasts in Thailand and their relationships with stock market valuation. Acc Bus Res 35:269–286
59. Mak Y (1989) The determinants of accuracy of management earning forecast: a New Zealand study. Int J Acc 24:267–280
60. Mak Y (1994) The voluntary review of earnings forecasts disclosed in IPO prospectuses. J Acc Public Polic 13:141–158
61. McGuinness PB (2016) Post-IPO performance and its association with subscription cascades and issuers' strategic-political importance. Rev Quant Financ Acc 46:291–333
62. Megginson W, Nash R, Netter J, Schwartz A (2000) The long-run return to investors in share issue privatizations. Financ Manag 29:67–77
63. Mitchell ML, Stafford E (2000) Managerial decisions and long-run stock price performance. J Bus 73:287–320
64. Pedwell K, Warsame H, Neu D (1994) The accuracy of Canadian and New Zealand earnings forecasts: a comparison of voluntary versus compulsory disclosures. J Int Acc Audit Tax 3:221–236
65. Ritter J (1984) The hot issue market of 1980. J Bus 57:215–240
66. Ritter J (1991) The long performance of initial public offerings. J Financ 46:3–28
67. Selva M, Ma A, Wa J (1994) The reliability of prospectus-based profit forecasts in Hong Kong. Working paper, Hong Kong City University
68. Shen Z, Coakley J, Instefjord N (2014) Earnings management and IPO anomalies in China. Rev Quant Financ Acc 42:69–93
69. Skinner K (1994) Why firms voluntarily disclose bad news. J Acc Res 32:38–60
70. Tan Q, Dimovski W, Fang V (2015) The underpricing of infrastructure IPOs: evidence from China. Rev Pac Basin Financ Mark Polic 18:1–31
71. Teoh S, Welch I, Wong T (1998a) Earnings management and the long-run market performance of initial public offerings. J Financ 53:1935–1974
72. Teoh S, Welch I, Wong T (1998b) Earnings management and the underperformance of seasoned equity offerings. J Financ Econ 50:63–99
73. Thomadakis S, Gounopoulos D, Nounis C (2012) Long term performance of greek IPOs. Eur Financ Manag 17:117–141
74. Thomadakis S, Gounopoulos D, Nounis C, Riginos M (2016) Innovation and upheaval: early growth in the Greek Capital Market listing and IPOs from 1880 to World War II in the Athens Stock Exchange. Econ Hist Rev (forthcoming)
75. Thomadakis S, Gounopoulos D, Nounis C, Merikas A (2016) Collateral regulation and IPO specific rebelarization: the case or price limits in the Athens stock exchange. Eur Financ Manag 22:1–32
76. Wu C (2014) Underpricing of homecoming A-share IPOs by Chinese firms already listed abroad. Rev Quant Financ Acc 43:627–649
77. Yi J-H (2001) Pre-offering earnings and the long-run performance of IPOs. Int Rev Financ Anal 10:53–67
78. Yung C, Colak G, Wang W (2008) Cycles in the IPO market. J Financ Econ 89:192–208

CHAPTER 4

Stock Price Reaction to Profit Warnings: The Role of Time-Varying Betas

Shuxing Yin[1], Khelifa Mazouz[2], Abdelhafid Benamraoui[3*], Brahim Saadouni[4]

[1.] *Sheffield Management School Sheffield University, Sheffield, UK*

[2.] *Cardiff Business School Cardiff University, Cardiff, UK*

[3.] *Westminster Business SchoolWestminster University, London, UK*

[4.] *Manchester Business School, Manchester University, Manchester, UK*

ABSTRACT

This study investigates the role of time-varying betas, event-induced variance and conditional heteroskedasticity in the estimation of abnormal returns around important news announcements. Our analysis is based on the stock price reaction to profit warnings issued by a sample of firms listed on the Hong Kong Stock Exchange. The standard event study methodology indicates the presence of price reversal patterns following both positive and negative warnings. However, incorporating time-varying betas, event-induced variance and conditional heteroskedasticity in the modelling process results in post-negative-warning price patterns that are consistent with the predictions of the efficient market hypothesis. These adjustments also cause the statistical significance of some post-positive-warning cumulative abnormal returns to disappear and their magnitude to drop to an extent that minor transaction costs would eliminate the profitability of the contrarian strategy.

Keywords: Profit warnings, Market efficiency, Overreaction, Time-varying betas, Event-induced variance

JEL Classification G14

1 INTRODUCTION

This study investigates the market reaction to profit warnings[1] in the Hong Kong Stock Exchange (HKEx). Empirical evidence on the price reaction to profit warnings is limited to a few markets, such as the US, the UK and mainland China. Clare (2001) shows that investors in the UK tend to overreact more to negative warnings than to positive ones. Using Chinese data, Lui et al. (2009) find that stock prices overreact to the negative news contained in profit warnings. Specifically, they report a significant price drop of about −3% over the [−1, +1] window and a significant price increase of 7.81% over the [+2, +60] window around negative warnings. Consistent with the overreaction hypothesis, Tucker (2004) documents that investors react more negatively to firms that warn, when they anticipate negative earnings news, than those that do not warn. Jackson and Madura (2003) examine stock price behavior around profit warnings containing negative news in the US market. They find a significant price drop of −21.7% over the 11-day period ending 5 days after the announcement. However, they find no evidence of reversal after this period and conclude that the market reaction to negative warnings is not excessive.

Previous studies on the stock price reaction to profit warnings are based on the standard event study methodology, which ignores the potential impact of news on stock betas and the residual variance (Brown et al. 1988; Corrado and Jordan 1997; Cyree and DeGennaro 2002; Lui et al. 2009; Savickas 2003; Zolotoy 2011; Cam and Ramiah 2014), a deficiency that may hinder the testing of the efficient market hypothesis (EMH).[2] Specifically, if a stock's beta changes after the event date, the use of the pre-event beta may result in inaccurate abnormal return estimates. Furthermore, several studies argue that ignoring the heteroskedastic nature of volatility and event-induced variance will lead to biased market model parameter estimates and inconsistent test statistics (e.g., Akgiray 1989; Corhay and Tourani-Rad 1994).

It is widely documented in the literature that equity betas are not constant over time. For example, Klemkosky and Martin (1975) and Bollerslev et al. (1988) argue that investors' expected returns are conditional on the information available at any particular point in time and their consistent reestimation of factor returns causes the betas of risky assets to vary considerably over time. Consistent with this prediction, several studies find that estimated betas exhibit statistically significant time variation (see, e.g., Harvey 1989; Ferson and Korajczyk 1995; Faff et al. 2000). Many researchers also show that the variation in betas is more pronounced around important news announcements. Zolotoy (2011), for instance, argues that, as equity value tends to rise (fall) following the arrival of good (bad) news, the weight attached to debt falls (rises). As such, the release of positive (negative) news decreases (increases) the riskiness of equity investments. Similarly, Lui et al. (2009) find that the systematic risk of individual stocks reacts asymmetrically to the positive and negative news contained in the analyst reports. However, Brown et al. (1988) argue that, as surprises increase uncertainty, stock betas should increase following both favourable and unfavourable surprises. Specifically, they claim that investors tend to set prices before the full ramifications of a dramatic financial event become known, and the arrival of news immediately causes risk-averse investors to set stock prices significantly below their conditional expected values. Using a sample of the 200 largest S&P firms over the period 1962–1985, Brown et al. (1988) show that the variance of returns, the residual variance and the coefficient of systematic risk exhibit significant increases following daily residual returns in excess of 2.5 (sign ignored). Kalay and Lowenstein (1985) argue that, since events convey information to the market, stock price volatility should also increase in the post-announcement period. They also maintain that the incremental risk above that on a random day may not be fully diversifiable. Using a sample of 302 US firms for the period 1962–1980, they show that stock betas exhibit a significant increase in the aftermath of dividend announcements. More recently, Patton and Verardo (2012) propose a simple learning model, which suggest that investors use firm-specific news to extract information on the aggregate economy. They argue that since earnings are affected by both market-wide and firm-specific conditions, investors can use earnings of the announcing firms to revise their expectations about the profitability of other firms in the economy. This process of learning across stocks is expected to drive up the comovement between announcing stocks and other stocks and yield an increase in the market beta of the announcing stocks. Consistent with the predictions of the learning model, Patton and Verardo show that the betas of the constituents of the S&P 500 increase significantly around earnings announcements, regardless of whether the news is good or bad.

The extant literature has recognised the importance of adjusting for stochastic volatility around the event period. For example, Brown and Warner (1980) argue that the failure to adjust for the event-induced variance may result in biased estimates of the traditional test statistics, and that the power of these tests can be improved by appropriately modelling the volatility process. To control for the event-induced variance, Boehmer et al. (1991) propose a statistical test, generated by first standardising event-period returns by the estimation-period standard deviation and then dividing the cross-sectional mean of the standardised returns by their cross-sectional standard deviation. Savickas (2003) argues that a limitation of Boehmer et al.'s approach stems from the implicit assumption that the event-induced variance is the same across all sample stocks. Brockett et al. (1999) model the volatility process around the events using a market model with GARCH effects and time-varying betas. However, their approach ignores the importance of event-induced variance. Savickas (2003) addresses the conditional heteroskedastic behavior of volatility and the event-induced residual variance in a single model. Nevertheless, his model ignores the importance of time-varying betas, a phenomenon that may be particularly important around dramatic financial events (see, e.g., Brown et al. 1988; Zolotoy 2011; Lui et al. 2009).

In this study, we investigate the market reaction to profit warnings using an event study methodology that addresses the conditional heteroskedastic behavior of volatility, the time-varying betas and the event-induced variance simultaneously. Our contribution to the literature is twofold. First, we examine whether the nature of profit warnings affects a stock's beta. Brown et al. (1988) shows that stock betas increase following daily price shocks (daily residual returns) of ±2.5% or more. However, we argue that large daily price shocks may be caused by the behavior of noise traders rather than the arrival of news. Thus, using profit warnings as the events provides a cleaner test of the impact of the arrival of news on time-varying betas. Second, we propose an event study methodology that adjusts the abnormal return estimates and statistical tests for the time-varying betas and the event-induced residual variances as well as the conditionally heteroskedastic behavior of volatility. Specifically, we improve on Savickas' (2003) model by allowing the beta parameter of the market model to vary over time, while adjusting for the event-induced variance and conditional heteroskedasticity. We argue that this improvement is important not only because a stock beta varies systematically over time, but also because its variation is likely to be more pronounced around news events (Brown et al. 1988; Zolotoy 2011; Lui et al. 2009).

The results of our study can be summarised briefly as follows. Firstly, we observe the highest positive (negative) abnormal returns on the day on which positive (negative) warnings are released. We also report strong price reversal patterns following both positive and negative warnings. Secondly, consistent with Brown et al.'s (1988) view that surprises increase uncertainty, the average beta of event stocks increases significantly after both positive and negative warnings. Thirdly, we show that the statistical significance of the cumulative abnormal returns (CARs) following negative warning announcements disappears completely after we account for the time-varying betas. We also find that the time-varying beta adjustments reduce the magnitude and the statistical significance of the price reversal patterns following positive warnings. Fourthly, the average abnormal returns on the first 2 days following warnings containing positive news lose their significance after the event-induced variance and conditional heteroskedasticity are accounted for, while the subsequent CARs become much smaller. Finally, we show that adjusting for time-varying betas, event-induced variance and conditional heteroskedasticity simultaneously yields even smaller post-positive-warning CARs. We argue that, although many of the post-positive-warning CARs remain significant after the incorporation of time-varying betas, event-induced variance and conditional heteroskedasticity in the modelling process, these patterns should not be used as evidence against the EMH for at least three reasons. First, the results from the pooled ordinary least square (OLS) regressions indicate that the post-positive-warning CARs are not related to the abnormal returns on the announcement days. This finding is not consistent with the prediction of the overreaction hypothesis, which suggests that greater overreaction would lead to greater correction (see, e.g., Cox and Peterson 1994; Choi and Jayaraman 2009). Second, the magnitude and the statistical significance of the post-positive-warning CARs are highly sensitive to the way in which abnormal returns are measured. Finally, incorporating time-varying betas, event-induced variance and conditional heteroskedasticity into the modelling process yields post-positive-warning CARs that are too small to cover the transaction costs that one would incur in pursuing a contrarian strategy.

The remainder of the paper is organised as follows. Section 2 describes the dataset. Section 3presents the empirical tests and results. Section 4 provides some additional results and robustness checks, and Sect. 5 concludes.

2 DATA

The statements of profit warnings used in this study are obtained from the HKEx website. The exchange began publishing these statements in July 2007. Our initial sample includes all such statements issued during the period from July 2007 to July 2012. Daily closing prices of the issuing firms and the Hang Seng index are obtained from DataStream. Our final sample constitutes a total of 1723 profit warnings, of which 1238 contain negative news and 485 contain positive news about firms' earnings prospects. The dominance of warnings containing negative news in our sample may be attributed to the fact that only bad news is emphasised in the "listing rules" in Hong Kong (see, e.g., Wang and Zhang 2011). Furthermore, several studies suggest that warnings are more likely to be issued prior to negative than positive news (see, e.g., Barmber and Cheon 1998; Libby and Tan 1999; Wang and Zhang 2011).

3 TESTS AND RESULTS

To estimate the abnormal returns around profit warnings, we initially use the standard event study methodology. Then, we relax the market model assumption that stock betas and residual variance are constant over time and not affected by the arrival of news. Specifically, we use a bivariate form of Engle and Kroner's (1995) BEKK GARCH model to account for time-varying betas. Then, we employ a model developed by Savickas (2003) to control for the effect of event-induced variance and the heteroskedastic behavior of volatility on the abnormal return estimates. Finally, we improve on Savickas' approach by allowing stock betas to vary over time.

3.1 Standard event study

The approach most commonly used to estimate expected returns is the market model[3]:

$$R_{it} = \alpha_i + \beta_i R_{mt} + \varepsilon_{it}, \quad E(\varepsilon_{it}) = 0, var(\varepsilon_{it}) = \sigma_i^2, \tag{1}$$

where $R_{i,t}$ and $R_{m,t}$ are day t's continuously compounded returns of stock i and the market portfolio m, respectively; αi and βi are the parameters of the market model and εit is the error term, which is assumed to have a zero mean and a constant variance, $\sigma i2$. Equation (1) is estimated over the [−200, −15] window prior to profit warnings. The abnormal return of stock i on day t, or $AR_{i,t}$, is then estimated as

$$AR_{it} = R_{it} - \left(\widehat{\alpha_i} + \widehat{\beta}_i R_{mt}\right). \tag{2}$$

The price effect of profit warnings is measured using the daily average abnormal return $(\overline{AR_t})$ and the cumulative average abnormal return $(\overline{CAR_s})$. The average abnormal return on day t is computed as

$$\overline{AR_t} = \frac{\sum_{i=1}^{N} AR_{i,t}}{N}. \tag{3}$$

The cumulative abnormal return of stock i over a window of S days starting one day after the event is given as

$$CAR_{is} = \sum_{t=1}^{S} AR_{it}. \tag{4}$$

The average cumulative abnormal return over a window of S days beginning 1 day after the event and across N stocks, is computed as

$$\overline{CAR_s} = \frac{1}{N}\sum_{i=1}^{N} CAR_{is}. \tag{5}$$

Two test statistics, T_1 and T_2, are used to assess the statistical significance of $\overline{AR_t}$ and $\overline{CAR_s}$, respectively, and are specified as follows:

$$T_1 = \frac{\overline{AR_t}}{\widehat{\sigma_{AR}}/\sqrt{N}} \text{ and } T_2 = \frac{\overline{CAR_s}}{\widehat{\sigma_{CAR}}/\sqrt{N}}, \tag{6}$$

where $\widehat{\sigma_{AR}}$ and $\widehat{\sigma_{CAR}}$ are the standard deviations of $AR_{i,t}$ and $CAR_{i,s}$, respectively, and are estimated as

$$\widehat{\sigma_{AR}} = \sqrt{\frac{1}{N-1}\sum_{i=1}^{N}\left(AR_{i,t} - \overline{AR_t}\right)^2} \text{ and } \widehat{\sigma_{CAR}} = \sqrt{\frac{1}{N-1}\sum_{i=1}^{N}\left(CAR_{i,s} - \overline{CAR_s}\right)^2}.$$

When standard assumptions hold, T_1 and T_2 follow Student t-distributions with $N - 1$ degrees of freedom.[4]

Table 1 reports the average daily ARs and the CARs around profit warnings. Panel A reports the CARs following negative warnings. The average AR on the announcement date is −3.46% and is highly significant (with a t-value of −16.73). The average CARs following the announcements of bad news are positive and significant, except for day 1, with values increasing monotonically from 0.01% on day 1 to 1.25% for the [+1, +10] window. This finding indicates that, on average, an investor can earn a significant abnormal return of 1.25% by holding a stock for a period of 10 days starting 1 day after a negative warning.

Table 1 Price reaction to profit warnings: standard event study approach

	Panel A: Negative warnings		Panel B: Positive warnings	
	Value (%)	t-statistic	Value (%)	t-statistic
$\overline{AR_0}$	−3.46***	−16.73	4.16***	12.46
$\overline{AR_1}$	−0.01	0.04	−0.90***	−5.00
$\overline{CAR_{[1,2]}}$	0.41*	1.76	−1.27***	−4.68
$\overline{CAR_{[1,3]}}$	0.49*	1.89	−1.30***	−4.31
$\overline{CAR_{[1,4]}}$	0.55*	1.87	−1.43***	−4.30
$\overline{CAR_{[1,5]}}$	0.74**	2.22	−1.73***	−4.89
$\overline{CAR_{[1,6]}}$	0.90**	2.45	−1.98***	−5.04
$\overline{CAR_{[1,7]}}$	1.02***	2.58	−2.15***	−5.10
$\overline{CAR_{[1,8]}}$	1.09***	2.66	−2.40***	−5.51
$\overline{CAR_{[1,9]}}$	1.28***	2.97	−2.71***	−5.99
$\overline{CAR_{[1,10]}}$	1.25***	2.73	−2.82***	−5.87
N	1238		485	

***, ** and * Significance at 1, 5 and 10%, respectively

The abnormal return of stock *i* on day *t*, or $AR_{i,t}$, is estimated using the parameters of the standard market model over the [−200, −15] window prior to profit warnings. The price effect of profit warnings is measured using the daily average abnormal return $(\overline{AR_t})$ and the cumulative average abnormal return $(\overline{CAR_s})$. The average abnormal return on day t is given by

$$CAR_{i,s} = \sum_{t=1}^{S} AR_{i,t}$$

The cumulative abnormal return of stock *i* over a window of *S* days starting 1 day after the event is given by $CAR_{i,s} = \sum_{t=1}^{S} AR_{i,t}$

The average cumulative abnormal return over a window of *S* days beginning 1 day after the shock and across *N* stocks is given by

$$\overline{CAR_s} = \frac{1}{N}\sum_{i=1}^{N} CAR_{i,s}$$

A standard *t* test is used to check the statistical significance of $(\overline{AR_t})$ and $(\overline{CAR_s})$

Panel B presents the results on the price reaction to positive warnings. The average AR on the announcement date is positive (with a value of

3.9%) and is statistically significant at less than the 1% level. The average CARs following positive news are negative and highly significant, indicating strong price reversal following the arrival of good news. This finding suggests that, on average, an abnormal return of 2.82% can be earned by short selling a stock 1 day after a positive warning and repurchasing it 10 days later. Our evidence is similar to that of Lui et al. (2009), who find strong share price reversal following negative warnings in the Chinese main market. Price reversal patterns are also documented by several studies on the price reaction to shocks. For instance, Atkins and Dyl (1990) report significantly negative (positive) average CARs from following the three largest winners (losers) on 500 randomly selected trading days from 1975 to 1984. Bremer and Sweeney (1991) examine stock price behavior following a 1-day price change of −10% or less, and their results confirm the overreaction hypothesis.

We argue that one weakness of the existing literature on the stock price reaction to profit warnings stems from the explicit assumption that the stock beta is constant over time and is not affected by the arrival of news. In this study, we adopt the bivariate form of Engle and Kroner's (1995) BEKK GARCH model to estimate time-varying systematic risk and the conditional abnormal returns.

3.2 Time-varying betas

Following Tsui and Yu (1999), Choudhry (2005) and Choudhry et al. (2010), we use the bivariate form of Engle and Kroner's (1995) BEKK GARCH to estimate the conditional betas of the market model. Under this model, the conditional variance–covariance matrix is specified as follows:

$$y_t = \upsilon + \Xi_t;\ \Xi_t|\Psi_{t-1} \sim N(0, H), \tag{7}$$

$$H_t = C'C + A'\Xi_{t-1}\Xi'_{t-1}A + B'H_{t-1}B, \tag{8}$$

here yt=(Rit,Rm,t)yt=(Rit,Rm,t) is a 2 × 1 vector containing the continuously compounded returns of stock *i* and the market portfolio *m*; υυ is a 2 × 1 vector of constants; *Ht* is a 2 × 2 conditional variance–covariance matrix, which depends on the elements of the information set Ψ_{t-1} of the past value of the error term Ξ_t; *C*, *A* and B are 2 × 2 matrices of parameters.

Engle and Kroner (1995) argue that the BEKK model is sufficiently general as it includes all positive definite diagonal representations and nearly all positive definite vector representations. It has also been suggested that the BEKK model addresses an important weakness of the general specification of the multivariate GARCH by ensuring that the *Ht* matrix is always positive definite (see, e.g., Bollerslev et al. 1994).

Equations (7) and (8) are estimated separately for the [−201, −1] and [0, +200] windows around warning dates, and the time-varying beta of stock *i* (*βi,t*) is computed as

$$\widehat{\beta_{it}} = \frac{h_{12,t}}{h_{22,t}}, \tag{9}$$

where *h12,t* and *h22,t* are elements of the matrix *Ht* and are defined as the conditional covariance between stock *i*'s returns and the market portfolio returns, and the conditional variance of the market portfolio returns, respectively.

The standard paired *t* test and the non-parametric Wilcoxon signed rank test (WSRT) are used to judge whether the average pre-warning beta, measured over the [−201, −1] window prior to the announcement date, is significantly different from the average post-warning beta, measured over a window of *S* days beginning 1 day after the event.

To account for the potential effect of changes in beta on the abnormal returns around profit warnings, we estimate conditional beta-adjusted abnormal returns as follows[5]:

$$AR_{it} = R_{it} - \left(\widehat{\alpha_i} + \widehat{\beta_{it}} R_{mt}\right). \tag{10}$$

The average conditional beta-adjusted average ARs and CARs are calculated using Eqs. (2) and (5), respectively. The standard *t* test is also used to gauge their statistical significance.

Table 2 reports the changes in conditional betas following profit warnings. Panel A presents the changes in conditional betas following negative warnings. The mean (median) pre-warning conditional beta is 0.609 (0.579). The results suggest that the post-warning betas are higher than the pre-warning betas, across all post-event windows. Both the paired *t* test and the WSRT suggest that the difference between the pre- and post-warning betas is statistically significant. Panel B presents the conditional betas around positive warnings. The mean (median) conditional beta over the [−201, −1] window prior to warnings containing positive news is 0.679 (0.685). Both the paired *t* test and the WSRT indicate that stock betas increase significantly following the arrival of positive news.

Table 2 Conditional betas following profit warnings

	Mean	t-statistic	Median	t-statistic
Panel A: Conditional betas around negative warnings				
$\beta_{[-201,-1]}$	0.609		0.579	
$\beta_{[0,+1]}$	0.655***	4.14	0.625***	4.30
$\beta_{[0,+2]}$	0.654***	3.98	0.623***	4.14
$\beta_{[0,+3]}$	0.653***	3.88	0.623***	4.17
$\beta_{[0,+4]}$	0.652***	3.61	0.622***	4.12
$\beta_{[0,+5]}$	0.652***	3.69	0.622***	4.12
$\beta_{[0,+6]}$	0.652***	3.74	0.619***	4.08
$\beta_{[0,+7]}$	0.652***	3.78	0.619***	4.08
$\beta_{[0,+8]}$	0.653***	3.82	0.619***	4.08
$\beta_{[0,+9]}$	0.653***	3.89	0.619***	4.10
$\beta_{[0,+10]}$	0.654***	3.93	0.617***	4.11
N	1238			
Panel B: Conditional betas around positive warnings				
$\beta_{[-201,-1]}$	0.679		0.685	
$\beta_{[0,+1]}$	0.729***	2.93	0.707***	2.96
$\beta_{[0,+2]}$	0.726***	2.76	0.707***	2.86
$\beta_{[0,+3]}$	0.727***	2.78	0.709***	2.86
$\beta_{[0,+4]}$	0.728***	2.88	0.709***	2.93
$\beta_{[0,+5]}$	0.728***	2.89	0.709***	2.90
$\beta_{[0,+6]}$	0.728***	2.88	0.711***	2.88
$\beta_{[0,+7]}$	0.728***	2.90	0.711***	2.90
$\beta_{[0,+8]}$	0.728***	2.89	0.711***	2.89
$\beta_{[0,+9]}$	0.728***	2.90	0.711***	2.91
$\beta_{[0,+10]}$	0.729***	2.93	0.711***	2.94
N	485			

***, ** and * Significance at 1, 5 and 10%, respectively

We use the bivariate form of Engle and Kroner's (1995) BEKK GARCH model to estimate the conditional betas. The model is estimated separately for the [-201, −1] and [0, +200] windows around announcement dates. The standard paired *t* test and the non-parametric Wilcoxon signed rank test (WSRT) are used to judge whether the average pre-warning announcement beta, measured over the [−201, −1] window prior to the announcement date, is significantly different from the average post-warning beta, measured over a window of *S* days beginning 1 day after the event

Figures 1 and 2 indicate the presence of a structural break in average daily betas following both positive and negative warning announcements. Specifically, the figures show that betas are relatively stable before warning announcements, increase substantially on the announcement dates and stabilise thereafter. In contrast to Patton and Verardo (2012), our results suggest that betas do not revert back to the pre-announcement average over the 10-day window following warning announcements.

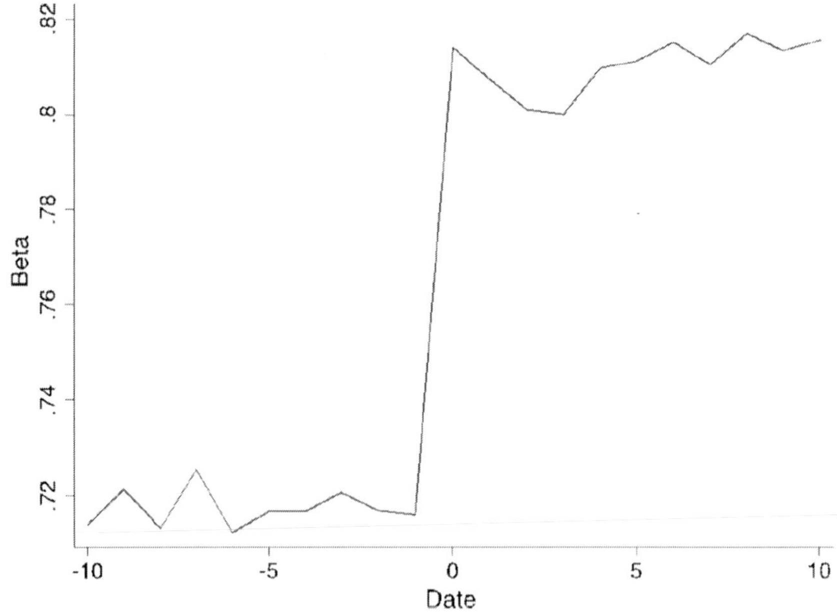

Figure. 1 Changes in beta around positive warnings

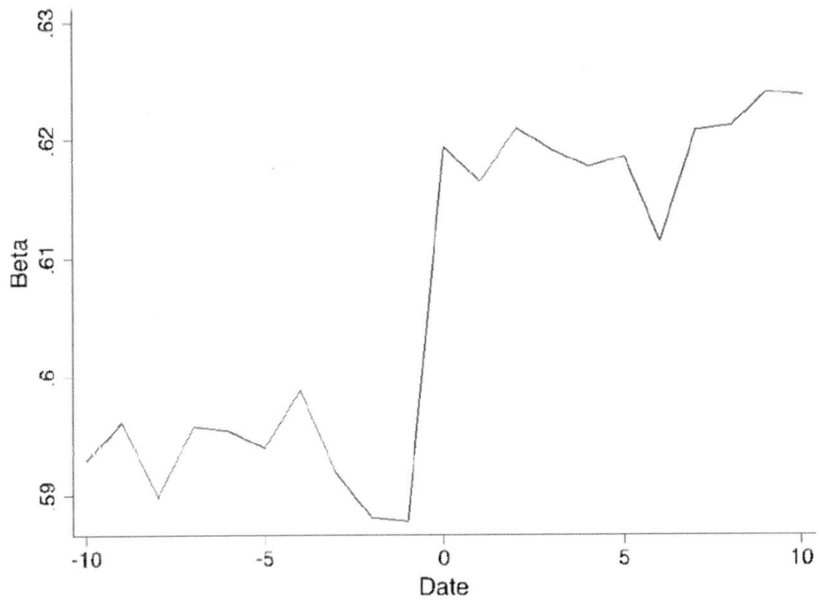

Figure. 2 Changes in beta around negative warnings

We conduct two additional tests to investigate whether the changes in betas are indeed caused by the arrival of news. First, for each event stock, we select a date randomly from its pre-warning window. We treat the randomly selected dates as if they were the event dates. The paired *t* test and WSRT suggest that the changes in beta following these pretended event dates are not significantly different from zero (the table is omitted to conserve space). Second, we match event stock with another stock from the same industry, with the closest market capitalization and with no price sensitive news within the [−10, +10] window around the warning announcement dates. In untabulated results, we find that the changes in the betas of the control stocks around the warning events are not significantly different from zero, implying that the increase in the post-warning betas is more likely to be caused by the arrival of news than the systematic time-varying nature of stock betas. This evidence is consistent with Kalay and Lowenstein (1985), Brown et al. (1988) and Patton and Verardo (2012), who show that, as surprises increase uncertainty, systematic risk increases significantly in the aftermath of both positive and negative news. Hence, ignoring the time-varying nature of the betas may result in biased abnormal return estimates.

Table 3 reports the conditional beta-adjusted average ARs and conditional beta-adjusted average CARs following the release of profit warnings.

Table 3 Price reaction to profit warnings: The conditional market model approach

	Conditional versus unconditional			
	Value (%)	t-statistic	t-statistic	Z-score
Panel A: Negative warnings				
$\overline{AR}_0$	−3.54	−16.14***	−4.97***	−5.49***
$\overline{AR}_1$	−0.16	−0.92	−4.73***	−4.82***
$\overline{CAR}_{[1,2]}$	0.04	0.18	−6.52***	−5.80***
$\overline{CAR}_{[1,3]}$	−0.02	−0.07	−8.27***	−7.15***
$\overline{CAR}_{[1,4]}$	−0.03	−0.10	−8.14***	−7.25***
$\overline{CAR}_{[1,5]}$	−0.01	−0.04	−8.68***	−7.76***
$\overline{CAR}_{[1,6]}$	0.01	0.03	−8.60***	−7.81***
$\overline{CAR}_{[1,7]}$	−0.06	−0.18	−8.74***	−7.92***
$\overline{CAR}_{[1,8]}$	−0.16	−0.44	−8.97***	−8.18***
$\overline{CAR}_{[1,9]}$	−0.17	−0.44	−9.27***	−8.39***
$\overline{CAR}_{[1,10]}$	−0.32	−0.79	−9.05***	−8.08***
N	1238			
Panel B: Positive warnings				
$\overline{AR}_0$	4.11	12.87***	4.43***	5.73***
$\overline{AR}_1$	−0.67	−3.96***	5.59***	5.68***
$\overline{CAR}_{[1,2]}$	−0.75	−3.10***	7.20***	7.21***
$\overline{CAR}_{[1,3]}$	−0.67	−2.37***	7.76***	8.15***
$\overline{CAR}_{[1,4]}$	−0.67	−2.23***	7.79***	7.89***
$\overline{CAR}_{[1,5]}$	−0.85	−2.60***	7.41***	8.29***
$\overline{CAR}_{[1,6]}$	−0.97	−2.70***	8.12***	8.24***
$\overline{CAR}_{[1,7]}$	−1.05	−2.73***	8.44***	8.51***
$\overline{CAR}_{[1,8]}$	−1.12	−2.80***	8.31***	8.43***
$\overline{CAR}_{[1,9]}$	−1.27	−3.06***	8.14***	8.35***
$\overline{CAR}_{[1,10]}$	−1.19	−2.73***	8.11***	8.38***
N	485			

***, ** and * Significance at 1, 5 and 10%, respectively

Panel A focuses on the negative warnings. Our results indicate that the post-warning CARs carry a positive sign. However, none of these post-warning CARs is significantly different from zero. This evidence indicates that individual stocks in Hong Kong react efficiently to the negative news contained in profit warnings. Both the paired *t* test and WSRT suggest that the values of time-varying-beta-adjusted CARs are significantly smaller than their corresponding market model CARs in Table 1. Panel B also presents the conditional beta-adjusted average ARs and CARs following positive warnings. Stock prices respond positively to the arrival of good news. The average conditional beta-adjusted abnormal return on the announcement date is 4.11%. This figure is statistically significant at less than the 1% level. All post-warning conditional beta-adjusted CARs are significantly negative, ranging from $CAR_1 = -0.67\%$ to $CAR_9 = -1.27\%$. Both the standard *t* test and the WSRT indicate a significant decline in the values of post-positive-warning CARs following time-varying beta adjustment.

To account for the potential effect of changes in beta on the abnormal returns around profit warnings, we estimate conditional beta-adjusted abnormal returns, $AR^c_{i,t}$, as follows:

$$AR_{i,t} = R_{i,t} - \left(\widehat{\alpha_h} + \widehat{\beta_{i,t}} R_{m,t}\right)$$

where βii,ttβii,tt is a time-varying beta estimated using the bivariate form of Engle and Kroner's (1995) BEKK GARCH model, which is estimated separately for the [−201, −1] and [0, +200] windows around the announcement dates. The average conditional beta-adjusted abnormal returns, $\overline{AR^c_t}$, are calculated by $\overline{AR_t} = \frac{\sum_{i=1}^{N} AR_{i,t}}{N}$

The time-varying beta-adjusted cumulative abnormal return of stock *i* over a window of *S* days starting 1 day after the event is given by

$$CAR_{i,s} = \sum_{t=1}^{S} AR_{i,t}$$

The average time-varying beta-adjusted cumulative abnormal return over a window of *S* days beginning 1 day after the shock and across *N* stocks is computed by

$$\overline{CAR_s} = \frac{1}{N}\sum_{i=1}^{N} CAR_{i,s}$$

The standard *t* test is used to assess the statistical significance of the price effect

In short, the findings in this section suggest that the price reversal patterns following negative profit warnings are more likely to be the outcome of model misspecification than investors' overreaction. However, while the negative CARs following positive warnings become small in magnitude, they remain highly significant even after the systematic variations in stock betas are accounted for.

3.3 Conditional heteroskedasticity and event-induced variance

As discussed earlier, the extant literature suggests that accounting for the conditionally heteroskedastic behavior of volatility and event-induced variance improves the market model parameter estimates and the power of the statistical tests. Following Savickas (2003), we estimate the abnormal returns as follows:

$$\begin{cases} R_{it} = \alpha_i + \beta_{it}R_{mt} + \gamma_{in}D_{in} + \varepsilon_{it}; \varepsilon_{it}|\Omega_{it-1} \sim N\left(0, h_{it}^2\right) \\ h_{it}^2 = \varphi_i + \psi_i h_{it-1}^2 + \theta_i \varepsilon_{it-1}^2 + \phi_{in}D_{in} \end{cases} \tag{11}$$

where εitεit in Eq. (11) is a residual term, with mean of zero and time-varying variance h2ithit2assumed to follow a GARCH (1,1) process.[6] The subscript n∈[1,+N] of the variable *Din*denotes the number of days after the event day t. *Di1*, *Di2*, ..., *DiN* are dummy variables with a value of unity if *t*∈[0,+1], *t*∈[0,+2], ..., *t*∈[0,+*N*], respectively, and zero otherwise. *βin* is the change in the systematic risk measured over the window of length *n* after the profit warning. The rest of the variables are as defined previously. The parameter *γin* in Eq. (11) reflects the abnormal returns around the event and the parameter *ϕin* in Eq. (11) captures the event-induced residual variance.

Savickas (2003) argues that the increase in variance around events may result in the misspecification of the traditional test statistics, and proposes the following test to account for the stochastic behavior of volatility during both event and non-event periods:

$$test = \sum_{i=1}^{N} \frac{S_{it}}{N} \Bigg/ \sqrt{\frac{1}{N(N-1)} \sum_{i=1}^{N} \left(S_{i,t} - \sum_{j=1}^{N} S_{j,t}/N \right)^2} \tag{12}$$

where

$$S_{it} = \frac{\widehat{\gamma_{in}}}{\sqrt{\widehat{h_{in}^2}}}$$

where $\widehat{\gamma_{in}}$ is the estimated mean abnormal return of security *i* on a given day or window and $\widehat{h_{in}^2}$ is the estimated standard deviation of the abnormal return on a given day (we use the average of $\widehat{h_{it}^2}$ when abnormal returns are estimated over a window). The test statistic presented in Eq. (12) follows the Student t-distribution with $N - 1$ degrees of freedom. Savickas argues that his test has more power as it allows the event-induced volatility effect to differ across securities and each security's variance to be stochastic outside the event period.

Table 4 presents the abnormal return estimates obtained using Savickas' (2003) approach. Panel A shows that the abnormal returns following negative warnings are small and not significantly different from zero. The paired *t* test suggests that, except on day 1, the CARs from Savickas' model are significantly smaller than the CARs obtained using the standard event methodology approach (see Panel A of Table 1). The WSRT also indicates that the difference between the ARs and CARs from Savickas' approach and the market model are significant for days 0 and +1 and for the window [+1, +9] after the negative warning. The ARs and CARs reported in Panel A are also smaller than their time-varying-beta-adjusted counterparts (see Panel A of Table 3). Overall, the results in Panel A are consistent with the predictions of the EMH, which posits that stock prices adjust instantly and accurately to the arrival of news.

We use Savickas' (2003) model to estimate the abnormal returns

$$\begin{cases} R_{it} = \alpha_i + \beta_{it} R_{mt} + \gamma_{in} D_{in} + \varepsilon_{it};\ \varepsilon_{it} \mid \Omega_{it-1} \sim N\left(0, h_{it}^2\right) \\ h_{it}^2 = \varphi_i + \psi_i h_{it-1}^2 + \theta_i \varepsilon_{it-1}^2 + \phi_{in} D_{in} \end{cases}$$

where εit is a residual term, with a mean of zero and a time-varying variance h_{it}^2 assumed to follow a GARCH (1,1) process. The rest of the variables are defined as previously. The parameter γin reflects the abnormal returns around the event and the parameter ϕin captures the event-induced residual variance. We use the following statistics to account for the stochastic behavior of volatility during both event and non-event periods:

$$test = \sum_{i=1}^{N} \frac{S_{it}}{N} \Big/ \sqrt{\frac{1}{N(N-1)} \sum_{i=1}^{N} \left(S_{i,t} - \sum_{j=1}^{N} S_{j,t}/N \right)^2}, \text{ where } S_{it} = \frac{\widehat{\gamma_{in}}}{\sqrt{\widehat{h_{in}^2}}}$$

Table 4 Price reaction to profit warnings: the role of event-induced variance and conditional heteroskedasticity

	Conditional versus unconditional			
	Value (%)	t-statistic	t-statistic	Z-score
Panel A: Negative warnings				
$\overline{AR}_0$	−3.48	−7.66***	0.18	4.93***
$\overline{AR}_1$	0.00	−0.50	0.01	2.19**
$\overline{CAR}_{[1,2]}$	0.03	−1.16	1.89*	0.94
$\overline{CAR}_{[1,3]}$	−0.04	−1.30	2.49**	1.30
$\overline{CAR}_{[1,4]}$	−0.10	−0.28	2.55**	0.75
$\overline{CAR}_{[1,5]}$	0.02	−0.78	2.61***	1.32
$\overline{CAR}_{[1,6]}$	0.12	0.87	2.48**	0.88
$\overline{CAR}_{[1,7]}$	0.22	0.11	2.25**	1.31
$\overline{CAR}_{[1,8]}$	0.10	0.33	2.64***	1.04
$\overline{CAR}_{[1,9]}$	0.04	0.10	3.21***	1.97**
$\overline{CAR}_{[1,10]}$	0.14	0.95	2.62***	1.52
N	1238			
	Conditional versus unconditional			
	Mean (%)	t-statistic	t-statistic	Z-score
Panel B: Positive warnings				
$\overline{AR}_0$	3.92	4.26***	−0.06	−4.05***
$\overline{AR}_1$	−0.85	−1.27	0.14	−4.76***
$\overline{CAR}_{[1,2]}$	−0.39	−1.33	−5.13***	−6.13***
$\overline{CAR}_{[1,3]}$	−0.27	−2.31**	−4.76***	−6.05***
$\overline{CAR}_{[1,4]}$	−0.19	−2.96***	−4.67***	−5.06***
$\overline{CAR}_{[1,5]}$	−0.15	−2.33**	−5.34***	−5.73***
$\overline{CAR}_{[1,6]}$	−0.25	−4.04***	−5.20***	−6.28***
$\overline{CAR}_{[1,7]}$	−0.14	−3.71***	−5.39***	−6.39***
$\overline{CAR}_{[1,8]}$	−0.17	−3.74***	−5.78***	−6.54***
$\overline{CAR}_{[1,9]}$	−0.18	−3.26***	−6.22***	−6.86***
$\overline{CAR}_{[1,10]}$	−0.17	−3.72***	−6.01***	−6.72***
	Conditional versus unconditional			
	Mean (%)	t-statistic	t-statistic	Z-score
N	485			

***, ** and * Significance at 1, 5 and 10%, respectively

where $\widehat{\gamma_{in}}$ is the estimated mean abnormal return of security *i* on a given day or window and $\widehat{h_{in}^2}$ is the estimated standard deviation of the abnormal return on a given day (we use the average of $\widehat{h_{it}^2}$ when abnormal returns are estimated over a window). The test statistic presented in Eq. (14) follows the Student t-distribution with $N - 1$ degrees of freedom

Panel B of Table 4 shows that the average ARs and CARs from Savickas' model, following positive warnings, are negative but much smaller than those from the standard market model (Panel B of Table 1) and from the market model with time-varying beta (Panel B of Table 3). The AR_1 and CAR_2 after warnings containing positive news are not significantly different from zero. The subsequent CARs are significantly negative, but small in magnitude, ranging from −0.27% (CAR_3) to −0.14% (CAR_7). While the abnormal returns immediately after the events are not significant, the results suggest that, on average, short selling a stock immediately after a positive news announcement and repurchasing it 10 days later would generate a return of 0.17%. However, the average (median) relative bid-ask spread associated with our sample stocks during the study period is 3.06% (1.92%). Thus, such price patterns may not yield profits in excess of transaction costs and therefore cannot be used as evidence against the EMH.

3.4 Conditional heteroskedasticity, event-induced variance and time-varying betas

Savickas (2003) explicitly assumes that stock betas are constant over time and not affected by the arrival of news. To address this potential limitation, we replace the constant βi parameter in Eq. (11) with the time-varying betas generated by the BEKK GARCH model. We argue that allowing betas to be stochastic both within and outside of the event period should yield more accurate estimates of the abnormal return and residual variance and should, therefore, improve the power and accuracy of the cross-sectional test statistic presented in Eq. (12).

The results of this analysis are reported in Table 5. Panel A shows that the average ARs and CARs following negative warnings are positive but not statistically significant (except for CAR_6). The paired *t* test and the non-parametric WSRT indicate that the ARs and CARs in Panel A of Table 5 are in many cases significantly smaller in magnitude than their market model counterparts (Panel A of Table 1). Panel B of Table 5 shows that the ARs and CARs following positive warnings are small, but remain negatively significant. The paired *t* test and the WSRT also suggest that the ARs and CARs in Panel B of Table 5 are significantly smaller than those generated by the standard market model. Furthermore, the ARs and CARs in Panel B of Table 5are very similar in magnitude to those obtained using Savickas' model (Panel B of Table 4), implying that profits in excess of transaction costs cannot be earned from the contrarian strategy.

We allow the beta parameter in Savickas' (2003) model to vary over time. Specifically, we estimate the abnormal returns as follows:

$$\begin{cases} R_{it} = \alpha_i + \beta_{it} R_{mt} + \gamma_{in} D_{in} + \varepsilon_{it};\ \varepsilon_{it} | \Omega_{it-1} \sim N\left(0, h_{it}^2\right) \\ h_{it}^2 = \varphi_i + \psi_i h_{it-1}^2 + \theta_i \varepsilon_{it-1}^2 + \phi_{in} D_{in} \end{cases}$$

where εit is a residual term, with a mean of zero and a time-varying variance h_{it}^2 assumed to follow a GARCH (1,1) process, βit is the time-varying beta estimated from the BEKK GARCH model, and the rest of the variables are defined as previously. The parameter γin reflects the abnormal returns around the event and the parameter ϕin captures the event-induced residual variance. We use the following statistics to account for the stochastic behavior of volatility during both event and non-event periods:

$$test = \sum_{i=1}^{N} \frac{S_{it}}{N} \Big/ \sqrt{\frac{1}{N(N-1)} \sum_{i=1}^{N} \left(S_{i,t} - \sum_{j=1}^{N} S_{j,t}/N \right)^2},\ \text{where } S_{it} = \frac{\widehat{\gamma_{in}}}{\sqrt{\widehat{h_{in}^2}}}$$

where $\widehat{\gamma_{in}}$ is the estimated mean abnormal return of security *i* on a given day or window and $\widehat{h_{in}^2}$ is the estimated standard deviation of the abnormal return on a given day (we use the average of h_{it}^2 when abnormal returns are estimated over a window). The test statistic presented in Eq. (14) follows the Student t-distribution with *N* – 1 degrees of freedom

Table 5 Price reaction to profit warnings: the role of event-induced variance, conditional heteroskedasticity and time-varying betas

	Conditional versus unconditional			
	Value (%)	t-statistic	t-statistic	Z-score
Panel A: Negative warnings				
$\overline{AR}_0$	−3.46	−5.95	0.25	1.89*
$\overline{AR}_1$	0.44	1.11	−1.11	1.22
$\overline{CAR}_{[1,2]}$	0.28	−0.63	0.72	−0.12
$\overline{CAR}_{[1,3]}$	−0.04	−1.64	2.57**	1.32
$\overline{CAR}_{[1,4]}$	0.32	1.09	0.76	0.09
$\overline{CAR}_{[1,5]}$	0.00	−0.53	2.63***	1.38
$\overline{CAR}_{[1,6]}$	0.23	2.14**	2.07**	0.82
$\overline{CAR}_{[1,7]}$	0.12	1.16	2.59***	1.35
$\overline{CAR}_{[1,8]}$	0.16	0.10	2.53**	1.00
$\overline{CAR}_{[1,9]}$	0.10	1.16	3.00***	1.72*
$\overline{CAR}_{[1,10]}$	0.05	0.30	2.89***	1.75*
N	1238			
Panel B: Positive warnings				
$\overline{AR}_0$	3.88	4.19	0.14	−3.47***
$\overline{AR}_1$	−0.92	−1.79*	0.54	−4.00***
$\overline{CAR}_{[1,2]}$	−0.57	−2.23**	−2.71***	−5.43***
$\overline{CAR}_{[1,3]}$	−0.34	−3.04***	−4.26***	−5.50***
$\overline{CAR}_{[1,4]}$	−0.15	−3.47***	−4.59***	−4.71***
$\overline{CAR}_{[1,5]}$	−0.09	−2.40**	−5.36***	−5.74***
$\overline{CAR}_{[1,6]}$	−0.19	−3.77***	−5.30***	−6.27***
$\overline{CAR}_{[1,7]}$	−0.25	−4.96***	−5.17***	−6.12***
$\overline{CAR}_{[1,8]}$	−0.16	−3.73***	−5.74***	−6.56***
$\overline{CAR}_{[1,9]}$	−0.12	−3.14***	−6.29***	−6.93***
$\overline{CAR}_{[1,10]}$	−0.16	−4.35***	−6.07***	−6.76***
	Conditional versus unconditional			
	Value (%)	t-statistic	t-statistic	Z-score
N	485			

***, ** and * Significance at 1, 5 and 10%, respectively

In summary, the market model CARs suggest that investors overreact to both positive and negative warnings. However, adjusting for the time-varying risk and the event-induced variance causes the overreaction patterns following negative news to disappear completely. These adjustments also cause abnormal returns on days +1 and +2 after positive warnings to lose their statistical significant (see Sect. 3.3) and the magnitudes of the remaining post-positive-warning CARs to decline to such an extent that a contrarian strategy does not earn a profit in excess of transaction costs. These findings are consistent with the large body of literature that posits that ignoring time-varying betas, the conditional heteroskedastic behavior of the residual variance, and the event-induced volatility may yield biased market model parameter estimates and inconsistent test statistics (e.g., Corhay and Tourani-Rad 1994; Brown et al. 1988; Savickas 2003; Kolari and Pynnönen 2010).

3.5 Multivariate analysis

So far, we have shown that incorporating the time-varying betas, event-induced variance and conditional heteroskedasticity into the modelling process causes the overreaction patterns after negative warnings to disappear completely. These adjustments also yield smaller, and in some cases statistically insignificant, post-positive-warning CARs. However, the presence of some significantly negative, albeit small, post-positive-warning CARs may still be used as an argument for the overreaction hypothesis. To verify the validity of this claim, we estimate the following pooled OLS regression:

$$CAR_{is} = \propto_0 + \propto_1 AR_{0i} + \sum_k \propto_{ik} Control_i^k + IndustryDummies + YearDummies + \varepsilon, \quad (13)$$

where *CARis* is firm *i*'s average cumulative abnormal return over a window of *S* days beginning 1 day after a positive warning; *AR0i* is firm *i*'s average event-day abnormal return (both *AR0i* and *CARis* are adjusted for the time-varying betas, event-adjusted variance and conditional heteroskedasticity); $Control_i^k$ is a vector of control variables, which includes a set of firm-specific variables that are deemed to influence stock returns. These variables include the book-to-market ratio (*BTMVi*), the natural logarithm of stock *i*'s market capitalization measure (*lnMVi*) and the average turnover—i.e., the daily trading volume divided by the number of shares outstanding (*AvgTOi*). *BTMVi* and *lnMVi* are measured 11 days prior to the positive warning, while *AvgTOi* is computed over the [−105, −6] window prior to the positive warning. Industry and year dummies are included to control for the industry and year fixed effects. We also correct standard errors for firm-level clustering.

If the overreaction hypothesis holds, the parameter *α1* in Eq. (13) will be negative and significant, as the greater overreaction would lead to a greater correction (see, e.g., Cox and Peterson, 1994; Choi and Jayaraman, 2009). Fama and French (1993, 1996) show that *BTMVi* and *lnMVi* are amongst the key determinants of cross-sectional stock returns. If smaller stocks reverse more than larger stocks, ∝2 is expected to be negative and significant (see, e.g., Bremer and Sweeney 1991). The parameter ∝3 is expected to be positive, as investors may overreact to a greater extent when positive warnings are made by mature firms than when they are made by growth firms.[7] Finally, if the price reversal process is caused by illiquidity, ∝4 is expected to be negative and significant (see, e.g., Choi and Jayaraman 2009).

The results of the pooled OLS estimates are shown in Table 6. The coefficient α1α1 is negative for all post-event CARs, but only significant in the case of CAR_3. This evidence contradicts the predictions of the overreaction hypothesis, which suggests that greater overreaction causes greater correction. The results indicate that the post-positive-warning price patterns are unlikely to be driven by the size effect, as the coefficient ∝2 carries a positive, rather than the predicted negative, sign for all post-event CARs except for AR_1 (significant at the 5% level) and CAR_4(statistically insignificant). The coefficient ∝3 is generally negative, but only significant in the case of CAR_4, which is too far from the event date. This finding suggests that growth opportunities do not play a major role in determining the post-positive-warning price reversal patterns. Finally, ∝4 is negative, but not statistically significant, indicating that the post-positive-warning price patterns cannot be attributed to the illiquidity effect.

Table 6 Multivariate regressions

	AR_1	$CAR_{[1,2]}$	$CAR_{[1,3]}$	$CAR_{[1,4]}$	$CAR_{[1,5]}$	$CAR_{[1,6]}$	$CAR_{[1,7]}$	$CAR_{[1,8]}$	$CAR_{[1,9]}$	$CAR_{[1,10]}$
Constant	−4.57**	−4.61**	−1.05	−0.74	−0.93	−0.67	−0.9	−0.6	−0.96*	−0.51
	(−2.32)	(−2.01)	(−1.05)	(−0.56)	(−0.89)	(−0.92)	(−1.30)	(−1.13)	(−1.82)	(−1.04)
AR_0	−1.94	−5.99	−4.65**	−3.59	−3.25	0.1	−0.94	−0.77	−0.7	−0.76
	(−0.48)	(−1.64)	(−2.46)	(−1.49)	(−1.37)	(0.08)	(−0.75)	(−0.83)	(−0.74)	(−0.91)
lnMV	0.54**	0.322	0.07	−0.05	0.05	0.08	0.03	0.03	0.06	0.04
	(2.33)	(1.46)	(0.74)	(−0.43)	(0.49)	(1.09)	(0.55)	(0.57)	(1.12)	(0.85)
BTMV	−0.015	0.023	−0.00	−0.01**	−0.02	−0.00	0.01	−0.01	−0.01	−0.01
	(−0.51)	−0.74	(−1.33)	(−2.50)	(−1.33)	(−0.12)	(0.56)	(−1.21)	(−1.36)	(−0.55)
AvgTO	−1.77	−0.243	−0.442	−0.35	−0.14	0.241	−0.457	0.157	−0.182	−0.22
	(−1.05)	(−0.16)	(−0.58)	(−0.48)	(−0.25)	−0.42	(−1.00)	(0.33)	(−0.46)	(−0.49)
Year dummies	Yes	Yes	Yes	Yes	Yes	Yes	Yes	Yes	Yes	Yes
Industry dummies	Yes	Yes	Yes	Yes	Yes	Yes	Yes	Yes	Yes	Yes
N	484	484	484	484	484	484	484	484	484	484
Adjusted R^2	0.008	0.019	0.016	0.004	0.005	0.007	0.012	0.011	0.012	0.001

***, ** and * Significance at 1, 5 and 10%, respectively

This table reports the results of the following pooled OLS regression:

$CARis=\propto 0+\propto 1AR0i+\sum k\propto ikControlki+IndustryDummies+YearDummies+\varepsilon$,

where *CARis* is firm *i*'s average cumulative abnormal return over windows of *S* days following positive warnings; *AR0i* is firm *i*'s average abnormal return on the event date (both *AR0i* and *CARis* are adjusted for the time-varying betas, event-adjusted variance and conditional heteroskedasticity); $Control_i^k$ is a vector of control variables, which include a set of firm-specific variables that are deemed to influence stock returns. These variables include the book-to-market ratio (*BTMVi*), the natural logarithm of stock *i*'s market capitalization measure (*lnMVi*) and the average turnover - i.e., the daily trading volume divided by the number of shares outstanding (*AvgTOi*). *BTMVi* and *lnMVi* are measured 11 days prior to positive warnings, while *AvgTOi* is computed over the [−105, −6] windows prior to the positive warnings. Industry and year dummies are included to control for the industry and year fixed effects. We also correct standard errors for firm-level clustering

The figures in parentheses are the t-statistics

The regression coefficients are multiplied by 10^2

4 ROBUSTNESS CHECKS

4.1 Alternative estimations of time-varying betas

hile the BEKK model has been widely used to estimate the conditional betas of the market model (Tsui and Yu 1999; Choudhry 2005; Choudhry et al. 2010), a number of other techniques have emerged for modelling and estimating time-varying betas.[8] For robustness purposes, we also estimate conditional betas using other popular members of the multivariate GARCH (M-GARCH) model family. Following Faff et al. (2000), we estimate a time series of conditional betas, β_{it}^G, for a stock *i* as

$$\beta_{it}^G = \frac{\rho_{im}\sigma_{it}}{\sigma_{mt}} \tag{14}$$

where *σit* and *σmt* are the time-varying conditional variances for the stock and the market, respectively, and *ρim* is the covariance between the stock and the market returns, which is assumed constant to overcome the onerous computational burden (see, e.g., Pagan 1996). The time-varying conditional variances (*σit* and *σmt*) are initially modelled using the standard GARCH (1,1) specification. Then, the Exponential GARCH (EGARCH) of Nelson (1991) and the Threshold GARCH (TGARCH) of Glosten et al. (1993) are employed to model the asymmetry in the stock price volatility reaction to positive and negative shocks. Following Faff et al. (2000), we also use the in-sample mean squared error of forecasts, or $MSE_i = \sum (R^*_{it} - R_{it})^2 / T$, to assess the accuracy of each forecasted beta series (including those obtained using the BEKK model). Here, R^*_{it} are the predicted values of stock i's return series generated from the single index model with time-varying betas generated from Eq. (14), and *T* is the number of days in the estimation period.

The GARCH-, TGARCH- and EGARCH-based time-varying beta estimates are provided in Panel A of Table 7. The time-varying betas from the standard GARCH model are similar to those obtained from the BEKK model (see Table 2). Specifically, the GARCH model indicates that stock betas increase in the aftermath of news, but the increase is more pronounced following good than bad news. The TGARCH estimates suggest that stock betas increase significantly after positive warnings, but exhibit no significant changes following warnings containing negative news. The EGARCH model suggests that stock betas do not exhibit any significant changes in the post-warning periods, irrespective of the news contained in the warnings. Panel A of Table 7 also reports the average mean squared errors (MSEs) associated with the different conditional beta estimates. It shows that the post-negative-warning (post-positive-warning) average MSEs associated with the GARCH- and TGARCH-based estimates are 12.54×10^{-4} and 12.53×10^{-4} (23.93×10^{-4} and 23.94×10^{-4}), respectively. However, the EGARCH estimates generate average MSEs of 50.10×10^{-4} and 50.47×10^{-4} for the cases of negative and positive warnings, respectively. This evidence suggests that GARCH and TGARCH estimates provide relatively more accurate forecasts of time-varying betas than EGARCH.[9] It also implies that stock betas are more likely to exhibit significant increases after good news than bad news.

Following Faff et al. (2000), we estimate a time series of conditional betas, β^G_{it}, for a stock *i* as

$$\beta^G_{it} = \frac{\rho_{im}\sigma_{it}}{\sigma_{mt}}$$

Table 7 Alternative time-varying beta estimates

	GARCH				TGARCH				EGARCH			
	Negative warn		Positive warn		Negative warn		Positive warn		Negative warn		Positive warn	
	Mean (%)	t-stat	Mean (%)	t-stat	Mean (%)	t-stat	Mean (%)	t-stat	Mean (%)	t-stat	Mean (%)	t-stat
Panel A: Time-varying betas												
$\beta_{[-200,-1]}$	0.64		0.73		0.64		0.73		0.63		0.66	
$\beta_{[0,+1]}$	0.67**	2.44	0.89***	7.96	0.66	1.53	0.88***	7.59	0.62	−0.19	0.71**	2.03
$\beta_{[0,+2]}$	0.67**	2.43	0.88***	8.27	0.66	1.33	0.87***	7.81	0.63	0.15	0.70*	1.81
$\beta_{[0,+3]}$	0.66**	2.19	0.87***	8.22	0.65	0.98	0.87***	7.82	0.62	−0.29	0.69	1.51
$\beta_{[0,+4]}$	0.66*	1.82	0.87***	8.08	0.65	0.66	0.86***	7.75	0.62	−0.48	0.69	1.25
$\beta_{[0,+5]}$	0.66	1.58	0.86***	7.90	0.65	0.40	0.86***	7.63	0.62	−0.63	0.68	1.06
$\beta_{[0,+6]}$	0.65	1.36	0.85***	7.69	0.64	0.21	0.85***	7.49	0.62	−0.76	0.68	0.92
$\beta_{[0,+7]}$	0.65	1.18	0.85***	7.42	0.64	0.09	0.85***	7.23	0.62	−0.81	0.68	0.80
$\beta_{[0,+8]}$	0.65	1.02	0.84***	7.18	0.64	0.00	0.84***	6.98	0.62	−0.81	0.67	0.69
$\beta_{[0,+9]}$	0.65	0.88	0.84***	7.02	0.64	0.18	0.84***	6.82	0.61	−0.97	0.67	0.59
$\beta_{[0,+10]}$	0.65	0.81	0.84***	6.90	0.64	0.32	0.84***	6.70	0.61	−1.11	0.67	0.52
MSE	12.54×10^{-4}		23.92×10^{-4}		12.53×10^{-4}		23.94×10^{-4}		50.10×10^{-4}		50.47×10^{-4}	
	GARCH				TGARCH				EGARCH			
	Negative warn		Positive warn		Negative warn		Positive warn		Negative warn		Positive warn	
	Value (%)	t-stat	Value (%)	t-stat	Value (%)	t-stat	Value (%)	t-stat	Value (%)	t-stat	Value (%)	t-stat
Panel B: Time-varying beta-adjusted ARs and CARs												
$\overline{AR}_0$	−3.45***	−16.76	3.93***	12.47	−3.45***	−16.76	3.93***	12.47	−3.46***	−16.21	3.98***	12.04
$\overline{AR}_1$	0.01	0.04	−0.83***	−5.00	0.01	0.04	−0.83***	−5.00	−0.03	−0.16	−0.85***	−5.06
$\overline{CAR}_{[1,2]}$	0.40*	1.74	−1.14***	−4.68	0.40*	1.74	−1.14***	−4.68	0.35	1.51	−1.12***	−4.79
	GARCH				TGARCH				EGARCH			
	Negative warn		Positive warn		Negative warn		Positive warn		Negative warn		Positive warn	
	Value (%)	t-stat	Value (%)	t-stat	Value (%)	t-stat	Value (%)	t-stat	Value (%)	t-stat	Value (%)	t-stat
$\overline{CAR}_{[1,3]}$	0.49*	1.89	−1.22***	−4.31	0.49*	1.89	−1.22***	−4.31	0.45*	1.71	−1.15***	−4.08
$\overline{CAR}_{[1,4]}$	0.54*	1.86	−1.32***	−4.30	0.54*	1.86	−1.32***	−4.30	0.55*	1.89	−1.27***	−4.12
$\overline{CAR}_{[1,5]}$	0.72**	2.21	−1.63***	−4.89	0.72**	2.21	−1.63***	−4.89	0.71**	2.14	−1.58***	−4.73
$\overline{CAR}_{[1,6]}$	0.87**	2.45	−1.89***	−5.04	0.87**	2.45	−1.89***	−5.04	0.90**	2.48	−1.80***	−4.73
$\overline{CAR}_{[1,7]}$	0.98***	2.58	−2.05***	−5.10	0.98**	2.58	−2.05***	−5.10	0.96**	2.46	−1.98***	−4.83
$\overline{CAR}_{[1,8]}$	1.04***	2.66	−2.33***	−5.51	1.04***	2.66	−2.33***	−5.51	1.04***	2.59	−2.26***	−5.22
$\overline{CAR}_{[1,9]}$	1.24***	2.97	−2.61***	−5.99	1.24***	2.97	−2.61***	−5.99	1.20***	2.82	−2.57***	−5.75
$\overline{CAR}_{[1,10]}$	1.18***	2.73	−2.68***	−5.87	1.18***	2.73	−2.68***	−5.87	1.13***	2.54	−2.54***	−5.46

***, ** and * Significance at 1, 5 and 10%, respectively

where σit and σmt are the time-varying conditional variances of the stock and the market, respectively, and ρim is the covariance between the stock and market returns, which is assumed to be constant. The time-varying conditional variances (σit and σmt) are modelled using the standard GARCH (1,1), TGARCH(1,1) and EGARCH(1,1). We use the in-sample mean squared error of forecasts, or $MSE_i = \sum\left(R_{it}^{*} - R_{it}\right)^2/T$, to assess the accuracy of each forecasted beta series (including those obtained from the BEKK model). Here, R∗itRit∗ are the predicted values of stock i's return series generated from the single index model with timing-varying betas, and T is the number of days in the estimation period

To gain further insight into whether the price patterns following warnings depend on the way the conditional betas are estimated, we reestimate the ARs and CARs reported in Table 3 using the procedure described in

Sect. 3.2, but with time-varying betas generated using the GARCH, TGARCH and EGARCH models, respectively. The results in Panel B of Table 7 indicate that, despite some variations in the magnitude and the statistical significance of the post-warning CARs, the general price patterns do not seem to depend on how the conditional betas are estimated and are consistent with those outlined earlier (see Sect. 3.2). Specifically, Panel B of Table 7 shows that the price reversal patterns following negative profit warnings disappear (almost) completely when the betas are allowed to vary over time. It also shows that, while the CARs following warnings containing good news remain significantly negative, their values decline considerably once adjusted for the time-varying betas.[10]

4.2 Subperiod analysis

1999) suggesting that the post-warning price patterns may also vary over time. To test this prediction, we subdivide the sample into two subperiods: July 2007 to July 2009 and August 2009 to July 2012.[11] Table 8 presents the ARs and CARs from the standard market model (Sect. 3.1), the market model with time-varying betas (Sect. 3.2), the Savikas model (Sect. 3.3) and the Savikas model with time-varying betas (Sect. 3.4). Panel A of Table 8presents the post-negative-warning and post-positive-warning ARs and CARs for the period from July 2007 to July 2009. The market model produces price patterns that are consistent with the predictions of the overreaction hypothesis. Specifically, the market model CARs following negative shocks are positive and significant, with an average CAR of 2.4% to be earned over the 10-day window following a negative warning. Similarly, the post-positive-warning CARs are significantly negative, with a contrarian strategy earning an average CAR of −4.1% over the 10-day window after the announcement date. However, alternative abnormal return estimation methods imply that the post-warning price patterns are likely to be due to a bad model problem. Specifically, adjusting for the time-varying betas causes the post-negative-warning price patterns to disappear completely and the magnitude of the post-positive-warning CARs to decline considerably. The post-warning CARs become even smaller after the stochastic behavior of volatility during both event and non-event periods have been accounted for. In particular, CARs from Savickas' (2003) model are not significantly different from zero following warnings containing bad news, and are very small, and in many cases insignificant, after warnings containing good news. The simultaneous adjustment for the time-varying beta, event-induced variance and conditional heteroskedasticity yields slightly smaller CARs than the Savickas model.

Table 8 Subperiod analysis

	Market model		Time-varying beta adjustment (BEKK)		Savickas (2003) model		Savickas (2003) with time-varying betas	
	Negative warning	Positive warning	Negative warning	Positive warning	Negative warning	Positive warning	Negative warning	Positive warning
Panel A: Price reaction to profit warnings: July 2007 to July 2009								
$\overline{AR}_0$	−3.31***	6.04***	−3.47***	6.08***	−2.94***	6.79***	−3.44***	6.01***
$\overline{AR}_1$	0.14	−1.40***	0.02	−1.19***	0.39	−1.34	−0.01	−1.05**
$\overline{CAR}_{[1,2]}$	0.85**	−2.28***	0.60*	−1.95***	0.07	−0.87	−0.03**	−0.76
$\overline{CAR}_{[1,3]}$	1.06***	−2.40***	0.60	−1.91***	0.06	−0.80**	−0.06	−0.78**
$\overline{CAR}_{[1,4]}$	1.16***	−2.63***	0.54	−2.10**	−0.13	−0.68**	0.15	−0.81***
$\overline{CAR}_{[1,5]}$	1.49***	−2.37***	0.69	−1.84**	0.01	−0.18	−0.17	−0.27
$\overline{CAR}_{[1,6]}$	1.79***	−2.99***	0.87*	−2.31**	0.19	−0.32**	0.10	−0.27**
$\overline{CAR}_{[1,7]}$	2.00***	−3.14***	0.92*	−2.34**	0.31	−0.26**	0.07	−0.30**
$\overline{CAR}_{[1,8]}$	2.21***	−3.99***	1.03*	−3.04***	0.13	−0.34**	0.12	−0.35**
$\overline{CAR}_{[1,9]}$	2.45***	−4.14***	1.10*	−3.27***	0.12	−0.39**	0.00	−0.27*
$\overline{CAR}_{[1,10]}$	2.37***	−4.09***	0.86	−3.17***	0.20	−0.30**	0.08	−0.23*
N	718	98	718	98	718	98	718	98
Panel B: Price reaction to profit warnings: August 2009 to July 2012								
$\overline{AR}_0$	−3.61***	3.39***	−3.53***	3.44***	−4.27***	3.20***	−3.44***	3.36***
$\overline{AR}_1$	−0.18	−0.68***	−0.03	−0.56***	−0.53	−0.72	−0.01	−0.91
$\overline{CAR}_{[1,2]}$	−0.24	−0.83***	−0.06	−0.72***	−0.02	−0.27**	−0.03**	−0.52*
$\overline{CAR}_{[1,3]}$	−0.32	−0.91***	−0.13	−0.74**	−0.19	−0.14	−0.06	−0.23**
$\overline{CAR}_{[1,4]}$	−0.37	−0.98***	−0.07	−0.75**	−0.07	−0.06**	0.15	0.01**
$\overline{CAR}_{[1,5]}$	−0.38	−1.42***	−0.09	−1.14***	0.01	−0.14**	−0.17*	−0.04**
$\overline{CAR}_{[1,6]}$	−0.43	−1.60***	−0.06	−1.25***	0.01	−0.23***	0.10	−0.18***
	Market model		Time-varying beta adjustment (BEKK)		Savickas (2003) model		Savickas (2003) with time-varying betas	
	Negative warning	Positive warning	Negative warning	Positive warning	Negative warning	Positive warning	Negative warning	Positive warning
$\overline{CAR}_{[1,7]}$	−0.49	−1.77***	−0.08	−1.32***	0.08	−0.11***	0.07	−0.23***
$\overline{CAR}_{[1,8]}$	−0.61	−1.90***	−0.15	−1.37***	0.07	−0.13***	0.12	−0.11***
$\overline{CAR}_{[1,9]}$	−0.49	−2.21***	0.00	−1.62***	−0.08	−0.13***	0.00	−0.09***
$\overline{CAR}_{[1,10]}$	−0.53	−2.32***	0.02	−1.67***	0.06	−0.14***	0.08	−0.14***
N	522	387	522	387	522	387	522	387

***, ** and * Significance at 1, 5 and 10%, respectively

This table reports the average abnormal returns $(\overline{AR}_i)$ and the average cumulative abnormal returns $(\overline{CAR}_s)$ from the standard market model (Sect. 3.1), the market model with time-varying betas (Sect. 3.2), the Savikas model (Sect. 3.3) and the Savikas model with time-varying betas (Sect. 3.4), separately for the periods July 2007 to July 2009 and August 2009 to July 2012

The post-negative-warning and post-positive-warning ARs and CARs for the period from August 2009 to July 2012 are presented in Panel B of Table 8. The results indicate that the price patterns following warnings with bad news are consistent with the EMH, irrespective of the abnormal return estimates. However, the post-positive-warning CARs seem to depend largely on the estimation method. The market model CARs suggest that investors overreact to warnings containing good news. Specifically, the post-positive-warning market model CARs are significantly negative and increase monotonically from $CAR_1 = -0.7\%$ to $CAR_{10} = -2.3\%$. The market model with time-varying betas generates similar results, but with slightly smaller post-positive-warning CARs ($CAR_1 = -0.6\%$ and $CAR_{10} = -1.7\%$). The Savickas (2003) approach breaks the monotonic abnormal return patterns, causing a large decline in the magnitude and the disappearance of the statistical significance of some of the post-positive-warning CARs. Similar results are obtained when the beta parameter in the Savickas model is allowed to vary systematically over time. Specifically, on average, shorting a stock 1 day after a positive warning and repurchasing it 10 days later would only yield an abnormal profit of 0.1%. This figure is considerably smaller than the average (median) relative bid-ask spread of 2.75% (1.71%) associated with the sample stocks during this period, implying that the contrarian strategy is not profitable after accounting for transaction costs.[12]

5 CONCLUSION

This study examines the impact of announcements containing profit warnings on the risk and return characteristics of the underlying stocks. Our main purpose is to test whether the price reaction anomalies around profit warnings survive adjustments for the time-varying beta, event-induced variance and conditional heteroskedasticity. The standard event study methodology indicates the presence of strong price reversal patterns following both positive and negative warnings. Specifically, the abnormal returns are positive (negative) on the days on which positive (negative) warnings are released and negative (positive) on the subsequent days. These results are consistent with the findings of Lui et al. (2009), Bremer and Sweeney (1991) and Atkins and Dyl (1990).

Several studies show that the betas and residual variances of individual stocks are affected by the arrival of news (e.g., Brown et al. 1988; Grullon et al. 2005; DeAngelo et al. 2006). Others find that the residual variance of the standard market model varies systematically over time (see, e.g., Patell and Wolfson 1979; Kalay and Lowenstein 1985; Kolari and Pynnönen 2010). This study shows that relaxing these assumptions results in post-warning abnormal returns patterns that are largely consistent with the predictions of the EMH.

In conclusion, the results of this study suggest that, while the literature seems to produce several price anomalies, the efficient market hypothesis should only be rejected if the abnormal returns estimates survive stringent tests. Our conclusion is consistent with Park (1995), who finds that the overreaction to large price change disappears after bid-ask spread bounces are accounted for, and Fama (1998), who finds that most of the return anomalies around corporate announcement events are mitigated with reasonable changes to the way abnormal returns are measured.

REFERENCES

1. Akgiray V (1989) Conditional heteroscedasticity in time series of stock returns: evidence and forecasts. J Bus 62(1):55–80
2. Armitage S (1995) Event study methods and evidence on their performance. J Econ Surv 9(1):25–52
3. Atkins AB, Dyl EA (1990) Price reversals, bid-ask spreads, and market efficiency. J Financ Quant Anal 25:535–547
4. Barmber LS, Cheon YS (1998) Discretionary management earnings forecast disclosures: antecedents and outcomes associated with forecast specificity choices. J Account Res 36:167–190
5. Boehmer E, Musumeci J, Poulsen A (1991) Event study methodology under conditions of event induced variance. J Financ Econ 30:253–272
6. Bollerslev T, Engle R, Wooldridge JM (1988) A capital asset pricing model with time-varying covariances. J Polit Econ 96:116–131
7. Bollerslev T, Engle R, Nelson D (1994) ARCH models. In: Engle R, McFadden D (eds) Handbook of econometrics, vol 4. Elsevier Science, New York, pp 2960–3038
8. Bremer M, Sweeney RJ (1991) The reversal of large stock-price decreases. J Finance 46:747–754
9. Brockett P, Chen HM, Garven JR (1999) A new stochastically flexible event methodology with application to Proposition 103. Insur Math Econ 25:197–217
10. Brown S, Warner J (1980) Measuring security price performance. J Financ Econ 8:205–258
11. Brown K, Harlow WV, Tinic SM (1988) Risk aversion, uncertain information, and market efficiency. J Financ Econ 22:355–385
12. Cam MA, Ramiah V (2014) The influence of systematic risk factors and econometric adjustments in catastrophic event studies. Rev Quant Finance Account 42:171–189
13. Campbell JY, Lo A, MacKinlay AC (1997) The econometrics of financial markets. Princeton University Press, Princeton
14. Choi HS, Jayaraman N (2009) Is reversal of large stock-price declines caused by overreaction or information asymmetry? Evidence from stock and option markets. J Futures Mark 26(3):348–376
15. Choudhry T (2005) Time-varying beta and the Asian financial crisis: investigating the Malaysian and Taiwanese firms. Pac Basin Finance J 13:93–118
16. Choudhry T, Lu L, Peng K (2010) Time-varying beta and the Asian financial crisis: evidence from the Asian industrial sectors. Jpn World Econ 22:228–234
17. Clare M (2001) The information in UK company profit warnings. Bank Engl Q Bull Spring 104–109
18. Corhay A, Tourani-Rad A (1994) Expected returns and volatility in European stock markets. Int Rev Econ Finance 3(1):45–56
19. Corrado CJ, Jordan BD (1997) Risk aversion, uncertain information, and market efficiency. Rev Quant Finance Account 8:51–68
20. Cox DR, Peterson DR (1994) Stock returns following large one-day declines: evidence on short-term reversals and longer-term performance. J Finance 49:255–267
21. Cyree KB, DeGennaro RP (2002) A generalized method for detecting abnormal returns and changes in systematic risk. Rev Quant Finance Account 19:399–416

22. DeAngelo H, DeAngelo L, Stulz RM (2006) Dividend policy and the earned/contributed capital mix: a test of life-cycle theory. J Financ Econ 81:227–254
23. Dyckman T, Philbrick D, Stephan J (1984) A comparison of event study methodologies using daily stock returns: a simulation approach. J Account Res 22:1–33
24. Engle R, Kroner K (1995) Multivariate simultaneous generalised ARCH. Econom Theor 11:122–150
25. Faff DW, Hillier D, Hillier J (2000) Time varying beta risk: an analysis of alternative modelling techniques. J Bus Financ Account 27:523–555
26. Fama EF (1998) Market efficiency, long-term returns, and behavioral finance. J Financ Econ 49:283–306
27. Fama EF, French KR (1993) Common risk factors in the returns on stocks and bonds. J Financ Econ 33:3–56
28. Fama EF, French KR (1996) Multifactor explanations of asset pricing anomalies. J Finance 51:55–84
29. Ferson W, Korajczyk R (1995) Do arbitrage pricing models explain the predictability of stock returns? J Bus 68:309–349
30. Glosten LR, Jagannathan R, Runkle D (1993) On the relation between the expected value and the volatility of the nominal excess return on stocks. J Finance 48:1779–1801
31. Grullon GR, Benartzi MS, Thaler RH (2005) Dividend changes do not signal changes in future profitability. J Bus 78(5):1659–1682
32. Harvey C (1989) Time-varying conditional covariances in tests of asset pricing models. J Financ Econ 24:289–317
33. Jackson D, Madura J (2003) Profit warnings and timing. Financ Rev 38.497–513
34. Kalay A, Lowenstein U (1985) Predictable events and excess returns: the case of dividend announcements. J Financ Econ 14:423–449
35. Klemkosky RC, Martin JD (1975) The adjustment of beta forecasts. J Finance 30(4):1123–1128
36. Kolari JW, Pynnönen S (2010) Event study testing with cross-sectional correlation of abnormal returns. J Financ Res 23:3996–4025
37. Libby R, Tan H (1999) Analysts' reaction to warnings of negative earnings surprises. J Account Res 37(2):415–435
38. Lui D, Markov S, Tamayo A (2009) The information content of analysts' risk rating. Working paper, http://www.ssrn.com. Accessed 20 Nov 2013
39. Nelson DB (1991) Conditional heteroskedasticity in asset returns: a new approach. Econometrica 59:347–370
40. Pagan A (1996) The econometrics of financial markets. J Empir Finance 3:15–102
41. Park J (1995) A market microstructure explanation for predictable variations in stock returns following large price changes. J Financ Quant Anal 30:241–256
42. Patell J, Wolfson M (1979) Anticipated information releases reflected in call option prices. J Account Econ 1:117–140
43. Patton AJ, Verardo M (2012) Does beta move with news? Firm-specific information flows and leaning about profitability. Rev Financ Stud 25(9):2789–2839
44. Savickas R (2003) Event-induced volatility and tests for abnormal performance. J Financ Res 16:165–178
45. Sullivan R, Timmermann A, White H (1999) Data-snooping, technical trading rule performance, and the bootstrap. J Finance 54(5):1647–1691
46. Tsui AK, Yu Q (1999) Constant conditional correlation in a bivariate GARCH model: evidence from the stock markets of China. Math Comput Simul 48:503–509
47. Tucker X (2004) Does disclosure reputation explain earnings warnings? PhD thesis, New York University

48. Wang R, Zhang Y (2011) The spillover effect of disclosure rules and materiality thresholds: evidence from profit warnings issued in Hong Kong market. China J Account Res 4:63–80
49. Yen G, Lee CF (2008) Efficient market hypothesis (EMH): past, present and future. Rev Pac Basin Financ Mark Policies 11:305–329
50. Zolotoy L (2011) Earnings news and market risk: Is the magnitude of the post-earnings announcement drift underestimated? J Financ Res 34:523–535

END NOTES

[1] Profit warnings are announcements made by publicly traded companies, prior to the issuance of their

formal financial reports, to warn investors that their earnings will differ from previously expected levels.

Profit warnings can be either positive or negative.

[2] See Yen and Lee (2008) for a thorough review of the literature on efficiency market hypothesis.

[3] Dyckman et al. (1984) show that the market model performs significantly better than other models, such as the index or average return models. Similarly, Armitage (1995: 25) argues that "...the market model is most commonly used to generate expected returns and no better alternative has yet been found despite the weak relationship between beta and actual returns...".

[4] For robustness purposes, we also use Newey-West heteroskedasticity, the serial correlation consistent estimator and the J-statistic of Campbell et al. (1997) and our conclusions remain unchanged. Details of these results are available upon request.

[5] The parameter ais generated by regressing the variable ðRi;t _ bi;tRm;tÞ against a constant term.

[6] The results from GJR-GARCH (1,1) and E-GARCH (1,1) are very similar to those reported here. Further details are available upon request.

[7] The earnings of mature firms tend to be more stable and predictable than those of growth firms. Therefore, investors' surprise is likely to be greater when warnings are issued by mature firms.

[8] See, for example, Faff et al. (2000) for a detailed review of these techniques.

[9] The average MSEs from the BEKK model are 12.58 9 10-4 and 24.06 9 10-4 for the negative and positive warnings, respectively. These figures are very close to those produced by the GARCH and TGARCH methods, but deviate considerably from those generated by EGARCH.

[10] Note that we also repeat the analysis in Sect. 3.4 using time-varying beta estimates from GARCH, TGARCH and EGARCH and our conclusions remain largely unchanged. Further details are available upon request.

[11] Subdividing the sample into three and four subperiods does not alter our main conclusions and the results are available upon request.

[12] We also stratify our sample stocks into different size and industry subsamples and our results remain unchanged. Specifically, we show that the magnitude and statistical significance

of CARs are highly sensitive to the estimation method. While the standard market model results are in many cases consistent with the predictions of the overreaction hypotheses, the price patterns obtained after accounting for the timevarying betas, event-induced variance and conditional heteroskedasticity are generally consistent with the predictions of the EMH. Further details of these results are available upon request.

CHAPTER 5

On an Efficient Multiple Time Step Monte Carlo Simulation of The SABR Model

Álvaro Leitao*†‡, Lech A. Grzelak†§ And Cornelisw. Oosterlee†‡

†TU Delft, Delft Institute of Applied Mathematics, Delft, The Netherlands

‡CentrumWiskunde & Informatica (CWI), Amsterdam, The Netherlands

§ING, Quantitative Analytics, Amsterdam, The Netherlands

ABSTRACT

In this paper, we will present a multiple time step Monte Carlo simulation technique for pricing options under the *Stochastic Alpha Beta Rho* model. The proposed method is an extension of the one time step Monte Carlo method that we proposed in an accompanying paper Leitao *et al.* [*Appl. Math. Comput.* 2017, 293, 461–479], for pricing European options in the context of the model calibration. A highly efficient method results, with many very interesting and nontrivial components, like Fourier inversion for the sum of log-normals, stochastic collocation, Gumbel copula, correlation approximation, that are not yet seen in combination within a Monte Carlo simulation. The present multiple time step Monte Carlo method is especially useful for long-term options and for exotic options.

Keywords: SABR model, Exact simulation, Monte Carlo methods, Copulas, Stochastic collocation, Fourier techniques, Exotic options
JEL Classifications: C15, C63

1. INTRODUCTION

The Stochastic Alpha Beta Rho (SABR) model (Hagan *et al.* 2002) is an established stochastic differential equation (SDE) model which, in practice, is often used for interest rates and foreign-exchange (FX) modelling. It is based on a parametric local volatility component in terms of a model parameter, β, and reads,

$$\begin{aligned} dS(t) &= \sigma(t)S^{\beta}(t)dW_S(t), \quad & S(0) &= S_0 \exp(rT), \\ d\sigma(t) &= \alpha\sigma(t)dW_{\sigma}(t), \quad & \sigma(0) &= \sigma_0. \end{aligned} \tag{1}$$

Here, $S(t) = \bar{S}(t)\exp(r(T-t))$ denotes the forward price of the underlying asset $\bar{S}(t)$, with r an interest rate, S_0 the spot price and T the maturity. Further, $\sigma(t)$ represents a stochastic volatility process, with $\sigma(0) = \sigma_0$, $W_S(t)$ and $W_{\sigma}(t)$ are correlated Brownian motions with constant correlation coefficient ρ (i.e. $W_S W_{\sigma} = \rho t$). The parameters of the SABR model are $\alpha > 0$ (the volatility of volatility, *vol-vol*), $0 \leq \beta \leq 1$ (the variance elasticity) and ρ (the correlation coefficient). Because of a practically very useful closed-form expression for the implied volatility, Hagan *et al.* (2002) and Obloj (2008), the model has gained its popularity. However, since this expression is derived by perturbation theory, the formula is not always accurate for small strike values or for long time to maturity. By the one time step SABR method in Leitao *et al.* (2017), we can efficiently calculate accurate implied volatilities for any strike, however, only for short times to maturity (less than two years). This fits well to the context of FX markets.

Here, we extend the one time step Monte Carlo method from Leitao *et al.* (2017), and propose *a multiple time step Monte Carlo simulation technique* for pricing options under the SABR dynamics. We call it the *mSABR method.* With this new simulation approach, we can price options with longer maturities, payoffs relying on the transitional distributions and exotic options (like path-dependent options) under the SABR dynamics, overcoming the limitations of Hagan's formula in the mentioned situations.

Because a robust and efficient SABR simulation scheme may be involved and quite expensive, we aim for using as few time steps as possible. This fact, together with the long-time horizon assumption, implies that we focus on larger time steps. In this context, the important research line is called *exact simulation*, where, rather than Taylor-based simulation techniques, the probability density of the SDE under consideration is highly accurately approximated. Point-of-departure is Broadie and Kaya's exact simulation for the Heston stochastic volatility model (Broadie and Kaya, 2006). That method is, however, time-consuming because of multiple evaluations of Bessel functions and the use of Fourier inversion techniques on each Monte Carlo path. Inspired by this approach, in Smith (2007) and van Hasstrecht and Pelsser (2010), computational improvements were proposed. Furthermore, Andersen (2008) presented two efficient alternatives to the Broadie and Kaya scheme, the so-called Truncated Gaussian and the Quadratic Exponential (QE) schemes.

For the SABR model, exact Monte Carlo simulation is nontrivial because the forward process in (1) is governed by *constant elasticity of variance* (CEV) dynamics (Cox 1996 Schroder 1989). In Islah (2009), the connection between the CEV and a squared Bessel processes was explained, and an analytic approximation for the SABR conditional distribution based on the non-central χ^2 distribution was presented. The expression relies on the stochastic volatility dynamics, but also on the *integrated variance* process (which itself also depends on the volatility).

An (approximately) exact SABR simulation can then be subdivided in to the simulation of the volatility process, the simulation of the integrated variance (conditional on the volatility process) and the simulation of the underlying CEV process (conditional on the volatility and the integrated variance processes). Based on this, Chen *et al.* (2012) proposed a low-biased Monte Carlo scheme with moment-matching (following the techniques of the QE scheme by Andersen (2008) and a direct inversion approach. For the simulation of the integrated variance, they also employed moment-matching based on conditional moments that were approximated by a small disturbance expansion. In Cai *et al.* (forthcoming), the authors employed direct inversion for the forward distribution and used a Laplace transform inversion technique to simulate the integrated variance. Kennedy and Pham (2014) provided the moments of the integrated variance for specific SABR models (like the normal, log-normal and displaced diffusion SABR models) and derived an approximation of the SABR distribution. In Lord and Farebrother (2014), a summary of different approaches was presented.

In this work, we approximate the distribution of the integrated variance conditional on the volatility dynamics by joining the two involved marginal distributions, i.e. the stochastic volatility and integrated variance distributions, by means of copula techniques. With both marginal distributions available, we use the Archimedean Gumbel copula to define a multivariate distribution which approximates the conditional distribution of the integrated variance given the stochastic volatility.

An approximation of the cumulative distribution function (CDF) of the integrated variance can be obtained by a recursive procedure, as described in Zhang and Oosterlee (2013), originally employed to price arithmetic Asian options. This iterative technique is based on the derivation of the characteristic function of the integrated variance process and Fourier inversion to recover the probability density function (PDF).

The fact that we need to apply recursion and compute the corresponding characteristic function, PDF and CDF of the integrated variance for each Monte Carlo sample at each time step, makes this approach however computationally very expensive (even with the improvements already made in Leitao *et al.* (2017), like the Heston exact simulation. In order to reduce the use of this procedure as much as possible, highly efficient sampling from the CDF of the integrated variance is proposed, which is based on the so-called *Stochastic Collocation Monte Carlo sampler* (SCMC), see Grzelak *et al.* (2015). The technique employs a sophisticated interpolation (based on Lagrange polynomials and collocation points) of the CDF under consideration.

As will be seen, the resulting 'almost exact' Monte Carlo SABR simulation method that we present here (the mSABR method), contains several interesting (and not commonly used) components, like the Gumbel copula, a recursion plus Fourier inversion to approximate the CDF of the integrated variance and efficient interpolation by means of the SCMC sampler. The proposed mSABR scheme allows for fast and accurate option pricing under SABR dynamics, providing a better ratio of accuracy and computational cost than Taylor-based simulation schemes. Compared to the other approaches mentioned in this introduction, we provide a highly accurate SABR simulation scheme, based on only a few time steps.

The paper is organized as follows. In section 2, SABR model simulation is introduced. In section 3, the different parts of the multiple time step copula-based technique are described, including the derivation of the marginal distributions and the application of the copula. Some numerical experiments are presented in section 4. We conclude in section 5. In the derivation of the CDF of the integrated variance, we need to perform several approximations, which comes with approximation errors. In appendix 1, we list and discuss these errors.

2. 'ALMOST EXACT' SABR SIMULATION

The SABR model with dependent Brownian motions was already given in (1). The multiple time step Monte Carlo SABR simulation is based on the corresponding SDE system with independent Brownian motions $d\hat{W}_S(t)$ and $d\hat{W}_\sigma(t)$, i.e.

$$
\begin{aligned}
dS(t) &= \sigma(t)S^{\beta}(t)\left(\rho d\hat{W}_\sigma(t) + \sqrt{1-\rho^2}d\hat{W}_S(t)\right),\\
S(0) &= S_0 \exp(rT),\\
d\sigma(t) &= \alpha\sigma(t)d\hat{W}_\sigma(t),\\
\sigma(0) &= \sigma_0.
\end{aligned} \tag{2}
$$

The forward dynamics are governed by a CEV process. Based on this fact and the work by Schroder (1989) and Islah (2009), an analytic approximation for the CDF of the SABR conditional process has been obtained. For some generic time interval $[s, t]$, $0 < s < t \leq T$, assuming $S(s) > 0$, the conditional cumulative distribution for forward $S(t)$ with an absorbing boundary at $S(t) = 0$, given $\sigma(s), \sigma(t)$, and $\int_s^t \sigma^2(z)dz$, is given by

$$
\begin{aligned}
&\Pr\left(S(t) \leq K | S(s) > 0, \sigma(s), \sigma(t), \int_s^t \sigma^2(z)dz\right)\\
&\quad = 1 - \chi^2(a; b, c),
\end{aligned} \tag{3}
$$

Where

$$
\begin{aligned}
a &= \frac{1}{\nu(t)}\left(\frac{S(s)^{1-\beta}}{(1-\beta)} + \frac{\rho}{\alpha}(\sigma(t) - \sigma(s))\right)^2,\\
c &= \frac{K^{2(1-\beta)}}{(1-\beta)^2\nu(t)},\\
b &= 2 - \frac{1-2\beta-\rho^2(1-\beta)}{(1-\beta)(1-\rho^2)},\\
\nu(t) &= (1-\rho^2)\int_s^t \sigma^2(z)dz,
\end{aligned}
$$

and $\chi^2(x; \delta, \lambda)$ is the non-central chi-square CDF.

This formula is exact in the case of $\rho = 0$ and constitutes an *approximation* otherwise (because the CEV process is approximated by a shifted process with an approximated initial condition for $\rho \neq 0$ in the derivation). Based on equation (3), an *approximately* exact simulation of SABR model is feasible by inverting the conditional SABR cumulative distribution when the conditional integrated variance is known, see section 2.1.

2.1. SABR Monte Carlo simulation

In order to apply an 'almost exact' multiple time step Monte Carlo method for the SABR model, several steps need to be performed, that are described in the following:

- *Simulation of the SABR volatility process,* $\sigma(t)$ given $\sigma(s)$. By equation (2), the stochastic volatility process of the SABR model exhibits a log-normal distribution. The solution is a *geometric Brownian motion,* i.e. the exact simulation of $\sigma(t)|\sigma(s)$ reads
$$\sigma(t) \sim \sigma(s) \exp\left(\alpha \hat{W}_\sigma(t) - \frac{1}{2}\alpha^2 t\right) \tag{4}$$
- *Simulation of the SABR integrated variance process,* $\int_s^t \sigma^2(z)\mathrm{d}z|\sigma(s), \sigma(t)$. This conditional distribution is not available in closed form. We will therefore derive an approximation of the conditional distribution of the SABR integrated variance given $\sigma(t)$ and $\sigma(s)$. The integrated variance sampling can be done by simply inverting it.
- *Simulation of the SABR forward price process.* The forward price $S(t)$ can be simulated by inverting the CDF in equation (3). By this, we avoid negative forward prices in the simulation, as an absorbing boundary at zero is considered. There is no analytic expression for the inverse distribution and therefore this inversion has to be computed by means of some numerical approximation.

The multiple time step Monte Carlo simulation for the SABR model is thus summarized in algorithm 1. Input parameters are the initial conditions (S_0 and σ_0), the maturity time, T, the SABR parameters α, β, and ρ, the number of Monte Carlo samples (n) and the number of time steps (m).

Algorithm 1: Multiple time step SABR Monte Carlo simulation.

Data: $S_0, \sigma_0, T, \alpha, \beta, \rho, n, m$.
$S(0) = S_0 \exp(rT)$, $\sigma(0) = \sigma_0$
for *Path* $p = 1 \ldots n$ **do**
  for *Step* $k = 1 \ldots m$ **do**
    $s = (k-1)\frac{T}{m}$
    $t = k\frac{T}{m}$
    Sampling $\sigma(t)$ given $\sigma(s)$.
    Sampling $\int_s^t \sigma^2(z)\mathrm{d}z$ given $\sigma(s)$ and $\sigma(t)$.
    Sampling $S(t)$ given $S(s)$, $\sigma(s)$, $\sigma(t)$ and $\int_s^t \sigma^2(z)\mathrm{d}z$.

In section 3.3, we propose a procedure to sample $\int_s^t \sigma^2(z)\mathrm{d}z|\sigma(s), \sigma(t)$ based on the Gumbel copula. For this, the CDF of the integrated variance given the initial volatility, $\sigma(s)$, (as a marginal distribution) must be derived. In section 3.1, a recursive technique to obtain an approximation of this CDF is presented. Because we need to apply this recursion to approximate the characteristic function, the PDF and the CDF of $\int_s^t \sigma^2(z)\mathrm{d}z|\sigma(s)$ for each sample of $\sigma(s)$ at each time step, this approach is expensive in terms of computational cost. To overcome this drawback, an efficient alternative will be employed here, based on Lagrange interpolation, as in the SCMC (Grzelak *et al.*, 2015). In section 2.2, this methodology is briefly described.

2.2. Stochastic collocation Monte Carlo sampler

In a multiple time step exact simulation Monte Carlo method we need to perform a significant number of computations each time step on each Monte Carlo path. This often hampers the applicability of the exact simulation for systems of SDEs. One of the important components of the multiple time step Monte Carlo SABR simulation is therefore an accurate and highly efficient interpolation, as proposed in the SCMC sampler in Grzelak *et al.* (2015).

In this section, we will introduce the SCMC technique. The details in the SABR conditional integrated variance context will be presented in section 3.2.

The SCMC technique relies on the property that a CDF of a distribution (if it exists) is uniformly distributed. A well-known standard approach to sample from a given distribution, Y, with CDF F_Y reads

$$F_Y(Y) \stackrel{d}{=} U \quad \text{thus} \quad y_n = F_Y^{-1}(u_n),$$

where $\stackrel{d}{=}$ means equality in distributional sense, $U \sim \mathcal{U}([0,1])$ and u_n are samples from $\mathcal{U}([0,1])$. The computational cost of this approach highly depends on the cost of the inversion F_Y^{-1}, which is assumed to be expensive.

We therefore consider another, 'cheap', random variable X, whose inversion, F_X^{-1}, is computationally much less expensive. In this framework, the following relation holds

$$F_Y(Y) \stackrel{d}{=} U \stackrel{d}{=} F_X(X).$$

The samples y_n of Y, and x_n of X, are thus related via the following inversion,

$$y_n = F_Y^{-1}(F_X(x_n)).$$

However, this does not yet imply any improvement since the number of expensive inversions F_Y^{-1} remains the same. The goal is to compute y_n using a function $g(\cdot) = F_Y^{-1}(F_X(\cdot))$, such that

$$F_X(x) = F_Y(g(x)) \quad \text{and} \quad Y \stackrel{d}{=} g(X),$$

where evaluations of function $g(\cdot)$ do not require many inversions $F_Y^{-1}(\cdot)$.

In Grzelak *et al.* (2015), function $g(\cdot)$ is approximated by means of Lagrange interpolation, which is a well-known interpolation also used in the Uncertainty Quantification (UQ) context. The result is a polynomial, $g_{N_Y}(\cdot)$, which approximates function $g(\cdot) = F_Y^{-1}(F_X(\cdot))$, and the samples y_n can be obtained by employing $g_{N_Y}(\cdot)$ as

$$y_n \approx g_{N_Y}(x_n) = \sum_{i=1}^{N_Y} y_i \ell_i(x_n), \quad \ell_i(x_n) = \prod_{j=1, j\neq i}^{N_Y} \frac{x_n - \bar{x}_j}{\bar{x}_i - \bar{x}_j},$$

where x_n is a vector of samples from X and $\bar{x}_i$ and $\bar{x}_j$ are so-called *collocation points.* N_Y represents the number of collocation points and y_i the exact inversion value of F_Y at the collocation point $\bar{x}_i$, i.e. $y_i = F_Y^{-1}(F_X(\bar{x}_i))$. By applying this interpolation, the number of inversions is reduced and, with only N_Y expensive inversions $F_Y^{-1}(F_X(\bar{x}_i))$, we can generate any number of samples by evaluating the polynomial $g_{N_Y}(x_n)$. This constitutes the 1D version of SCMC sampler in Grzelak *et al.* (2015).

According to the definition of the SCMC technique, a crucial aspect for the computational cost is parameter N_Y: the fewer collocation points are required, the more efficient will be the sampling procedure. The collocation points must be *optimally* chosen in a way to minimize their number. For any random variable X, the optimal collocation points will be based on the moments of X. The optimal collocation points are here chosen to be Gauss quadrature points that are defined as the zeros of the related orthogonal polynomial. This approach leads to a stable interpolation under the probability distribution of X. The complete procedure to compute the collocation points is described in Grzelak *et al.*(2015).

In section 3.2, the application of SCMC technique to generate samples from the CDF of $\int_s^t \sigma^2(z)\mathrm{d}z|\sigma(s)$ is presented. Since we deal with a conditional distribution, the 2D version of SCMC needs to be used.

3. COMPONENTS OF THE MSABR METHOD

In this section, we will discuss the different components of the mSABR method. For simplicity, hereafter, we denote the SABR's integrated variance process by $Y(s,t) := \int_s^t \sigma^2(z)\mathrm{d}z$. We will explain how to efficiently sample the integrated variance given the initial and the final volatility, i.e. $Y(s,t)|\sigma(t)\sigma(s)$, as well as its use in a complete SABR simulation. Since the distribution is not available in closed form, some approximations need to be made. In section 3.3, we propose an accurate sampling method based on copula theory (Sklar, 1959), which is employed to approximate the required conditional distributions. The copula relies on the availability of the *marginal* distributions to simulate the *joint* distribution. As the marginal distributions, $Y(s,t)|\sigma(s)$ and $\sigma(t)|\sigma(s)$ appear as the natural choices. In section 3.1, a procedure to recover the CDF of the integrated variance process given the initial volatility is presented.

3.1. CDF of $\int_s^t \sigma^2(z)\mathrm{d}z|\sigma(s)$ using the COS method

We present a technique to approximate the CDF of $Y(s,t)|\sigma(s)$, i.e. $F_{Y|\sigma(s)}$. We will work in the log-space, so an approximated CDF of $\log Y(s,t)|\log\sigma(s)$, $F_{\log Y|\log\sigma(s)}$, will be derived. We have employed this technique in the one time step SABR method (Leitao *et al.* 2017), but the multiple time step version is more involved. We approximate *Y*(*s*, *t*) by its discrete equivalent, i.e.

$$Y(s,t) := \int_s^t \sigma^2(z)\mathrm{d}z \approx \sum_{j=1}^{M} \Delta t \sigma^2(t_j) =: \hat{Y}(s,t) \tag{5}$$

where *M* is the number of intermediate or discrete time points, † $\Delta t = \frac{t-s}{M}$ and $t_j = s + j\Delta t,\ j = 1,\dots,M$. $\hat{Y}(T)$ is subsequently transformed to the logarithmic domain, with

$$F_{\log\hat{Y}|\log\sigma(s)}(x) = \int_{-\infty}^{x} f_{\log\hat{Y}|\log\sigma(s)}(y)\mathrm{d}y \quad ,(6)$$

and $f_{\log\hat{Y}|\log\sigma(s)}$ the PDF of $\log\hat{Y}(s,t)|\log\sigma(s)$.

Density $f_{\log\hat{Y}|\log\sigma(s)}$ is, in turn, found by approximating the associated characteristic function, $\phi_{\log\hat{Y}|\log\sigma(s)}$, and applying a Fourier inversion procedure. The characteristic function and the desired PDF of $\log\hat{Y}(T)|\log\sigma(s)$ form a *Fourier pair*. Based on the work in Benhamou (2002) and Zhang and Oosterlee (2013), we can define a recursive procedure to recover the characteristic function of $f_{\log\hat{Y}|\log\sigma(s)}$.

3.1.1. Recursive procedure to recover $\phi_{\log \hat{Y} | \log \sigma(s)}$

We start by defining the sequence,

$$R_j = \log\left(\frac{\sigma^2(t_j)}{\sigma^2(t_{j-1})}\right), \quad j = 1, \ldots, M \tag{7}$$

where R_j is the logarithmic increment of $\sigma^2(t)$ between t_j and t_{j-1}, $j = 1, \ldots, M$. As the volatility process follows log-normal dynamics, and increments of Brownian motion are independent and identically distributed, the R_j are also independent and identically distributed, i.e. $R_j \overset{d}{=} R$. In addition, the characteristic function of R_j is well known and reads, $\forall u, j$,

$$\phi_{R_j}(u) = \phi_R(u) = \exp(-iu\alpha^2 \Delta t - 2u^2\alpha^2 \Delta t) \tag{8}$$

with α as in (1) and $i = \sqrt{-1}$ the imaginary unit. By the definition of R_j in equation (7), we write $\sigma^2(t_j)$ as

$$\sigma^2(t_j) = \sigma^2(t_0)\exp(R_1 + R_2 + \cdots + R_j) \tag{9}$$

At this point, a backward recursion procedure in terms of R_j will be defined by which we can recover $\phi_{\log \hat{Y} | \log \sigma(s)}(u)$. We define

$$Y_1 = R_M, \quad Y_j = R_{M+1-j} + Z_{j-1}, \quad j = 2, \ldots, M \tag{10}$$

with $Z_j = \log(1 + \exp(Y_j))$.

By equations (9) and (10), the discrete integrated variance can be expressed as follows:

$$\hat{Y}(s,t) = \sum_{i=1}^{M} \sigma^2(t_i)\Delta t = \Delta t \sigma^2(s)\exp(Y_M) \tag{11}$$

From equation (11) and by applying the definition of characteristic function, we determine $\phi_{\log \hat{Y} | \log \sigma(s)}$, as follows:

$$\begin{aligned}
&\phi_{\log \hat{Y} | \log \sigma(s)}(u) \\
&\quad = \mathbb{E}[\exp(iu \log \hat{Y}(s,t)) | \log \sigma(s)] \\
&\quad = \exp\left(iu \log(\Delta t \sigma^2(s))\right) \mathbb{E}[\exp(iuY_M) | \log \sigma(s)] \\
&\quad = \exp\left(iu \log(\Delta t \sigma^2(s))\right) \phi_{Y_M}(u).
\end{aligned} \tag{12}$$

We have reduced the computation of $\phi_{\log \hat{Y} | \log \sigma(s)}$ to the computation of ϕ_{Y_M}. As Y_M is defined recursively, its characteristic function can be obtained by a recursion as well. According to the definition of the (backward) sequence Y_j in equation (10), the initial and recursive characteristic functions are given by the following expressions,

$$\begin{aligned}
\phi_{Y_1}(u) &= \phi_{R_M}(u) = \phi_R(u) = \exp(-iu\alpha^2 \Delta t - 2u^2\alpha^2 \Delta t), \\
\phi_{Y_j}(u) &= \phi_{R_{M+1-j}}(u)\phi_{Z_{j-1}}(u) = \phi_R(u)\phi_{Z_{j-1}}(u),
\end{aligned} \tag{13}$$

where equation (8) is used in both expressions and the independence of R_{M+1-j} and Z_{j-1} is also employed. By definition, the characteristic function of Z_{j-1} reads

$$\phi_{Z_{j-1}}(u) := \int_{-\infty}^{\infty} (\exp(x) + 1)^{iu} f_{Y_{j-1}}(x) \mathrm{d}x$$

PDF $f_{Y_{j-1}}$c is not known. To approximate it, the Fourier cosine series expansion on $f_{Y_{j-1}}$ is applied. We first truncate the integration range to $[a, b]$. The calculation of integration boundaries a and b follows the cumulant-based approach as described in Zhang and Oosterlee (2013) and Fang and Oosterlee (2008). After truncation of the integral and by applying a cosine series expansion to $f_{Y_{j-1}}$, we have

$$\phi_{Z_{j-1}}(u) \approx \frac{2}{b-a} \sum_{l=0}^{N-1}{}' B_l \int_a^b (\exp(x)+1)^{iu} \times \cos\left((x-a)\frac{l\pi}{b-a}\right) dx =: \hat{\phi}_{Z_{j-1}}(u),$$
$$B_l = \Re\left(\hat{\phi}_{Y_{j-1}}\left(\frac{l\pi}{b-a}\right)\exp\left(-ia\frac{l\pi}{b-a}\right)\right), \tag{14}$$

with N the number of cosine expansion elements, and where

$$\hat{\phi}_{Y_1}(u) := \phi_R(u),$$
$$\hat{\phi}_{Y_j}(u) := \phi_R(u)\hat{\phi}_{Z_{j-1}}(u).$$

We wish to compute $\hat{\phi}_{Y_M}(\frac{k\pi}{b-a})$, for $k = 0, \ldots, N-1$. Considering equation (14), we rewrite $\hat{\phi}_{Y_j}(u) = \phi_R(u)\hat{\phi}_{Z_{j-1}}(u)$ in matrix–vector form, as follows:

$$\Phi_{j-1} = \mathcal{M} A_{j-1} \tag{15}$$

Where

$$\Phi_{j-1} = \left[\Phi_{j-1}(k)\right]_{k=0}^{N-1},$$
$$\Phi_{j-1}(k) = \hat{\phi}_{Z_{j-1}}(u_k),$$
$$\mathcal{M} = [\mathcal{M}(k,l)]_{k=0}^{N-1},$$
$$\mathcal{M}(k,l) = \int_a^b (\exp(x)+1)^{iu_k}\cos((x-a)u_l)\,dx,$$
$$A_j = \frac{2}{b-a}\left[A_j(l)\right]_{l=0}^{N-1},$$
$$A_j(l) = \Re\left(\hat{\phi}_{Y_{j-1}}(u_l)\exp(-iau_l)\right), \tag{16}$$

With

$$u_z = \frac{z\pi}{b-a} \tag{17}$$

By the recursion procedure in equation (15), we obtain the approximation $\hat{\phi}_{Y_M}$ of the characteristic function ϕ_{Y_M} of Y_M.

3.1.2. Numerical approximation of integral

For the approximation in equation (15), we must compute matrix $\mathcal{M}$ in equation (16) highly efficiently. The fast computation of the integral occurring for each element of $\mathcal{M}$ is a key aspect for the performance of the mSABR method. The number of elements in matrix $\mathcal{M}$ can be large, as it corresponds to the number of elements in the cosine series expansion, i.e. N^2. The efficiency also depends on the required accuracy. In this work, we take advantage of the performed analysis in Leitao *et al.* (2017). Among several options, the authors have shown that a numerical approximation based on a *piecewise linear approximation* provides a good balance between performance and accuracy. Following this approximation, we rewrite the considered integral as follows:

$$I_{\mathcal{M}} := \int_a^b \exp\left(iu_k h(x)\right) \cos\left((x-a)u_l\right) \mathrm{d}x,$$
$$h(x) = \log((\exp(x)+1)), \tag{18}$$

with u_k and u_l as in (17). Although function $h(x)$ is not linear, it is smooth and monotonic (see figure 1). Hence, the proposed numerical technique is to define sub-intervals of the integration range $[a, b]$ in which $h(x)$ is almost constant or linear, i.e.

$$I_{\mathcal{M}} = \sum_{j=1}^{L} \int_{a_j}^{b_j} \exp(iu_k h(x)) \cos\left((x-a)u_l\right) \mathrm{d}x$$
,

so that we can perform a first-order expansion in each sub-interval $[a_j, b_j]$, $j = 1, \ldots, L$. In order to guarantee continuity of the approximation, $h(x)$ is approximated by a linear function in each interval as follows:

$$h(x) \approx c_{1,j} + c_{2,j}x, \quad x \in [a_j, b_j].$$

This gives us an approximation $\hat{I}_{\mathcal{M}}$, as follows

$$\hat{I}_{\mathcal{M}} = \sum_{j=1}^{L} \exp(iu_k c_{1,j}) \underbrace{\int_{a_j}^{b_j} \exp(iu_k c_{2,j} x) \cos((x-a)u_l)\, \mathrm{d}x}_{I_j} \quad (19)$$

In each sub-interval $[a_j, b_j]$, we need to determine a simple integral, I_j, which can be computed analytically.

The optimal integration points a_j and b_j in equation (19) are found by differentiation of $h(x)$ w.r.t. x, i.e.

$$H(x) := \frac{\partial h(x)}{\partial x} = \frac{\exp(x)}{\exp(x)+1} = \frac{1}{1+\exp(-x)},$$

giving a relation between the derivative of $h(x)$ and the logistic distribution. This representation indicates that $H(x)$ is the CDF of the logistic distribution, with a so-called *location parameter* $\mu = 0$ and a *scale parameter* $\eta = 1$. In order to determine the optimal integration points so that their positions correspond to the logistic distribution, we need to compute the quantiles. The quantiles of the logistic distribution are analytically available and given by

$$q(p) = \log\left(\frac{p}{1-p}\right)$$

The algorithm for the optimal integration points, given integration interval $[a, b]$ and the number of intervals L, is then as follows:

- Determine an interval in the probability space: $P_h = [H(a), H(b)]$.
- Divide the interval P_h equally into L parts: $P_h^j = [p_j, p_{j+1}],\ j = 0 \ldots L-1$.
- Determine a_j and b_j by calculating the quantiles, $a_j = q(p_j)$ and $b_j = q(p_{j+1}),\ j = 0 \ldots L-1$.

The algorithm above ensures that integration points are optimally distributed (see figure 1(a)). However, for the particular problem, the integration interval can be rather large. As the integration points are typically concentrated near the function's curved part, the errors in the outer sub-intervals dominate the overall error and more points at the beginning and at the end of the interval are required to reduce this error. A more evenly distribution of the integration points can be achieved by introducing a scale parameter, $\eta \neq 1$, which implies that the integration points correspond to a CDF with increased variance. In figure 1(b), the case of $\eta = 3$ is presented. When $\eta \neq 1$, the expression for the quantiles reads,

$$q(p) = \eta \log\left(\frac{p}{1-p}\right)$$

Through the experiments in Leitao *et al.* (2017), we concluded that the choice $\eta = 3$ provided accurate results in this context.

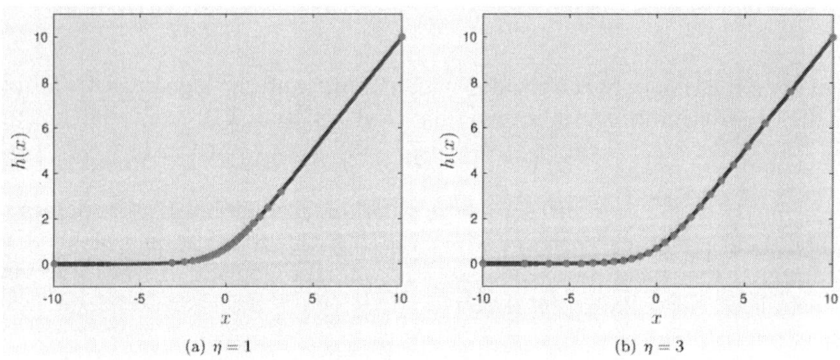

Figure 1. Optimal integration points. *L*=25.

3.1.3. Recovering $f_{\log \hat{Y} | \log \sigma(s)}$ by COS method

Once the approximation of ϕ_{Y_M}, $\hat{\phi}_{Y_M}$ has been efficiently derived, we can recover $f_{\log \hat{Y} | \log \sigma(s)}$ from $\phi_{\log \hat{Y} | \log \sigma(s)}$ by employing the COS method (Fang and Oosterlee 2008), as follows:

$$f_{\log \hat{Y} | \log \sigma(s)}(x) \approx \frac{2}{\hat{b}-\hat{a}} \sum_{k=0}^{N-1'} C_k \cos\left(\left(x-\hat{a}\right)\frac{k\pi}{\hat{b}-\hat{a}}\right) \quad (20)$$

With

$$C_k = \Re\left(\phi_{\log \hat{Y} | \log \sigma(s)}\left(\frac{k\pi}{\hat{b}-\hat{a}}\right)\exp\left(-i\frac{\hat{a}k\pi}{\hat{b}-\hat{a}}\right)\right),$$

and

$$\begin{aligned}\phi_{\log \hat{Y} | \log \sigma(s)}\left(\frac{k\pi}{\hat{b}-\hat{a}}\right) &= \exp\left(i\frac{k\pi}{\hat{b}-\hat{a}}\log\left(\Delta t\sigma^2(s)\right)\right)\phi_{Y_M}\left(\frac{k\pi}{\hat{b}-\hat{a}}\right) \\ &\approx \exp\left(i\frac{k\pi}{\hat{b}-\hat{a}}\log\left(\Delta t\sigma^2(s)\right)\right)\hat{\phi}_{Y_M}\left(\frac{k\pi}{\hat{b}-\hat{a}}\right),\end{aligned}$$

where N is the number of COS terms, $[\hat{a}, \hat{b}]$ is the support † of $\log \hat{Y}(s,t) | \log \sigma(s)$ and the prime′and $\Re$ symbols in equation (20) mean division of the first term in the summation by two and taking the real part of the complex-valued expressions in the brackets, respectively. CDF $F_{\log \hat{Y} | \log \sigma(s)}$ can be obtained by integration, plugging the approximated $f_{\log \hat{Y} | \log \sigma(s)}$ from equation (20) into equation (6).

The characteristic function $\phi_{\log \hat{Y} | \log \sigma(s)}$ and PDF $f_{\log \hat{Y} | \log \sigma(s)}$ need to be derived for each sample of $\log \sigma(s)$ and this makes the computation of $F_{\log \hat{Y} | \log \sigma(s)}$ (and also its subsequent inversion) very expensive. In order to overcome this issue, the SCMC technique (see section 2.2) will be employed approximating these rather expensive computations by means of an accurate interpolation.

3.2. Efficient sampling of $\log \hat{Y} | \log \sigma(s)$

By employing the SCMC technique, instead of directly computing $F_{\log \hat{Y} | \log \sigma(s)}$ for each sample of $\log \sigma(s)$, we only have to compute $F_{\log \hat{Y} | \log \sigma(s)}$ at the collocation points. In general, only a few collocation points are sufficient to obtain accurate approximations, which is a well-known fact from the UQ research field. This fact allows us to drastically reduce the computational cost of sampling the required distribution.

For the problem at hand, we require samples from the integrated variance conditional on the initial volatility, $\hat{Y}(s,t)|\log\sigma(s)$. Therefore, we need to make use of the *2D version of the SCMC technique*. Two levels of collocation points need to be chosen, one for each dimension. If we denote them by $N_{\hat{Y}}$ and N_{σ}, respectively, the resulting number of inversions equals $N_{\hat{Y}}.N_{\sigma}$. The formal definition of the 2D SCMC technique applied to our context reads

$$y_n|v_n \approx g_{N_{\hat{Y}},N_\sigma}(x_n)$$
$$= \sum_{i=1}^{N_{\hat{Y}}}\sum_{j=1}^{N_\sigma} F^{-1}_{\log\hat{Y}|\log\sigma(s)=\bar{v}_j}(F_X(\bar{x}_i))\ell_i(x_n)\ell_j(v_n) \quad , (21)$$

where x_n are the samples from $X \sim \mathcal{N}(0,1)$, which is used as the cheap variable, and u_n the samples of $\log\sigma(s)$; $\bar{x}_i$ and $\bar{v}_j$ are the collocation points for approximating variables $\log Y$ and $\log\sigma(s)$, respectively. The Lagrange polynomials ℓ_i and ℓ_j are defined by

$$\ell_i(x_n) = \prod_{k=1,k\neq i}^{N_{\hat{Y}}} \frac{x_n - \bar{x}_k}{\bar{x}_i - \bar{x}_k}, \quad \ell_j(v_n) = \prod_{k=1,k\neq j}^{N_\sigma} \frac{v_n - \bar{v}_k}{\bar{v}_i - \bar{v}_k}$$

The collocation points $\bar{x}_i$ and $\bar{v}_j$ are calculated based on the moments of the random variables. The two involved variables in the application of the 2D SCMC technique are the collocation variables, $X \sim \mathcal{N}(0,1)$, and $\log\sigma(s) \sim \mathcal{N}(\log\sigma_0 + \frac{1}{2}\alpha^2 s, \alpha\sqrt{s})$. The moments of a normal variable, $X \sim \mathcal{N}(\mu_X, s_X)$ are analytically available, and are given by

$$\mathbb{E}[X^p] = \begin{cases} 0 & \text{if } p \text{ is odd}, \\ s_X^p(p-1)!! & \text{if } p \text{ is even} \end{cases},$$

where *p* is the moment order and the expression !! represents the *double factorial*. In order to compute the optimal collocation points, $\bar{v}_j$, the precalculated collocation points by means of the $\log\sigma(s)$ moments need to be shifted according to the mean, i.e. $\log\sigma_0 + \frac{1}{2}\alpha^2 s$.

We first test numerically the accuracy of the SCMC technique for our problem. In figure 2, two experiments are presented. In figure 2(a), we compare the samples obtained by direct inversion and by the SCMC technique. The fit is highly accurate, already with only seven collocation points. In figure 2(b), the empirical CDFs are depicted, where the excellent match is confirmed.

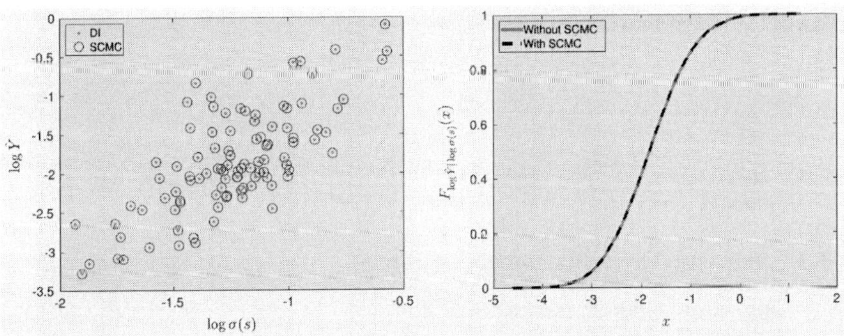

(a) $\log\hat{Y}|\log\sigma(s)$, direct inversion vs. SCMC. (b) $F_{\log\hat{Y}|\log\sigma(s)}(x)$.

Figure 2. (a) Points cloud $\log\hat{Y}|\log\sigma(s)$ by direct inversion (DI) sampling (red dots) and SCMC sampling (black circles). (b) Empirical CDFs of $\log Y|\log\sigma(s)$ with and without using SCMC.

In order to show the performance of SCMC technique for the simulation of $\log\hat{Y}|\log\sigma(s)$, the execution times of generating different numbers of samples are presented in table 1.

In appendix 1, the different sources of error due to the approximations made in the sections 3.1 and 3.2 are analysed.

3.3. Copula-based simulation of $\int_s^t \sigma^2(z)\mathrm{d}z|\sigma(t), \sigma(s)$

In this section, we present an algorithm for the simulation of the integrated variance given $\sigma(t)$ and $\sigma(s)$ by means of a copula. In order to obtain the joint distribution, we require the marginal distributions. In our case, the required CDFs are $F_{\log Y|\log\sigma(s)}$ and $F_{\log\sigma(t)|\log\sigma(s)}$. In section 3.1, an approximated CDF of $\log Y|\log\sigma(s)$, $F_{\log\hat{Y}|\log\sigma(s)}$, has been derived. Since $\sigma(t)$ follows a log-normal distribution, by definition, $\log\sigma(t)$ is normally distributed, and the conditional process $\log\sigma(t)$ given $\log\sigma(s)$, follows a normal distribution,

$$\log\sigma(t)|\log\sigma(s)\sim\mathcal{N}\left(\log\sigma(s)-\frac{1}{2}\alpha^2(t-s),\alpha\sqrt{t-s}\right) \quad (22)$$

Table 1. SCMC time in seconds.

Samples	Without SCMC	With SCMC $N_{\hat{Y}}=N_\sigma=3$	$N_{\hat{Y}}=N_\sigma=7$	$N_{\hat{Y}}=N_\sigma=11$
100	1.0695	0.0449	0.0466	0.0660
10000	16.3483	0.0518	0.0588	0.0798
1000000	1624.3019	0.2648	0.5882	1.0940

3.3.1. Pearson's correlation coefficient

For any copula some measure of the correlation between the marginal distributions is needed. We will employ the Pearson correlation coefficient for $\log Y(s,t)$ and $\log\sigma(t)$. For this quantity, an approximated analytic formula can be derived. By definition, we have

$$\mathcal{P}_{\log Y,\log\sigma(t)}=\text{corr}\left[\log\int_s^t\sigma^2(z)\mathrm{d}z,\log\sigma(t)\right]$$
$$=\frac{\text{cov}\left[\log\int_s^t\sigma^2(z)\mathrm{d}z,\log\sigma(t)\right]}{\sqrt{\text{var}\left[\log\int_s^t\sigma^2(z)\mathrm{d}z\right]\text{var}\left[\log\sigma(t)\right]}}.$$

We employ the following approximation

$$\log\int_s^t\sigma^2(z)\mathrm{d}z\approx\int_s^t\log\sigma^2(z)\mathrm{d}z=2\int_s^t\log\sigma(z)\mathrm{d}z$$

where the logarithm and the integral are interchanged. Since the log function is concave, this approximation forms a lower bound for the true value. This can be seen by applying Jensen's inequality, i.e.

$$\log \int_s^t \sigma^2(z)\mathrm{d}z \geq \int_s^t \log \sigma^2(z)\mathrm{d}z$$

It has been numerically shown that, under the model settings with an interval size *t-s*less than two years, the results based on this approximation are highly satisfactory (further details in Leitao *et al.* (2017). The correlation coefficient can then be approximated by

$$\mathcal{P}_{\log Y(T),\log\sigma(t)} \approx \frac{\mathrm{cov}\left[\int_s^t \log\sigma(z)\mathrm{d}z, \log\sigma(t)\right]}{\sqrt{\mathrm{var}\left[\int_s^t \log\sigma(z)\mathrm{d}z\right]\mathrm{var}\left[\log\sigma(t)\right]}}$$

To compute the covariance, we need $\mathbb{E}\left[\left(\int_s^t \log\sigma(z)\mathrm{d}z\right)\log\sigma(t)\right]$, $\mathbb{E}\left[\int_s^t \log\sigma(z)\mathrm{d}z\right]$ and $\mathbb{E}\left[\log\sigma(t)\right]$. From equation (4), the last expectation as well as $\mathrm{var}\left[\log\sigma(t)\right]$ are known. We find that

$$\log\sigma(t) = \log\sigma_0 - \frac{1}{2}\alpha^2 t + \alpha W(t) =: \eta(t) + \alpha W(t),$$

and

$$\begin{aligned}\int_s^t \log\sigma(z)\mathrm{d}z \\ &= \log\sigma_0(t-s) - \frac{1}{4}\alpha^2(t^2 - s^2) + \alpha\int_s^t W(z)\mathrm{d}z \\ &=: \gamma(s,t) + \alpha\left(tW(t) - sW(s) - \int_s^t z\mathrm{d}W(z)\right).\end{aligned}$$

Based on these equations, using

$$\mathbb{E}\left[\log\sigma(T)\right] = \eta(t), \quad \mathbb{E}\left[\int_s^t \log\sigma(z)\mathrm{d}z\right] = \gamma(s,t),$$

we obtain the following expressions for the remaining expectation,

$$\mathbb{E}\left[\left(\int_s^t \log\sigma(z)\mathrm{d}z\right)\log\sigma(t)\right]$$
$$= \mathbb{E}\left[\left(\gamma(s,t)+\alpha\left(tW(t)-sW(s)-\int_s^t z\mathrm{d}W(z)\right)\right)\right.$$
$$\left.\times\ (\eta(t)+\alpha W(t))\right]$$
$$= \eta(t)\gamma(s,t)+\alpha^2\left(t^2-s^2-\mathbb{E}\left[\int_s^t z\mathrm{d}W^2(z)\right]\right)$$
$$= \eta(t)\gamma(s,t)+\frac{1}{2}\alpha^2(t^2-s^2).$$

For the variance of $\int_s^t \log\sigma(z)\mathrm{d}z$, we first compute

$$\mathbb{E}\left[\left(\int_s^t \log\sigma(z)\mathrm{d}z\right)^2\right]$$
$$= \mathbb{E}\left[\left(\gamma(s,t)+\alpha\left(tW(t)-sW(s)-\int_s^t z\mathrm{d}W(z)\right)\right)^2\right]$$
$$= \gamma^2(s,t)+\alpha^2\mathbb{E}\left[(tW(t)-sW(s))^2\right]$$
$$+\alpha^2\mathbb{E}\left[\left(\int_s^t z\mathrm{d}W(z)\right)^2\right]$$
$$-2\alpha^2\mathbb{E}\left[(tW(t)-sW(s))\int_s^t z\mathrm{d}W(z)\right].$$

Where

$$
\begin{aligned}
&\mathbb{E}\left[(tW(t) - sW(s))^2\right] \\
&= t^2\mathbb{E}\left[(W(t))^2\right] + s^2\mathbb{E}\left[(W(s))^2\right] - 2st\mathbb{E}\left[W(s)W(t)\right] \\
&= t^3 + s^3 - 2s^2 t, \\
&\mathbb{E}\left[\left(\int_s^t z\mathrm{d}W(z)\right)^2\right] \\
&= \int_s^t z^2 \mathrm{d}z = \frac{1}{3}(t^3 - s^3), \\
&\mathbb{E}\left[(tW(t) - sW(s))\int_s^t z\mathrm{d}W(z)\right] \\
&= t\mathbb{E}\left[W(t)\int_s^t z\mathrm{d}W(z)\right] - s\mathbb{E}\left[W(s)\int_s^t z\mathrm{d}W(z)\right] \\
&= t\mathbb{E}\left[\int_s^t \mathrm{d}W(z)\int_s^t z\mathrm{d}W(z)\right] - s\mathbb{E}\left[\int_s^s \mathrm{d}W(z)\int_s^t z\mathrm{d}W(z)\right] \\
&= \frac{1}{2}t(t^2 - s^2),
\end{aligned}
$$

and the variance reads

$$
\begin{aligned}
&\mathrm{var}\left[\int_s^t \log\sigma(z)\mathrm{d}z\right] \\
&= \mathbb{E}\left[\left(\int_s^t \log\sigma(z)\mathrm{d}z\right)^2\right] - \left(\mathbb{E}\left[\int_s^t \log\sigma(z)\mathrm{d}z\right]\right)^2 \\
&= \alpha^2\left(t^3 + s^3 - 2s^2 t + \frac{1}{3}(t^3 - s^3) - t(t^2 - s^2)\right) \\
&= \alpha^2\left(\frac{1}{3}t^3 + \frac{2}{3}s^3 - s^2 t\right).
\end{aligned}
$$

An approximation of the Pearson correlation coefficient is then obtained as follows:

$$\mathcal{P}_{\log Y, \log \sigma(t)} \approx \frac{\eta(t)\gamma(s,t) + \frac{1}{2}\alpha^2(t^2 - s^2) - \eta(t)\gamma(s,t)}{\sqrt{\alpha^2\left(\frac{1}{3}t^3 + \frac{2}{3}s^3 - s^2 t\right)(\alpha^2 t)}}$$
$$= \frac{t^2 - s^2}{2\sqrt{\left(\frac{1}{3}t^4 + \frac{2}{3}ts^3 - t^2 s^2\right)}}. \quad (23)$$

Some copulas do not employ, in their definition, Pearson's coefficient as the correlation measure. For example, Archimedean copulas employ the Kendall's τ rank correlation coefficient. However, a relation between both Pearson's and Kendall's τ coefficients is available,

$$\mathcal{P} = \sin\left(\frac{\pi}{2}\tau\right).$$

3.3.2. Gumbel copula

In this work, we will use Archimedean Gumbel copula. The Gaussian copula may seem a natural choice, as $\sigma(t)$ follows a log-normal distribution and $Y(T) = \int_s^t \sigma^2(z)\mathrm{d}z$ can be seen as summation of squared log-normal processes. However, in Leitao *et al.* (2017), we have empirically shown that the Gumbel copula performs most accurately in this context for a variety of SABR parameter values. The formal definition of the Archimedean Gumbel copula, considering $F_1 \ldots, F_d \in [0,1]^d$ as the marginal distributions, reads

$$C_\theta(F_1 \ldots, F_d) = \exp\left(-\left(\sum_{i=1}^{d}(-\log(F_i))^\theta\right)^{1/\theta}\right) \quad (24)$$

where $(-\log(\cdot))^\theta, \theta > 0$ is the so-called *generator function* of the Gumbel copula. In order to calibrate parameter θ, a relation between θ and the rank correlation coefficient Kendall's τ can be employed, i.e. $\theta = 1/(1-\tau)$.

So, we have derived approximations for all the components required to apply the copula-based technique for the integrated variance simulation.

The algorithm to sample $\int_s^t \sigma^2(z)\mathrm{d}z$ given $\sigma(t)$ and $\sigma(s)$ then consists of the following steps:

(1) Determine *F*log *σ(t)*| log *σ(s)* by equation (22).

(2) Determine *F*log *Y*ˆ | log *σ(s)* by equation (6).

(3) Determine the correlation between log *Y (s, t)* and log *σ(t)* by equation (23).

(4) Generate correlated uniform samples, *U*log *σ(t)*| log *σ(s)* and *U*log *Y*ˆ | log *σ(s)* from the Gumbel copula by equation (24).

(5) From *U*log *σ(t)*| log *σ(s)*, invert the CDF *F*log *σ(t)*| log *σ(s)* to get the samples ˜*σn* of log *σ(t)*| log *σ(s)*. This procedure is straightforward since the normal distribution inversion is analytically available.

(6) From *U*log *Y*ˆ | log *σ(s)*, invert the CDF *F*log *Y*ˆ | log *σ(s)* to get the samples *y*˜*n* of log *Y*ˆ | log *σ(s)*. We propose an inversion procedure based on linear interpolation. First, we evaluate the CDF function at some discrete points. Then, the insight is that, by rotating the CDF under consideration, we can interpolate over probabilities. This is possible when the CDF function is a smoothly increasing function. The interpolation polynomial provides the quantiles of the original distribution from some given probabilities. Since *F*log *Y*ˆ | log *σ(s)* is indeed a smooth and increasing function, the interpolationbased inversion is definitely applicable. This procedure together with the use of 2DSCMCsampler (see section 3.2) results in a fast and accurate inversion.

(7) The samples *σn* of *σ(t)*|*σ(s)* and *yn* of *Y (s, t)* = $\int_s^t$ *t s σ2(z)*dz|*σ(t), σ(s)* are obtained by simply taking exponentials as follows:

$$\sigma_n = \exp(\tilde{\sigma}_n), \quad y_n = \exp(\tilde{y}_n).$$

3.4. Simulation of S(t) given S(s), $\sigma(s), \sigma(t)$ and $\int_s^t \sigma^2(z)\mathrm{d}z$

We complete the mSABR method by the conditional sampling of *S*(*t*). The most commonly used techniques can be classified in two categories: direct inversion of the SABR distribution function given in equation (3) and moment-matching approaches. The direct inversion procedure has a higher computational cost because of the evaluation of the non-central X^2 distribution. However, some recent developments make this computation affordable. Chen *et al.* (2012) proposed a forward asset simulation based on a combination of moment-matching (Quadratic Gaussian) and enhanced direct inversion procedures. We employ this technique also here. Note however that, for some specific values of β, the simulation of the conditional $S(t)$ given $S(s)$, $\sigma(s), \sigma(t)$ and $\int_s^t \sigma^2(z)\mathrm{d}z$ enables analytic expressions.

3.4.1. Case $\beta = 0$.

For $\beta = 0$, it is easily seen from equation (2) that the asset process (in integral form) becomes

$$S(t) = S(s) + \rho \int_s^t \sigma(z)\mathrm{d}\hat{W}_\sigma(z) + \sqrt{1-\rho^2} \int_s^t \sigma(z)\mathrm{d}\hat{W}_S(z).$$

Where

$$\int_s^t \sigma(z)\mathrm{d}\hat{W}_\sigma(z) = \frac{1}{\alpha}(\sigma(t) - \sigma(s)) \tag{25}$$

And

$$\int_s^t \sigma(z)\mathrm{d}\hat{W}_S(z)|\sigma(t), \sigma(s) \sim \mathcal{N}\left(0, \sqrt{\int_s^t \sigma^2(z)\mathrm{d}z}\right) \tag{26}$$

Since this allows negative asset prices, the simulation of $S(t)$ with $\beta = 0$ must be corrected. If the price at the initial time is less or equal to zero, i.e. $S(s) \leq 0$, the price $S(t)$ must remain zero for all time. Otherwise,

$$S(t) = \max(S(t), 0).$$

This correction is performed to be consistent within the whole simulation procedure, since the distribution of the 'exact' SABR simulation in equation (3) relies on the condition $S(s) > 0$. If negative values are permitted, this must be handled in a different way (see Hagan *et al.* (2014) or Antonov *et al.* (2015), for example).

3.4.2. Case $\beta = 1$

In the case of $\beta = 1$, the asset dynamics in equation (1) become log-normal and the solution is given by

$$S(t) = S(s)\exp\left(-\frac{1}{2}\int_s^t \sigma^2(z)\mathrm{d}z + \rho\int_s^t \sigma(z)\mathrm{d}\hat{W}_\sigma(z) + \sqrt{(1-\rho^2)}\int_s^t \sigma(z)\mathrm{d}\hat{W}_S(z)\right).$$

If we take the log-transform,

$$\log\left(\frac{S(t)}{S(s)}\right) = -\frac{1}{2}\int_s^t \sigma^2(z)\mathrm{d}z + \rho\int_s^t \sigma(z)\mathrm{d}\hat{W}_\sigma(z) + \sqrt{(1-\rho^2)}\int_s^t \sigma(z)\mathrm{d}\hat{W}_S(z),$$

and by considering again equations (25) and (26), we obtain the distribution of $\log\left(\frac{S(t)}{S(s)}\right) \mid \int_s^t \sigma^2(z)\mathrm{d}z, \sigma(t), \sigma(s)$, which reads

$$\mathcal{N}\left(-\frac{1}{2}\int_s^t \sigma^2(z)\mathrm{d}z + \frac{\rho}{\alpha}(\sigma(t) - \sigma(s)), \times\sqrt{(1-\rho^2)\int_s^t \sigma^2(z)\mathrm{d}z}\right). \quad (27)$$

By employing equation (27), the asset dynamics $S(t)$ can be sampled from

$$S(t) \sim S(s)\exp\left(-\frac{1}{2}\int_s^t \sigma^2(z)\mathrm{d}z + \frac{\rho}{\alpha}(\sigma(t) - \sigma(s)) + X\sqrt{(1-\rho^2)\int_s^t \sigma^2(z)\mathrm{d}z}\right), \quad (28)$$

where X is the standard normal.

3.4.3. Case $\beta \neq 0, \beta \neq 1$

As mentioned before, for the generic case of $\beta \in (0, 1)$, we employ the enhanced inversion of the SABR asset price distribution in equation (3) as introduced in Chen *et al.* (2012). We briefly summarize this approach. The asset simulation is performed either by a moment-matched quadratic Gaussian approximation or by an enhanced direct inversion. The authors in

Chen *et al.* (2012) proposed a threshold value to choose the suitable technique among these two. The moment-matching approach relies on the fact that, for $S(s) \gg 0$, the distribution function in equation (3) can be approximated by

$$\Pr\left(S(t) \leq K | S(s) > 0, \sigma(s), \sigma(t), \int_s^t \sigma^2(z)\mathrm{d}z\right)$$
$$\approx \chi^2(c; 2-b, a).$$

By definition, the mean, μ and the variance, κ^2, of a generic non-central chi-square
distribution $\chi^2(x; \delta, \lambda)$ are $\mu := \delta + \lambda$ and $\kappa^2 := 2(\delta + 2\lambda)$, respectively. A variable $Z \sim \chi^2(x; \delta, \lambda)$ is accurately approximated by a quadratic Gaussian distribution, as follows

$$Z \overset{\mathrm{d}}{\approx} d(e+X)^2, \quad X \sim \mathcal{N}(0,1),$$

With

$$\psi := \frac{\kappa^2}{\mu^2}, \quad e^2 = 2\psi^{-1} - 1 + \sqrt{2\psi^{-1}}\sqrt{2\psi^{-1} - 1} \geq 0,$$
$$d = \frac{\mu}{1+e^2}.$$

Applying this to the case in equation (3), we can sample the conditional asset dynamics, $0 \leq s < t$, by

$$S(t) = \left((1-\beta)^2(1-\rho^2)d(e+X)^2 \int_s^t \sigma^2(z)\mathrm{d}z\right)^{\frac{1}{2(1-\beta)}}$$

The moment-matching approximation is safely and accurately applicable when $0 < \psi \leq 2$ and $\mu \geq 0$ (Chen *et al.* 2012). Otherwise, a direct inversion procedure must be employed. A root-finding Newton method is then used. In order to reduce the number of Newton iterations (and the expensive evaluation of the non-central chi-square CDF), the first iterations are based on an approximated formula for the non-central chi-square distribution, which is based on the normal CDF and derived by Sankaran (1963). Then, the obtained value is employed as the initial guess for the Newton algorithm. The result is a significant reduction in the number of iterations and, hence, in the computational cost. Furthermore, this root-finding procedure consists of only basic operations, so that the whole procedure can be easily vectorized, leading to a further efficiency improvement.

As in the case of $\beta = 0$, the absorbing boundary at zero needs to be prescribed to avoid negative prices.

The resulting sampling procedure is robust and efficient.

3.4.4. Martingale correction

As already pointed by Andersen (2008) and Chen *et al.* (2012), the almost exact simulation of the asset price in some stochastic volatility models can produce a loss of the martingale property, due to the approximation of a continuous process by its discrete equivalent. This is especially seen, when the size of time step is large and for certain configurations of the SABR parameters, like small β values, close-to-zero initial asset values S_0, high *vol-vol* parameter α or large initial volatility σ_0. In order to overcome this issue, we incorporate to the mSABR method the use of a simple but effective numerical martingale correction, as follows:

$$
\begin{aligned}
S(t) &= S(t) - \frac{1}{n}\sum_{i=1}^{n} S_i(t) + \mathbb{E}[S(t)], \\
&= S(t) - \frac{1}{n}\sum_{i=1}^{n} S_i(t) + S_0,
\end{aligned}
$$

where $S_i(t)$ represents the ith Monte Carlo sample. Note, however, that, for practical SABR parameters, the martingale error is very small and can be easily controlled by increasing the number of overall time steps m.

4. NUMERICAL EXPERIMENTS

In this section, we benchmark the mSABR method by pricing options under the SABR dynamics. We will consider several parameter configurations, extracted from the literature (see table 2), a zero-correlation set as in Grzelak *et al.* (2015) (set I), a set under log-normal dynamics as in Chen *et al.* (2012) (set II), a medium long time to maturity set as in Antonov *et al.* (2013) (set III) and a more extreme set as in Antonov *et al.* (2015) (set IV). The strikes, K_i, are chosen following the expression

$$K_i(T) = S(0)\exp(0.1 \times T \times \delta_i),$$
$$\delta = \{-1.5, -1.0, -0.5, 0.0, 0.5, 1.0, 1.5\},$$

introduced by Piterbarg (2005).

Table 2. Data sets.

	S_0	σ_0	α	β	ρ	T
Set I (Grzelak *et al.* 2015)	0.5	0.5	0.4	0.5	0.0	4
Set II (Chen *et al.* 2012)	0.04	0.2	0.3	1.0	−0.5	5
Set III (Antonov *et al.* 2013)	1.0	0.25	0.3	0.6	−0.5	20
Set IV (Antonov *et al.* 2015)	0.0056	0.011	1.080	0.167	0.999	1

Other parameters include:

- Step size in equation (5): $\Delta t = 5 \times 10^{-5}$.
- Number of terms in equations (14) and (20): $N = 150$.
- Number of intervals for piecewise linear approximation

in equation (19): $L = 100$.

- Number of collocation points in equation (21): $N_{\hat{Y}} = N_\sigma = 3$.
- Number of samples: $n = 1.000.000$.

The experiments were performed on a computer with CPU Intel Core i7-4720HQ 2.6 GHz and RAM memory of 16 Gigabytes. The employed software package was Matlab r2015b.

As usual, the European option prices are obtained by averaging the maximum of the forward asset values at maturity minus the strike and zero, i.e. $(S(T) - K_i(T), 0)$ (call option). Subsequently, the Black–Scholes formula is inverted to determine the *implied volatilities*. Note that, once the forward asset is simulated by the mSABR method, we can price options at multiple strikes simultaneously.

4.1. Convergence test

We perform a convergence study in terms of the number of time steps, m and, subsequently, in terms of the number of samples, n. Set I is considered suitable for this experiment, since an analytic solution is available by Antonov *et al.* (2013) when $\rho = 0$, and it can be used as a reference. Also, as equation (3) is employed to sample the forward asset dynamics, the case $\rho = 0$ results in an *exact simulation*.

In table 3, the implied volatilities obtained by the mSABR method for several choices of the number of time steps m are presented. We observe by the errors in basis points (bp) that, by increasing the number of time steps, the technique converges very rapidly. Furthermore, only a few time steps are required to obtain accurate results: with only m=4 time steps (one per year) the error is already less than ten basis points and with m=4T (the time step size is a quarter of year), the error remains below one basis points. Compared to the *low-bias SABR simulation scheme* proposed in Chen *et al.* (2012), the mSABR method is more accurate, since it relies on the *exact* simulation of the integrated variance, whereas the low-bias scheme approximates the integrated variance by a small disturbance expansion and moment-matching techniques. That scheme performs worse when bigger time steps are considered.

Table 3. Implied volatility, increasing m: Antonov vs. mSABR. Set I.

Strikes	K_1	K_2	K_3	K_4	K_5	K_6	K_7
Antonov	73.34%	71.73%	70.17%	N/A	67.23%	65.87%	64.59%
$m = T/4$	73.13%	71.75%	70.41%	69.11%	67.05%	66.64%	65.48%
Error(bp)	−21.51	2.54	24.38	N/A	61.71	76.66	89.26
$m = T/2$	73.30%	71.78%	70.29%	68.86%	67.49%	66.17%	64.93%
Error(bp)	−4.12	4.94	12.71	N/A	25.48	30.40	34.73
$m = T$	73.25%	71.67%	70.14%	68.66%	67.24%	65.89%	64.62%
Error(bp)	−9.56	−5.93	−2.79	N/A	0.92	2.21	3.17
$m = 2T$	73.32%	71.71%	70.16%	68.65%	67.22%	65.85%	64.55%
Error(bp)	−2.08	−1.56	−1.20	N/A	−1.65	−2.35	−3.36
$m = 4T$	73.34%	71.73%	70.18%	68.67%	67.24%	65.87%	64.58%
Error(bp)	0.15	0.58	0.78	N/A	0.43	0.04	−0.48

Table 4. Implied volatility, increasing n: Antonov vs. mSABR. Set I.

Strikes	K_1	K_2	K_3	K_4	K_5	K_6	K_7
Antonov	73.34%	71.73%	70.17%	N/A	67.23%	65.87%	64.59%
$n = 10^2$	67.29%	65.55%	63.84%	62.20%	60.63%	59.01%	57.65%
RE	8.24×10^{-2}	8.61×10^{-2}	9.01×10^{-2}	N/A	9.82×10^{-2}	1.04×10^{-1}	1.07×10^{-1}
$n = 10^4$	73.41%	71.87%	70.36%	68.91%	67.51%	66.19%	64.94%
RE	9.65×10^{-4}	1.94×10^{-3}	2.75×10^{-3}	N/A	4.08×10^{-3}	4.93×10^{-3}	5.48×10^{-3}
$n = 10^6$	73.34%	71.73%	70.18%	68.67%	67.24%	65.87%	64.58%
RE	2.04×10^{-5}	8.08×10^{-5}	1.11×10^{-4}	N/A	6.39×10^{-5}	6.07×10^{-6}	7.43×10^{-5}

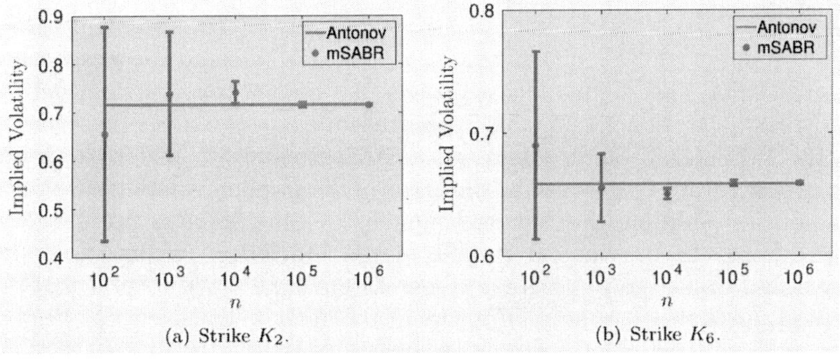

Figure 3. Convergence of the mSABR method.

PowerPoint slideOriginal jpg (97.00KB)Display full sizeWe also perform a convergence test in the number of samples, *n*. According to the previous experiment and in order to guarantee that the errors do not increase by the time discretization, we set ,m=4T. In table 4, the implied volatilities for increasing *n* are presented. By the *relative error* (RE), we can see that, as expected, the error is approximately reduced by a factor of $1/\sqrt{n}$. Furthermore, we wish to show how the number of samples affects the variance of the mSABR method. In figure 3, the standard deviations† for several choices of *n* are presented. Two strikes are evaluated, one *in-the-money* strike, K_2, and one *out-of-the-money* strike, ‡ K_6. An identical behaviour can be observed for all other strikes as well. Again, we see a fast convergence and variance reduction in terms of *n*.

4.2. Performance test

Next to the convergence of the mSABR method in terms of *m* and *n*, another important aspect is the computational cost. Specifically, we will measure the execution times of an option pricing experiment using different alternative Monte Carlo-based methods and compare them with the mSABR method. Again, parameter Set I is employed since a reference value is available for this case. We consider a plain Monte Carlo Euler (MC Euler) discretization scheme for the forward asset, $S(t)$. Note that the absorbing boundary at zero must be handled with care. Since our main contribution here is the exact simulation of the integrated variance, it is natural to also compare the performance with other, less involved, techniques. We therefore also present a comparison between the mSABR method and two techniques where the integrated variance is either approximated by $\int_s^t \sigma^2(z)\mathrm{d}z \approx \sigma^2(s)(t-s)$ (left rectangular rule) or $\int_s^t \sigma^2(z)\mathrm{d}z \approx \frac{1}{2}(\sigma^2(t)+\sigma^2(s))(t-s)$ (trapezoidal rule). We denote these alternatives as the *Y-Euler* and *Y-trpz* schemes, respectively. The simulation of $S(t)$ is, in both cases, carried out as explained in section 3.4. In table 5, we present execution times (in seconds) of the MC Euler, *Y-Euler*, *Y-trpz* and mSABR schemes, to achieve a certain level of accuracy (measured as the error of the implied volatilities in basis points). In addition, the required number of time steps is presented in parentheses to provide an insight in the efficiency. Furthermore, the speedup obtained by our method is included in table 6. We observe that the mSABR method is superior in terms of the ratio between computational cost and accuracy. The accuracy of *Y*-Euler or *Y*-trpz can be improved by adding intermediate time steps, but then also the computational cost increases.

Table 5. Execution times and time steps, m (parentheses).

Error	<100 bp	<50 bp	<25 bp	<10 bp
MC Euler	6.85(200)	10.71(300)	27.42(800)	42.90(1200)
Y-Euler	2.18(4)	6.55(16)	11.85(32)	45.12(128)
Y-trpz	2.17(3)	4.24(8)	7.25(16)	14.47(32)
mSABR	3.46(1)	2.98(2)	3.72(3)	4.89(4)

Table 6. Speedups provided by the mSABR method.

Error	<100 bp	<50 bp	<25 bp	<10 bp
MC Euler	1.98	3.59	7.37	8.77
Y-Euler	0.63	2.19	3.18	9.22
Y-trpz	0.62	1.42	1.94	2.95

4.3. 'Almost exact' SABR simulation by varying ρ

As a next experiment we test the stability of the mSABR method when correlation parameter ρ varies from negative to positive values. As stated before, equation (3) is only exact for ρ=0. For that, we use Set II since the conditional forward asset simulation enables a closed-form solution (given by equation (28)) for $\beta = 1$. The considered values are ρ= $\{-0.5, 0.0, 0.5\}$. Following the outcome of the previous experiment, we set m=4T. As a reference, a Monte Carlo (MC) simulation with a very fine Euler time discretization (1000T time steps) in the forward asset process is employed. In table 7, the resulting implied volatilities are provided. The differences between the Monte Carlo volatilities and the ones given by the mSABR method are within 10 basis points for all choices of ρ.

Table 7. Implied volatility, varying ρ: Euler Monte Carlo vs. mSABR. Set II.

Strikes	K_1	K_2	K_3	K_4	K_5	K_6	K_7
				$\rho = -0.5$			
MC	22.17%	21.25%	20.38%	19.57%	18.88%	18.33%	17.95%
mSABR	22.21%	21.28%	20.39%	19.58%	18.88%	18.32%	17.94%
Error(bp)	3.59	2.86	1.78	0.95	−0.19	−0.96	−1.10
				$\rho = 0.0$			
MC	21.35%	20.96%	20.71%	20.63%	20.71%	20.96%	21.34%
mSABR	21.35%	20.95%	20.69%	20.60%	20.68%	20.93%	21.32%
Error(bp)	0.04	−1.04	−2.51	−3.02	−3.33	−3.19	−2.56
				$\rho = 0.5$			
MC	19.66%	20.04%	20.61%	21.34%	22.20%	23.14%	24.16%
mSABR	19.59%	19.96%	20.54%	21.28%	22.15%	23.11%	24.11%
Error(bp)	−6.93	−7.36	−6.77	−5.53	−4.35	−3.76	−4.05

4.4. Pricing barrier options

We will price barrier options with the mSABR method. The *up-and-out* call option is considered here, with the barrier level, *B*, $B > S_0, B > K_i$. The price of this type of barrier option reads

$$V_i(K_i, B, T)$$
$$= \exp(-rT)\,\mathbb{E}\left[(S(T) - K_i)\mathbb{1}\left(\max_{0<t_k\leq T} S(t_k) > B\right)\right]$$

where $\mathbb{1}(\cdot)$ represents the indicator function and t_k are the predefined times where the barrier condition (whether or not it is hit) is checked.

As a reference, a Monte Carlo method with very fine Euler time discretization is employed, as in the previous experiment. The number of time steps for the mSABR scheme is again set to m=4T. In tables 8, 9 and 10, the obtained prices are presented for the parameter sets I, II and III, respectively. The prices are multiplied by a factor of 100. Also, the *mean squared error* (MSE) is shown. We define the MSE as

$$\text{MSE} = \frac{1}{7}\sum_{i=1}^{7}\left(V_i^{MC}(K_i, B, T) - V_i^{mSABR}(K_i, B, T)\right)^2$$

where $V_i^{MC}(K_i, B, T)$ and $V_i^{mSABR}(K_i, B, T)$ are the barrier option prices provided by Monte Carlo method and by the mSABR method, respectively.

Table 8. Pricing barrier options with mSABR: $V_i(K_i, B, T) \times 100$. Set I.

Strikes	K_1	K_2	K_3	K_4	K_5	K_6	K_7
				$B = 1.2$			
MC	6.2868	5.4577	4.6330	3.8263	3.0551	2.3358	1.6869
mSABR	6.3264	5.4890	4.6566	3.8436	3.0677	2.3452	1.6953
MSE	5.3196×10^{-8}						
				$B = 1.5$			
MC	10.2271	9.1568	8.0718	6.9856	5.9142	4.8753	3.8899
mSABR	10.2249	9.1507	8.0609	6.9702	5.8960	4.8572	3.8737
MSE	1.8884×10^{-8}						
				$B = 1.8$			
MC	13.5740	12.3571	11.1118	9.8513	8.5910	7.3477	6.1403
mSABR	13.5915	12.3686	11.1174	9.8507	8.5851	7.3380	6.1276
MSE	1.0851×10^{-8}						

Table 9. Pricing barrier options with mSABR: $V_i(K_i, B, T) \times 100$. Set II.

Strikes	K_1	K_2	K_3	K_4	K_5	K_6	K_7
				$B = 0.08$			
MC	1.1702	0.9465	0.7268	0.5215	0.3423	0.1996	0.0987
mSABR	1.1724	0.9486	0.7285	0.5226	0.3428	0.1997	0.0986
MSE	1.8910×10^{-10}						
				$B = 0.1$			
MC	1.3099	1.0766	0.8462	0.6290	0.4367	0.2794	0.1626
mSABR	1.3092	1.0761	0.8456	0.6282	0.4355	0.2782	0.1618
MSE	7.5542×10^{-11}						
				$B = 0.12$			
MC	1.3521	1.1168	0.8841	0.6644	0.4695	0.3093	0.1891
mSABR	1.3518	1.1166	0.8838	0.6639	0.4686	0.3080	0.1880
MSE	6.3648×10^{-11}						

Table 10. Pricing barrier options with mSABR: $V_i(K_i, B, T) \times 100$. Set III.

Strikes	K_1	K_2	K_3	K_4	K_5	K_6	K_7
				$B = 2.0$			
MC	29.1174	23.4804	17.2273	10.7825	5.0203	1.1750	0.0036
mSABR	29.2346	23.5828	17.3086	10.8327	5.0385	1.1805	0.0036
MSE	4.8146×10^{-7}						
				$B = 2.5$			
MC	41.3833	34.5497	26.8311	18.6089	10.7281	4.4893	0.9434
mSABR	41.3394	34.5097	26.7948	18.5747	10.6943	4.4546	0.9320
MSE	1.2131×10^{-7}						
				$B = 3.0$			
MC	48.5254	41.1652	32.7980	23.7807	14.9344	7.5364	2.6692
mSABR	48.5008	41.1515	32.7888	23.7655	14.9097	7.5117	2.6549
MSE	3.6201×10^{-8}						

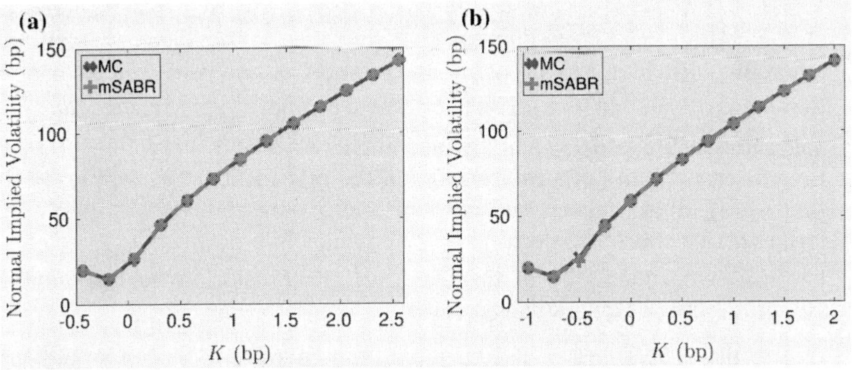

Figure 4. The mSABR method in the context of negative interest rates.

The resulting accuracy is very satisfactory. As expected, higher option prices are obtained for bigger barrier levels, B, and/or lower strikes, K_i.

4.5. Negative interest rates

The SABR model is very popular in the interest rate context. One of the most important current features in this market is the occurrence of negative interest rate values and strikes. Approaches dealing with this issue have appeared in the literature, like Antonov *et al.* (2015) or Hagan *et al.* (2014). To apply the mSABR scheme in the case of negative interest rates, we will use the *shifted SABR model* by Schlenkrich *et al.* (2014). The shifted SABR model is defined as follows:

$$\mathrm{d}S(t) = \sigma(t)(S(t)+\theta)^{\beta}\mathrm{d}W_S(t),$$
$$S(0) = (S_0+\theta)\exp(rT),$$

where $\theta > 0$ is a displacement, or shift, in the underlying. The volatility process, $\sigma(t)$, remains invariant (see equation (1)). Parameter θ can be seen as the maximum level of negativity that is expected in the rates. This generalization of the SABR model is widely used by market practitioners due to its simplicity, and, precisely, the advantage of keeping the existing simulation schemes. In mSABR, the assumption $S(t) > 0$ is required to employ the method.

In order to test the mSABR scheme in the context of negative interest rates, we employ the set IV. The SABR model parameters were obtained by Antonov *et al.* (2015) after a calibration process to real data† under the shifted SABR model. Note that this configuration corresponds to an extreme situation with high values of the correlation, ρ, and *vol-vol*, α. As a shift parameter, $\theta = 0.02$ is chosen. We compare the *normal implied volatilities*‡ obtained by employing Monte Carlo with very fine (1000T) time discretization (as a reference) and the mSABR method with m=4T. In figure 4(a), these curves are depicted and a perfect fit is observed. The strikes, K are chosen to permit negative values. We perform an even more extreme experiment, by setting $S_0 = 0$. The resulting normal implied volatilities are presented in figure 4(b), again with a excellent fit.

Due to the complexity of the negative interest rates experiment and this particular set of parameters $(\rho \approx 1)$, we wish to test the performance of our method under these conditions. In tables 11 and 12 the execution times obtained are presented. Again, the small number of time steps due to the 'almost' exact simulation of the mSABR method provides an important gain in terms of performance.

Table 11. Execution times and time steps, m (parentheses).

Error	< 25 bp	< 10 bp	< 5 bp	< 1 bp
Y-Euler	2.47(4)	9.64(16)	14.32(24)	19.25(32)
Y-trpz	1.32(2)	4.99(8)	9.75(16)	14.52(24)
mSABR	1.61(1)	2.57(2)	3.52(3)	4.57(4)

Table 12. Speedups provided by the mSABR method.

Error	< 25 bp	< 10 bp	< 5 bp	< 1 bp
Y-Euler	1.57	3.75	4.07	4.21
Y-trpz	0.82	1.94	2.77	3.17

5. CONCLUSIONS

In this work, we have proposed an accurate and robust multiple time step Monte Carlo method to simulate the SABR model with only a few time steps. The mSABR method employs many nontrivial components that have not been seen before in combination within a Monte Carlo simulation. A copula methodology to generate samples from the conditional SABR's integrated variance process has been introduced. The marginal distribution of the integrated variance has been derived by employing Fourier techniques. We have dealt with an increasing computational complexity, due to the multiple time steps and the almost exact simulation, by applying an interpolation based on stochastic collocation. This has resulted in an efficient SABR simulation scheme. We have numerically shown the convergence and the stability of the method. In terms of convergence, the mSABR method requires only very few time steps to achieve high accuracy, due to the focus on almost exact simulation (in contrast to the low-bias SABR scheme Chen *et al.* 2012). This fact impacts the performance, obtaining an excellent ratio between accuracy and computational cost. As a multiple time step method, it can be employed to price path-dependent options. As an example, a pricing experiment for barrier options has been carried out. Furthermore, we have shown that the mSABR scheme is also suitable in the context of negative interest rates, in combination with a shifted model.

Appendix 1. Error analysis of $\int_s^t \sigma^2(z)\mathrm{d}z|\sigma(s)$ simulation

A.1. Errors in the CDF of $\int_s^t \sigma^2(z)\mathrm{d}z|\sigma(s)$

In section 3.1, we have proposed a method to compute the approximated CDF for the conditional integrated variance, i.e. $Y(s,t)|\sigma(s)$. Here, we perform an error analysis, indicating how to reduce or bound the errors. The sources of error in the approximation include ϵ_D, the discretization of $Y(s, t)$ in equation (5), ϵ_T, the truncation in equations (14) and (20), ϵ_C, due to the cosine expansion in equations (14) and (20) and ϵ_M, the numerical computation of I_M in equation (19).

A.1.1. Error ϵ_D

This error occurs because SABR's time-integrated variance is approximated by its discrete equivalent, i.e. $Y(s,t) \approx \hat{Y}(s,t)$. Note that the process $Y(s,t) = \int_s^t \sigma^2(z)\mathrm{d}z$ is the solution of the SDE

$$\mathrm{d}Y(s,z) = \sigma^2(z)\mathrm{d}z, \quad Y(s,s) = 0,$$

which, by employing the same time discretization as for $\hat{Y}(s,t)$ in Equation (5), can be written as follows:

$$\bar{Y}(s,t_j) - \bar{Y}(s,t_{j-1}) = \sigma^2(t_{j-1})\Delta t.$$

This is essentially the Euler–Maruyama discretization scheme. Indeed, it is easy to see that, if we sum the right-hand side of the equality over the *M* discrete time points, we recover the discrete approximation $\hat{Y}(s,t)$, as follows:

$$\begin{aligned}\sum_{j=1}^{M} \sigma^2(t_{j-1})\Delta t &= \sum_{j=1}^{M} \left(\bar{Y}(s,t_j) - \bar{Y}(s,t_{j-1})\right)\\ &= \bar{Y}(s,t_M) - \bar{Y}(s,t_0) = \hat{Y}(s,t).\end{aligned}$$

The error (or convergence) of the Euler–Maruyama discretization scheme has been intensively studied in the literature (see, e.g. Kloeden and Platen 1992). Two errors can be defined when a stochastic process is discretized: *strong* and *weak*. It was proved by Kloeden and Platen (1992) that, under some conditions of regularity, the Euler–Maruyama scheme has strong convergence of order $\gamma = \frac{1}{2}$ and weak convergence of order $\gamma = 1$. We focus on the strong error between $Y(s, t)$ and $\hat{Y}(s, t)$. Then, error ϵ_D can be computed and bounded by employing the usual strong convergence expression as

$$\epsilon_D := \mathbb{E}\left[\left|\hat{Y}(s,t) - Y(s,t)\right|\right] \leq C\left(\frac{t-s}{M}\right)^{\gamma},$$

for some constant $C > 0$.

In the actual experiments, we consider $\log \hat{Y}(s, t)$ and $\log Y(s, t)$. This will reduce the error further. In Leitao *et al.* (2017), we have presented numerical results that confirm the expected behaviour of this error.

A.1.2. Error ϵ_T. The use of cosine expansions and the COS method implies a truncation of the integration range to interval $[a, b]$. This gives rise to a truncation error, which is defined by

$$\epsilon_T(H) := \int_{\mathbb{R}\setminus[a,b]} f_H(y)\mathrm{d}y,$$

where H corresponds to Y_j and $\log \hat{Y}(s, t)$, according to equations (14) and (20), respectively. By a sufficiently large integration range $[a, b]$, this error can be reduced and does not dominate the overall error.

A.1.3. Error ϵ_C The convergence due to the number of terms N in the cosine series expansion was derived in Fang and Oosterlee (2008). For any $f(y|x) \in \mathbf{C}^{\infty}[a, b]$, ϵ_C, can be bounded by

$$|\epsilon_C(N, [a, b])| \leq P^*(N)\exp(-(N-1)\nu),$$

being $\nu > 0$ a constant and $P^*(N)$ a term which varies less than exponentially with respect to N. As the integrated variance is a very smooth function, ϵ_C will decay exponentially with respect to N.

A.1.4. Error ϵ_M

When the piecewise approximation in equation (19) is employed, the convergence in terms of the number of sub-intervals, L, is numerically tested. In table A1, we observe a reduction of the error by a factor four, when going from L=25 to L=50 integration points. However, with L=100 the error reduces more drastically due to the choice of optimal integration points.

Table A1. MSE vs. L for $\hat{I}_{\mathcal{M}}$ in equation (19).

	$L = 25$	$L = 50$	$L = 75$	$L = 100$
Equidistant	2.4075×10^{-3}	2.6743×10^{-4}	3.1064×10^{-4}	2.9547×10^{-6}
Optimal ($\eta = 1$)	1.3273×10^{-2}	2.5343×10^{-3}	9.5959×10^{-4}	4.7953×10^{-4}
Optimal ($\eta = 3$)	3.4837×10^{-5}	1.1445×10^{-6}	2.1506×10^{-7}	6.7114×10^{-8}

A.2. Error in the application of SCMC to $\int_s^t \sigma^2(z)\mathrm{d}z|\sigma(s)$

In Grzelak *et al.* (2015), where SCMC sampler was proposed, an error analysis of the technique was carried out. Here we summarize the most important results related to our context and refer the reader to the original paper for further details.

To measure the error which results from the collocation method in a more general case (see section 2.2), we can consider either the difference between functions $g(X)$ and $g_N(X)$ (X the cheap variable) or the error associated with the approximated CDF. The first type of error is due to Lagrange interpolation, so the corresponding error estimate is well known. The error $\epsilon(\xi_n)$ with N_Y collocation points is given by the standard Lagrange interpolation reads

$$\epsilon_X(\xi_n) = \left|g(\xi_n) - g_{N_Y}(\xi_n)\right| = \left|\frac{1}{N_Y!}\frac{\partial^{N_Y} g(x)}{\partial x^{N_Y}}\Big|_{x=\xi}\prod_{i=1}^{N_Y}(\xi_n - \bar{x}_i)\right| \quad \text{(A1)}$$

where $\bar{x}_i$ are the collocation points, and $\hat{\xi} \in [\min(\mathbf{x}), \max(\mathbf{x})]$, $\mathbf{x} = (\bar{x}_1, \ldots, \bar{x}_{N_{\hat{Y}}})^T$. The error can be bounded by defining ξ as the point where $\left|\partial^{N_Y} g(x)/\partial x^{N_Y}\right|$ has its maximum. A small probability of large errors *in the tails* can be observed by deriving the error $\epsilon_U(\xi_n)$, by substituting the uniformly distributed random variable u_n in the previous equation, using $\xi_n = F_X^{-1}(u_n)$,

$$\begin{aligned}\epsilon_U(u_n) &= \left|g(F_X^{-1}(u_n)) - g_{N_Y}(F_X^{-1}(u_n))\right| \\ &= \left|\frac{1}{N_Y!}\frac{\partial^{N_Y} g(x)}{\partial x^{N_Y}}\bigg|_{x=\xi}\prod_{i=1}^{N_Y}(F_X^{-1}(u_n) - \bar{x}_i)\right|.\end{aligned}$$

A 'close-to-linear' relation between the involved stochastic variables, meaning $\partial^{N_Y} g(x)/\partial x^{N_Y}$ being small, gives a small approximation error. On the other hand, when approximating the CDF of the expensive variable Y, we have $F_Y(y) = F_Y(g(x)) \approx F_Y(g_{N_Y}(x))$, which is exact at the collocation points $\bar{x}_i$,

$$F_Y(y_i) = F_Y(g(\bar{x}_i)) = F_Y(g_{N_Y}(\bar{x}_i)).$$

A deeper analysis can be found in the original paper of SCMC (Grzelak *et al.*, 2015), including a theoretical convergence in L^2 and the convergence within the distribution tails.

We have chosen X to be standard normal, and $Y = \log \hat{Y} | \log \sigma(s)$ can be seen as the logarithm of a sum of log-normal variables. Therefore, it is a 'close-to-normal' distribution which can be efficiently approximated by a linear combination of standard normal distributions. In order to measure the error when SCMC is applied to the sampling of $\int_s^t \sigma^2(z)\mathrm{d}z | \sigma(s)$ (see section 3.2), we can consider the difference between the functions $g(x) = F^{-1}_{\log \hat{Y} | \log \sigma(s)}(F_X(x))$ and $g_{N_{\hat{Y}}, N_\sigma}(x)$ in equation (21). By following the same derivation as in equation (A1), we have

$$\begin{aligned}\epsilon_{\hat{Y}}(\xi_n) &= \left|g(\xi_n) - g_{N_{\hat{Y}}, N_\sigma}(\xi_n)\right| \\ &= \left|\frac{1}{N_{\hat{Y}}!}\frac{\partial^{N_{\hat{Y}}} g(x)}{\partial x^{N_{\hat{Y}}}}\bigg|_{x=\xi}\prod_{i=1}^{N_{\hat{Y}}}(\xi_n - \bar{x}_i)\prod_{j=1}^{N_\sigma}(v_n - \bar{v}_j)\right|,\end{aligned}$$

where $N_{\hat{Y}}$ and N_σ are the number of collocation points in each dimension and $\bar{v}_j$ are the collocation points corresponding to the conditional random variable $\log \sigma(s)$.

NOTES

1 So, we have the *m* large time steps, for the multiple time step Monte Carlo simulation, and we have *M* intermediate dates for the integrated variance, $M >> m$. We suggest to prescribe Δt, and use $M = \frac{t-s}{\Delta t}$, in order to avoid over-discretization issues. This makes the value of *M* dependent on the interval size, *t-s*.

2 It can be calculated given the support of Y_M, $[a, b]$.

3 Computed by performing 100 realizations.

4 We consider European call options.

5 1Y15Y Swiss franc swaption from 10 February 2015.

6 Determined using the *normal* Bachelier model instead of the *log-normal* Black–Scholes model.

REFERENCES

1. Andersen, L.B.G., Efficient simulation of the Heston stochastic volatility model. *J. Comput. Finance*, 2008, 11, 1–22.
2. Antonov, A., Konikov, M. and Spector, M., SABR spreads its wings. *Risk Mag.*, 2013, 58–63.
3. Antonov, A., Konikov, M. and Spector, M., The free boundary SABR: Natural extension to negative rates. *Risk Mag.*, August, 2015.
4. Benhamou, E., Fast Fourier transform for discrete Asian options. *J. Comput. Finance*, 2002, 6, 49–68.
5. Broadie, M. and Kaya, O., Exact simulation of stochastic volatility and other affine jump diffusion processes. *Oper. Res.*, 2006, 54, 217–231.
6. Cai, N., Song, Y. and Chen, N., Exact simulation of the SABR model. *Oper. Res.*, forthcoming.
7. Chen, B., Oosterlee, C.W. and van der Weide, H., A low-bias simulation scheme for the SABR stochastic volatility model. *Int. J. Theor. Appl. Finance*, 2012, 15, 1250016-1–1250016-37.
8. Cox, J.C., Note on option pricing I: Constant elasticity of variance diffusions. *J. Portfolio Manage.*, 1996, 22, 15–17.
9. Fang, F. and Oosterlee, C.W., A novel pricing method for European options based on Fourier-cosine series expansions. *SIAM J. Sci. Comput.*, 2008, 31, 826–848.
10. Grzelak, L.A., Witteveen, J.A.S., Suárez-Taboada, M. and Oosterlee, C.W., The stochastic collocation Monte Carlo sampler: Highly efficient sampling from "expensive" distributions. 2015. Available at SSRN: http://ssrn. com/abstract=2529691
11. Hagan, P.S., Kumar, D., Lesniewski, A.S. and Woodward, D.E., Managing smile risk. *Wilmott Mag.*, January, 2002, 84–108.
12. Hagan, P.S., Kumar, D., Lesniewski, A.S. and Woodward, D.E., Arbitrage-free SABR. *Wilmott Mag.*, January, 2014, 60–75.

13. Islah, O., Solving SABR in exact form and unifying it with LIBOR market model. 2009. Available at SSRN: http://ssrn.com/abstract=1489428
14. Kennedy, J.E. and Pham, D., On the Approximation of the SABR with mean reversion model: A probabilistic approach. *Appl. Math. Finance*, 2014, 21, 451–481.
15. Kloeden, P.E. and Platen, E., *Numerical solution of stochastic differential equations*, 1992 (Springer-Verlag: Berlin).
16. Leitao, A., Grzelak, L.A. and Oosterlee, C.W., On a one time-step Monte Carlo simulation approach of the SABR model: Application to European options. *Appl. Math. Comput.*, 2017, 293, 461–479.
17. Lord, R. and Farebrother, A., Fifty shades of SABR simulation. Presentation at 10th Fixed Income Conference, WBS Training, Barcelona, Spain, 2014. Available online at: http://www.rogerlord.com/fiftyshadessabrwbs.pdf
18. Obloj, J., Fine-tune your smile: Correction to Hagan et al. 2008. Available at arXiv: http://arxiv.org/abs/0708.0998v3
19. Piterbarg, V., A multi-currency model with FX volatility skew. 2005. Available at SSRN: http://ssrn.com/abstract=685084
20. Sankaran, M., Approximations to the non-central chi-square distribution. *Biometrika*, 1963, 50, 199–204.
21. Schlenkrich, S., Miemiec, A. and Wolff-Siemssen, T., Low strike extrapolation for SABR. *Wilmott Mag.*, November, 2014.
22. Schroder, M., Computing the constant elasticity of variance option pricing formula. *J. Finance*, 1989, 44, 211–219.
23. Sklar, A., Fonctions de répartition à *n* dimensions et leurs marges [N-dimensional distribution functions and their margins]. *Publ. Inst. Statist. Univ. Paris*, 1959, 8, 229–231.
24. Smith, R.D., An almost exact simulation method for the Heston model. *J. Comput. Finance*, 2007, 11, 115–125.
25. van Hasstrecht, A. and Pelsser, A.A., Efficient, almost exact simulation of the Heston stochastic volatility model. *Int. J. Theor. Appl. Finance*, 2010, 13, 1–43.
26. Zhang, B. and Oosterlee, C.W., Efficient pricing of European-style Asian options under exponential Lévy processes based on Fourier cosine expansions. *SIAM J. Financ. Math.*, 2013, 4, 399–426.

CHAPTER 6

Robust Stock Option Plans

Olaf Korn • Clemens Paschke • Marliese Uhrig-Homburg

Georg-August-Universita¨t Go¨ttingen and Centre for Financial Research Cologne (CFR), Platz der Go¨ttinger Sieben 3, 37073 Go¨ttingen, Germany

Hansaallee 136, 60320 Frankfurt, Germany

Karlsruhe Institute of Technology (KIT), 76049 Karlsruhe, Germany

ABSTRACT

The main purpose of this paper is to address the issue of robustness of stock option plans, which is essential for reliable accounting valuations. The introduction of the accounting standards SFAS 123R and IFRS 2 for executive stock options has led to an important change. As companies are now forced to value their stock options at grant date for accounting purposes, the robustness of prices against the choice of certain valuation models and input parameters has become a very important issue. We address this issue by first analyzing certain building blocks of existing stock option plans with regard to their robustness properties. Based on our analysis, we then show how robustness of stock option plans can be achieved. The resulting stock option plans are both transparent in structure and reasonable in respect to the incentives they provide in order to increase shareholder value. We therefore conclude that stock options can be reliably expensed, if the corresponding plans are properly designed.

Keywords: Stock option plans, Robustness, Accounting valuation, Corporate governance

1 INTRODUCTION

Accounting rules for executive stock options have changed substantially recently and moved towards a "fair value accounting" approach, both under US-GAAP and IFRS.[1] These changes have severe consequences. The former regulation in the US[2] allowed to expense just the option's intrinsic value and distinguished between plans with known exercise price (fixed plans) and plans with unknown exercise price, such as indexed options (variable plans). For fixed plans, expenses were only considered at the time when the plan was granted. Thus, the accounting treatment clearly favored plain vanilla at-the-money call options, where no expenses would ever occur. Under the new rules, however, there is no longer such a clear accounting-driven advantage for these types of plans. Hence, plan designers are expected to put more emphasis on the incentive effects of executive stock options, which might lead to more diversity and a more complex structure of plans. One indication of such a tendency has become apparent in Germany. Here, historically accounting rules did not favor a certain type of plan and we have already observed many different kinds of particularly complex plan designs. Apart from possible changes in the diversity and complexity of stock option plans, a major consequence of "fair value accounting" is the central importance of valuation models. As market prices of comparable option contracts are mostly unavailable,[3] model prices have to be used. The valuation of stock option plans raises a range of different problems. Firstly, standard no-arbitrage option pricing models might not be appropriate. Although we take the shareholders' perspective and interpret the option price as an opportunity cost that would arise if the options were traded in the market, we cannot ignore the behavior of managers who receive the options. Recipients of stock option plans are restricted from selling or hedging their options. Since these options usually allow for early exercise after some vesting period, such restrictions might cause an exercise behavior that violates the assumptions of standard no-arbitrage models.[4] As a result, the firm's "opportunity cost" is wrongly quantified according to standard models. Secondly, the increasing complexity of stock option plans might result in valuation problems that are difficult to solve, even within the standard no-arbitrage framework. In particular, one might need to rely on numerical valuation methods which are subject to approximation errors and make it more difficult to verify the results. Finally, even if valuation models provide reliable values in principle, these models need input parameters, such as volatility and future dividends, which are not directly observable and therefore must be estimated. Stock option plans usually have a long time to maturity, and therefore, estimation errors can have an important impact on the resulting value.

Each of the above-mentioned problems of valuation strongly raises the question as to how robust the actual model prices of stock option plans are. In this paper, we mainly concentrate on the third problem. However, we will see that all three problems are interrelated. For our purposes, we define robustness as a low sensitivity of an option's price with respect to the pricing model's input parameters that have to be estimated, such as the volatility, the future dividend yield, or the interest rate. If stock option plans are not robust, the reliability of their accounting values would be questionable. This would by itself call the entire approach to "fair value accounting" of stock option plans into question.

In addition, mispricing would not necessarily occur by chance, for example, due to unavoidable estimation errors. If a stock option plan is not robust, it provides opportunities for concise manipulation. It has been argued in the literature that certain design elements of stock option plans provide examples of managerial self-dealing.[5] Robustness is a feature of stock option plans that potentially protects shareholders against self-dealing by managers, as it makes it more difficult to camouflage a plan's true value. A better understanding of the robustness properties of stock option plans could provide a basis for empirical tests of such a hypothesis. Moreover, one could empirically investigate if firms with better governance structures grant more robust compensation packages.

Our paper is closely related to the literature on the valuation of stock option plans. Major issues of this literature are certain special features of executive stock options, namely, vesting periods, forfeiture, and early exercise. In particular, the problem of early exercise has received much attention. It has been dealt with by either integrating utility maximizing elements into the valuation model, or, by specifying an exogenous exercise threshold.[6] Another strand of the valuation literature has dealt with non-standard forms of executive stock options. Hemmer et al. (1998) and Dybvig and Loewenstein (2003) have considered reload options and Brenner et al. (2000) have dealt with resettable options.[7] Johnson and Tian (2000b) have analyzed indexed options and Câmara (2001) has looked at the valuation and the incentive effects of options which can only be exercised if an outperformance compared to a reference portfolio is achieved. In another paper, Johnson and Tian (2000a) analyze and compare six different types of non-standard stock option plans, namely premium options, performance-vested options, resettable options, purchased options, reload options, and indexed options.

The literature on the valuation of stock option plans has provided (sometimes as a byproduct) important results on the robustness issue: Ammann and Seiz (2004) consider the impact of model choice on an option's value, by comparing different valuation models. An important result of their study states that no-arbitrage option pricing models and utility maximizing models lead to very similar values if they are calibrated to the same expected lifetime of the option. Bettis et al. (2005), who calibrate different valuation models using an extensive data set of option exercises of corporate executives, arrived at the same conclusion, stating[8]: "For determining the costs of shareholders of granting ESOs, we show that simply adjusting the maturity of a tradable American option reflecting the expected time of exercise produces valuations that are close to those computed from the more elaborate utility based model". These results suggest that the values of stock options plans are quite robust with respect to the model choice. However, the effective time to maturity needs to be known to determine the value precisely.

Rubinstein (1995) also focuses strongly on the aspect of robustness, especially with respect to input parameters. Concentrating on an at-the-money call option (traditional or standard stock option plan), he analyzed the sensitivity of option prices to volatility, dividend yield, early exercise, delayed vesting, and the forfeiture rate, among other things. As his results show, a traditional stock option plan is strongly non-robust with respect to estimation errors, resulting in price differences of up to 60% for reasonable parameter variations.

Johnson and Tian (2000a) compare different plans with respect to their sensitivity to volatility changes, changes in the dividend yield, and changes in the stock price.[9] In particular, they have shown that traditional call options have lower deltas and vegas than indexed options, although they are more sensitive to dividend yield changes. Johnson's and Tian's selection of stock option plans was based on the Conference Board Report of 1992. New design elements have emerged, however, in practice recently, particularly combinations of both absolute and relative performance targets,[10] as well as barriers and caps. These elements are expected to have an important impact on a stock option plan's robustness.

The literature has provided important insights into the robustness of executive stock option plans, however, the focus has been on the characteristics of different plan designs that have existed in practice or which were suggested. Very little attention has been paid to the question of how different design elements can be combined in order to improve robustness of plans. This paper is the first that systematically deals with this important issue. Our main conclusion is that the construction of robust plans is indeed possible and that these plans possess reasonable incentives to increase shareholder value. The key to this result is the use of European style options instead of American or Bermudan style options. The former are able to resolve different problems related to the robustness issue. Firstly, the problem of determining the optimal exercise strategies of corporate executives disappears. Secondly, analytical pricing formulas become applicable, which facilitates the verification of model prices by outsiders and improves the transparency of the valuation process. Finally, robustness with respect to input parameters can be achieved more easily. In particular, the use of caps allows us to solve the problem of parameter robustness with respect to volatility, by explicitly constructing vega-neutral plans. In summary, our results suggest that "fair value accounting" introduces certain robustness problems, although at the same time, eliminating the bias towards a certain type of plan, it can offer the opportunity to design plans which actually minimize the robustness problem.

The remaining part of this paper is structured as follows: In Sect. 2 we begin by introducing some examples of real stock option plans. These show the variety of different features that constitute real plans and provide a first impression on the robustness of alternative plan structures with respect to variation of input parameters. In Sect. 3, that follows, we analyze different elements of stock option plans separately and quantify their effect on a plan's robustness. Based on this analysis, Sect. 4 discusses how robust plans could be constructed and analyzes some of their properties. Our conclusions are presented in Sect. 5.

2 EXAMPLES OF EXISTING STOCK OPTION PLANS

To exemplify the variety and complexity of stock option plans, a selection of stock option plans is presented in this section. All of these examples are taken from Germany, as historically, accounting rules in Germany did not favor stock option plans of a certain type of design (such as the fixed plans in the US). Thus, German stock option plans could provide prototypes for the kind of design we will see in the future.

With respect to their performance targets, stock option plans can be assigned to one of three groups: The first group sets only absolute performance targets compared to historical share prices. Most plans in the US belong to this group. The second group requires only relative performance targets, whereupon share price performance is measured relative to specific comparable shares or indices. Finally, the third group combines absolute and relative performance targets.

To illustrate different design principles of stock option plans and get an initial impression of the price sensitivity with respect to volatility, dividend, interest rate, and stock price changes, we provide concrete examples for each specific group of plans. All of these plans come from companies which are included in the DAX 30, Germany's major blue chip stock index[11] and contain the main design features associated with stock option plans in Germany.

Linde AG has started the Linde Management Incentive Program in 2002. As this plan specifies only an absolute performance target, it belongs to the first group. Here, call options with an exercise price of 20% above the stock price were issued, i.e., out-of-the-money calls. These options had an initial time to maturity of 7 years and early exercise was not allowed during a vesting period of 2 years. The Deutsche Boerse AG Option Plan is an example of the second group of plans, with purely relative performance targets. The options pay the difference in performance between the stock price and a benchmark index, which is the Dow Jones STOXX Technology Index. Hence, they are initially at-the-money exchange options. The initial time to maturity is 5 years with a vesting period of 3 years.

An example of a combination of relative and absolute performance targets which belongs to the third group is the BASF Stock Option Program 2005, issued by the chemical firm BASF. Within this plan recipients receive an equal amount of two sub-plans, called rights A and B. Right A is an at-the-money option with an absolute exercise threshold of 30% above the stock price at grant date and with a cap equal to 100% of the stock price at grant date. Right B is a dividend protected at-the-money exchange option against the MSCI Chemicals Index. It has an additional cap equal to the difference between the share price at exercise and the face value of the share. The options have an initial time to maturity of 8 years and a vesting period of 2 years. Rights A and B can only be exercised jointly and the total value is additionally capped at a level of 2.5 times the strike price. The BASF plan has substantial additional complexity, due to the joint exercise rule and the different caps. Such caps are indeed a common feature of stock option plans in Germany, which might be justified as a protection to shareholders against extraordinary and unforeseen events in the stock market.[12]

A further example of a combination of absolute and relative performance targets is the Group Incentive Plan of the insurance company Allianz AG. Unlike the BASF program, the relative performance is implemented through a barrier option rather than an exchange option. The plan contains an at-the-money call with 7 years to maturity and a vesting period of 2 years. The absolute performance target has a threshold at maturity that is 20% above the strike price. The plan also contains a cap that is 150% above the strike price. The relative performance target states that the option can only be exercised if a benchmark index (Dow Jones Europe STOXX Price Index) has been outperformed at least once for a period of five consecutive trading days.

Based on the above examples, we want to offer a first impression of input parameter robustness and incentive effects of various designs. To do so, we need appropriate robustness measures, valuation models, and base case input parameters.

Robustness refers to the sensitivity of a stock option plan's value (model price) to changes in the major "unknowns" of option pricing, volatility, dividend yield, and risk-free interest rate. Plans with a lower sensitivity to volatility, dividend yield, and risk-free interest rate are considered to be more robust. Incentive effects[13] are approximated by the sensitivity of a plan's value to a change in the stock price. Plans with higher sensitivities to changes in the stock price are considered to provide stronger incentives.

As a simple way to measure sensitivity we rely on the slope of a plan's value function with respect to the parameter under consideration. To ensure comparability across plans, we standardize their values to an amount of €100 by choosing the number of options granted for each plan appropriately. The reason for this is that firms can scale the number of options they grant so that they reach a specified total value of the plan. Therefore, different plans should be compared using this total value.

To obtain a value function and the corresponding sensitivity measures for each plan, we need a valuation model that can deal with the various absolute and relative performance targets of the four plans and the possible early exercise after the end of the vesting period. Our choice is a three-dimensional tree model[14] with 300 time steps. The advantage of this model lies in its relatively simple structure and its capability to price all different stock option plans within a unified framework. It is clearly a limitation of the model that it does not capture certain characteristics of stock returns, like large jumps or a stochastic volatility.[15] However, as the accounting valuation of stock option plans by German firms relies predominantly on simple closed form valuation formulas or tree models, it seems most relevant to base the robustness analysis on such a simple model.

Finally, we have to choose a set of reasonable base case input parameters, i.e., stock and index return volatilities, return correlations, dividend yields, and interest rates. To our knowledge, there is no industry standard to obtain parameter values. However, one practical rule is to estimate volatilities, correlations and dividend yields from historical data that reaches as far back into the past as the maturity date of the options lies in the future. For example, if a plan has a time to maturity of 7 years, like the plans by Linde and Allianz, the estimation period would consist of the 7 years preceding the valuation date. For most stock option plans, the valuation date is the last day of the year.[16] For the estimation of our reference parameters, we use the valuation date December 31st 2009 and the corresponding historical estimation periods. Volatilities and correlations are estimated from monthly returns. Dividend yields are the average yearly yields[17] over the estimation periods, and interest rates refer to the term structure of German government bonds at the end of 2009. The resulting parameter estimates are given in Table 1.

Table 1 Historical parameter estimates for four existing stock option plans

Parameter	Linde		Deutsche Boerse		Allianz		BASF	
	Stock	Index	Stock	Index	Stock	Index	Stock	Index
Return volatility	24%	–	34%	24%	41%	18%	26%	18%
Return correlation	–		0.50		0.75		0.80	
Dividend yield	2.3%	–	1.8%	1.7%	2.5%	2.8%	3.5%	–
Interest rate	3.1%		–		3.1%		3.3%	

This table shows historical parameter estimates for the example stock option plans. Return volatilities and return correlations are estimated from monthly data over the estimation periods. Dividend yields refer to average yearly yields. The estimation periods are January 2003 to December 2009 (Linde, Allianz), January 2002 to December 2009 (BASF), and January 2005 to December 2009 (Deutsche Boerse). Interest rates refer to the term structure of German government bonds at the end of 2009. For Linde and Allianz, we use the interest rate for a maturity of 7 years. For BASF, the interest rate for a maturity of 8 years is used

A first interesting robustness issue is how many parameters are needed to value the different plans. The more parameters we require, the more potential estimation problems can arise. Some particular features of the four example plans affect the number of input parameters. Most notably, the stock option plan by Linde has no relative performance target and therefore needs no information on index return volatility or dividends. The plan by Deutsche Boerse can formally be seen as an exchange option, which means that interest rates do not enter into the valuation.[18] Finally, Right B of the BASF plan is protected for dividend payments. Therefore, the dividend yield of the reference index is not needed. The dividend yield of the BASF stock is required to value Right A of the plan, however.

Table 2 provides the sensitivity measures based on the historical parameter estimates of Table 1. The slope values which are shown can be interpreted as follows: As an example, take the Linde stock option plan. If volatility increases from 24 to 25%, the option's price goes up from 100 to 105.11. Similarly, an increase in the dividend yield from 2.3 to 3.3%, leads to a price decline from 100 to 84.95. An increase in interest rates from 3.1 to 4.1% corresponds to a price increase from 100 to 109.98. Finally, a number of 2.62 for the incentive effect means that a 1% increase in the stock price would lead to a price increase of the plan from 100 to 102.62.

Table 2 Robustness and incentive effects of four existing stock option plans for historical parameter estimates

Plan	Robustness to changes in			Incentive effect
	Volatility	Dividend yield	Interest rate	
Linde	511	−1,505	998	2.62
Deutsche Boerse	233	−956	0	2.40
Allianz	157	−699	305	1.59
BASF	241	−445	202	2.60

This table shows the slopes of the plans' value functions with respect to volatility, dividend yield, interest rate, and stock price, respectively. Plan values are standardized at €100 for the historical parameter estimates of Table 1. The slope values have the following interpretation: If, for example, the return volatility of the Linde stock increases from 24 to 25%, the option's price goes up from 100 to 105.11. Similarly, an increase of the dividend yield from 2.3 to 3.3% leads to a price decline from 100 to 84.95. An increase in interest rates from 3.1 to 4.1% corresponds to a price increase from 100 to 109.98. Finally, a number of 2.62 for the incentive effect means that a 1% increase in the stock price would lead to a price increase of the plan from 100 to 102.62

The results of Table 2 show that the robustness of different plans can vary substantially. Some of these differences follow directly from certain design features of the plans. The plan by Deutsche Boerse shows no interest rate sensitivity, because interest rates do not enter the valuation model. The same holds for Right B of the BASF plan, which is another exchange option. Therefore, the interest rate sensitivity of the BASF plan stems exclusively from Right A, which explains the relatively small value of 202. The relatively low dividend yield sensitivity of the BASF plan follows from the dividend protection of Right B. Other major differences in robustness between plans, in particular with respect to the stock return volatility, need further examination.

The differences in Table 2 might not solely be due to differences in plan designs but could potentially be affected by the different parameter values that we used to value each plan. Therefore, we choose a common base case parameter set and recalculated all sensitivity measures based on this common set. Concretely, we choose a stock return volatility of 25% p.a., an index return volatility of 15%, a correlation of 0.7 between stock and index returns, a dividend yield of 2% p.a. for both the stock and the reference index, and a risk-free rate of 3% p.a. These parameter values lie in the range of the historical estimates. Note, however, that relatively low volatilities have been chosen, because the time of the financial crisis, which falls in our estimation period, led to exceptionally high historical estimates. The robustness results for the common base case parameter set are presented in Table 3. As a point of reference we also consider an at-the money call.

Table 3 Robustness and incentive effects of four existing stock option plans and an at-the-money call option for a common base case parameter set

Plan	Robustness to changes in			Incentive effect
	Volatility	Dividend yield	Interest rate	
At-the-money Call	341	−1,366	823	2.33
Linde	480	−1,513	984	2.53
Deutsche Boerse	445	−1,481	0	3.70
Allianz	344	−1,283	764	2.21
BASF	311	−437	217	2.64

This table shows the slopes of the plans' value functions with respect to volatility, dividend yield, interest rate, and stock price, respectively. Plan values are standardized at €100 for the base case scenario, characterized by a stock return volatility of 25% p.a., an index volatility of 15%, a correlation of 0.7 between stock and index returns, a dividend yield of 2% p.a. for the stock, a dividend yield of 2% p.a. for the reference index, and a risk-free rate of 3% p.a. The slope values have the following interpretation: If the return volatility increases from 25 to 26%, the price of the at-the-money call option, for example, goes up from 100 to 103.41, Similarly, an increase of the dividend yield from 2 to 3% leads to a price decline from 100 to 86.34. An increase in interest rates from 3 to 4% corresponds to a price increase from 100 to 108.23. Finally, a number of 2.33 for the incentive effect means that a 1% increase in the stock price would lead to a price increase of the plan from 100 to 102.33

The results of Table 3 confirm that stock option plans can be highly non-robust. Apparently, the reaction to volatility changes strongly depends on the design of the stock option program. The prices of the plans of Linde AG and Deutsche Boerse AG react much stronger to volatility changes than our reference point, the traditional at-the-money call. In contrast, the sensitivity of the plan of Allianz AG is less than that for the first two plans. The plan is, however, still less robust than the traditional call. Surprisingly, the plan of BASF AG is the most robust one with respect to volatility changes. This plan shows even less sensitivity than the at-the-money call. Besides the basic plan structure itself, possible explanations for the differences could be the role of caps or the joint exercise of sub-plans. The effect of theses design features will be further explored in the next section. In comparison with Table 2, the most striking result is that the robustness of the plans by Deutsche Boerse and Allianz is strongly reduced in general. This finding is due to the fact that for these stocks the base case volatility is clearly smaller than the historical volatility estimate used in Table 2. Therefore, the dependence of robustness on the level of volatility also deserves further investigations.

One might suspect that a low sensitivity of a plan with respect to volatility, dividend yield, and interest rate would render the plan unattractive from an incentives' perspective, as its sensitivity to stock price changes is reduced. The results of Table 3, however, show no clear relation between volatility, dividend, and interest rate sensitivities on the one hand and stock price sensitivity on the other. For example, although Linde AG's plan is most sensitive to volatility and dividend changes, it also shows a very similar stock price sensitivity to BASF AG's, which is the most robust one with respect to a mis-specification of either volatility or dividend yield.

In summary, we have seen that real plans can have a large set of different design features. However, there is no indication that these design features are generally used to make plans more robust. What we observe is that different plans show big differences in robustness. Therefore, it is worthwhile searching for certain properties of a plan that may strongly reduce or increase robustness. In addition, it is also evident that a higher complexity of a plan does not necessarily lead to a higher sensitivity, as can be seen from BASF AG's plan. Instead, we have to look more closely at the different design elements to understand which elements are useful to improve robustness and which ones just make the plan overly complex.

3 ROBUSTNESS OF STOCK OPTION PLANS

As a further step towards understanding the robustness of stock option plans, we investigate how single design elements of plans affect their robustness with regard to the input parameters volatility, dividend yield, and risk-free interest rate. For this purpose, we use the example plans presented in the previous section to pinpoint certain building blocks. We first differentiate between some basic plan structures and then add additional features, such as the joint exercise of sub-plans and caps.

3.1 Basic plan structures

Our investigation of existing stock option plans revealed five basic plan structures:

- *Traditional stock option plan*: an at-the-money call option that can be exercised at any time after the end of a vesting period. Thus, the strike price X equals simply the stock price S_0 at the grant date of that plan. This plan is widely used in the US and serves as the reference point for our analysis.

- *Premium option*: an out-of-the-money call, i.e., the strike price X lies above the stock price S_0. The difference between these two prices determines the performance target. One way to set the performance target would be to require an appropriate risk-adjusted hurdle rate per year.

- *Option with absolute exercise condition*: a plan with a payoff equal to a traditional option, if an absolute performance goal is reached, and zero otherwise. In other words, we have a portfolio consisting of a premium option and a digital option that pays out the difference between S_0 and X if the stock price at exercise exceeds X.
- *Indexed option*: an at-the-money option to exchange a reference portfolio (the index) for the stock of the company. The option is at the money if the index is normalized such that it equals the stock price at grant date.
- *Relative performance vested option*: a plan with a relative performance target, that is used as an exercise condition only. Formally, such a plan structure consists of an at-the-money up-and-in barrier call option with the price of a reference portfolio as stochastic barrier.

3.2 Robustness of basic plan structures

Figure 1 provides a first impression of the robustness of the different basic plans with respect to volatility ("vega", top of figure), the dividend yield ("dividend yield sensitivity", middle), and the risk-free interest rate ("rho", bottom). As in Sect. 2, the slope of a plan's value function is used to measure its parameter robustness. To construct Fig. 1, the base case parameters from Table 3 are used and we consider a time to maturity of 8 years and a vesting period of 2 years for all of the plans. In addition, for the premium option and the absolute performance vested option we assume that the strike price X is 30% above the stock price S_0.

The upper part of Fig. 1 provides an illustration of the robustness with respect to volatility. On the x-axis, we see different levels of volatilities, which might represent different kinds of stocks, such as growth stocks or value stocks. Thus, when multiplying the "vegas" on the y-axis of Fig. 1by 0.01, as an example, the result can be interpreted as the percentage change in value of the stock option plan corresponding to a one percentage point estimation error in volatility. First, the resulting values are rather high in magnitude, given that an estimation error of only one percentage point for volatility is considered. Second, we see that the premium option clearly has the highest vega independent of the volatility level. In contrast, the indexed option has the lowest vega for very small stock return volatilities but once volatilities approach a medium or even high level, its vega exceeds the vegas of the traditional option, the option with absolute exercise condition, and the relative performance vested option. Third, it should be noted that the vega strongly depends on the volatility level. With exception of the

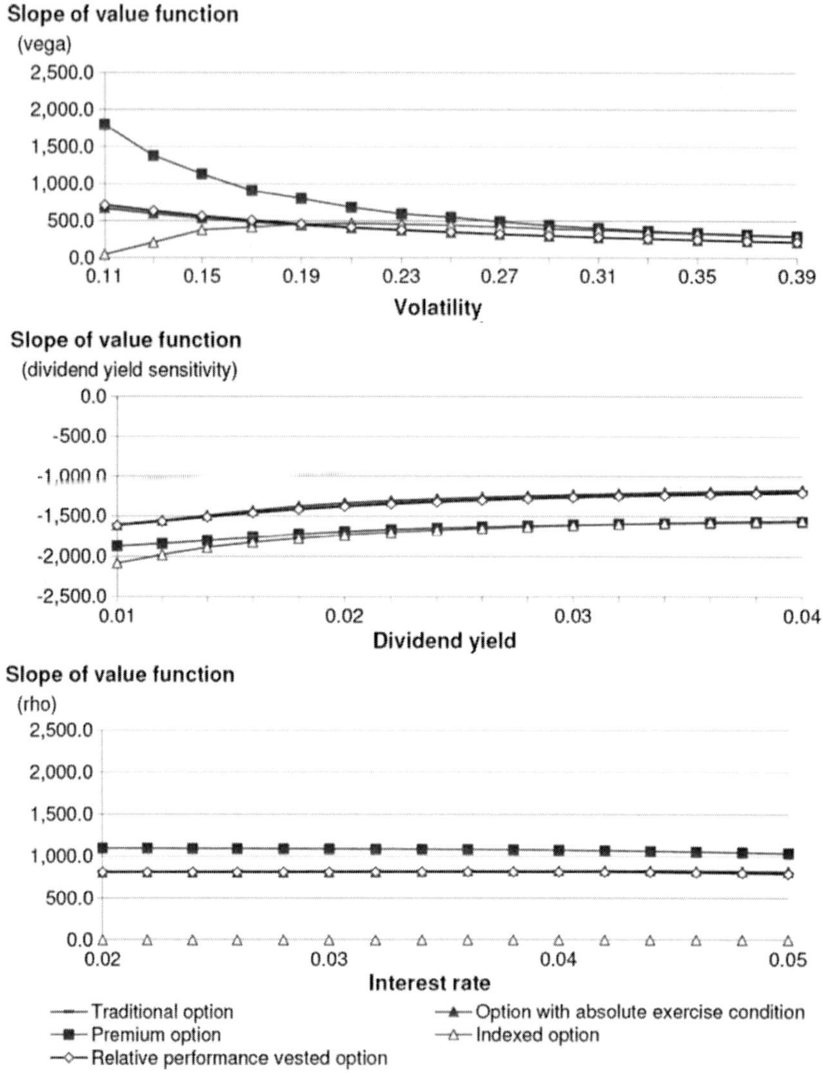

Figure 1 Slopes of the plans' value functions with respect to volatility, dividend yield, and interest rate. This figure shows the slopes of the plan's value functions with respect to volatility, dividend yield, and interest rate for different levels of volatility, dividend yield, and interest rate. Plan values are standardized at €100. The additional parameter values are a stock return volatility of 25% p.a., a dividend yield of 2% p.a. for the stock, a dividend yield of 2% p.a. for the reference index, an index volatility of 15% p.a., a risk-free rate of 3% p.a., and a correlation between stock and index returns of 0.7. The time to expiration equals 8 years and the vesting period equals 2 years

indexed option, all basic plan structures show that robustness improves monotonically with volatility (decreasing vega), which generalizes the observations from the previous section. This result is in line with theoretical considerations. For example, in the Black–Scholes model, the price of a call option converges to the stock price if the product of the return variance and the time to maturity goes to infinity. Therefore, if the volatility or the time to maturity becomes larger and larger, the option should behave more and more like the stock. Since the stock itself is completely robust, we would expect the option to become more and more robust. At first sight, the non-monotonic behavior of the indexed option's vega is puzzling. However, we need to consider that for an exchange option the relevant volatility measure is the square root of the sum of the stock return variance and the index return variance minus two times the covariance. This volatility measure is a non-monotonic function of the stock return volatility, i.e., it decreases with increasing stock return volatility for small volatilities and increases for large volatilities.

With respect to the dividend yield, the effect of a one percentage point prediction error is quite substantial, leading to decreases in value of between 10 and 20%. Moreover, we can distinguish between two groups of plans: the premium option and the exchange option fall under the first group, which is more sensitive to the dividend yield. The remaining three options, which are less sensitive to dividend yield changes, build the second group.

In addition, the interest-rate sensitivity "rho", given a one percentage point estimation error, varies heavily between plan structures, ranging from a value of zero to approximately 11%. The premium option is the most sensitive, followed by the traditional option, the option with an absolute exercise condition and the relative performance vested option. The last three of these plans behave very similarly. Finally, as already mentioned in the previous section, the indexed option shows complete robustness with respect to the interest rate because the risk-free interest rate does not enter valuation models for exchange options.

Different conclusions can be drawn from our analysis so far. Firstly, the non-robustness of stock option plans with respect to volatility, dividend yield, and risk-free interest rate can be strong. Secondly, there is no single plan that dominates all other plans in terms of robustness if we take all the three input parameters into account. Finally, Fig. 1 suggests that the traditional option, the option with an absolute exercise condition and the relative performance vested option have very similar robustness properties, i.e., the additional exercise conditions have very little impact on robustness. A sensitivity analysis shows that this result holds for a broad range of parameter values.[19] Thus, it is possible to concentrate on one of the three options, the natural choice being the traditional option, which is the simplest one.

3.3 Additional design elements and robustness

Existing option plans typically combine basic plan structures with additional elements such as caps, different vesting periods or joint-exercise requirements for sub-plans. How these elements affect the robustness of the plan structures is analyzed in the following subsections.

3.3.1 Impact of joint exercise of sub-plans

One design element of existing stock option plans is the joint exercise of sub-plans. Several German stock option plans such as those of BASF AG and Allianz AG combine options with absolute and relative performance targets into an options basket. A joint exercise condition forces the option's recipient to exercise options with absolute and relative performance targets simultaneously. If only one performance target is fulfilled at the time of exercise, the other option forfeits worthless, which reduces the likelihood of early exercise.

The existence of joint-exercise features raises the question of how this feature impacts robustness. To provide some evidence on this issue, we consider a combination of a traditional option and an indexed option. For a wide range of parameter values, sensitivities with respect to volatility, dividend yield, and interest rate were calculated.[20] These calculations show that the joint exercise of sub-plans has usually a small impact on robustness with respect to all sensitivity measures. Moreover, joint exercise even leads to higher sensitivities in most cases, i.e., it does not bring an improvement. For example, if we consider our base case parameters, joint exercise leads to a small increase of vega from 391 to 396. However, a decline in robustness can be worse. For a volatility of 15%, a dividend yield of 3%, and a risk-free interest rate of 3%, joint exercise leads to an increase of the dividend yield sensitivity from −2,142 to −2,360. Additional calculations for other combinations of sub-plans confirm these findings. If we combine, for example, an indexed option and an absolute performance vested option under the base case parameter scenario, all sensitivity measures increase when moving from separate exercise to joint exercise. If we consider in addition the real plan by BASF AG that involves additional complexity due to different caps on a sub-plan and the full plan, differences between joint exercise and separate exercise are very small. The only positive effect of joint exercise is a marginal reduction of vega from 319 to 311. In conclusion, our calculations show no evidence for a substantial and systematic improvement of robustness due to a joint-exercise requirement of sub-plans.

3.3.2 Impact of caps and vesting period

If we consider European options, the introduction of a cap can be seen as adding a short position in an out-of-the money call option to the original long call position. Since the sensitivities with respect to volatility, dividend yield, and interest rate of long and short positions have opposite signs, the introduction of caps could lead to a significant increase in the robustness of such stock option plans.

To provide some evidence of these effects, a cap (strike price of the shorted call option) equal to two times the stock price at grant date is introduced for the traditional option and the premium option. Since the relevant volatility for indexed options is lower than the volatility of the underlying stock itself, for comparison purposes, the cap of the indexed option is set equal to the cap of the traditional option in terms of distance to stock price at grant date, measured by standard deviations. As a result, in the case of the indexed option, the maximum outperformance of the stock was set as 0.72 times the stock price.

Using our base case parameters, a time to maturity of 8 years and a vesting period of 2 years, we observe that caps have only a small effect on the parameter robustness of the stock option plans, which is surprising at first sight. To this end, Fig. 2 shows the vegas for the case with caps. Compared to the vegas for the case without caps, as shown in Fig. 1, we see very little difference. If we take for example a volatility of 25%, the vega of the traditional option with cap equals 314, instead of 341 without a cap.

The reason for the rather weak impact of caps on robustness lies in the American feature of the stock option plans. If stock prices rise, the cap will trigger early exercise and the short position in the call can never be in the money. Therefore, the shorted call position will never develop its full potential to improve the overall robustness of the plan, compared to European options. As an example, using our base case parameters, compare an at-the-money European option without a cap with the same option plus an additional cap at two times the current stock price at grant date. Such a cap reduces the option's value from 26.06 to 21.81, or by approximately 18%. In comparison, the otherwise identical American option suffers a price decrease of only 3%.

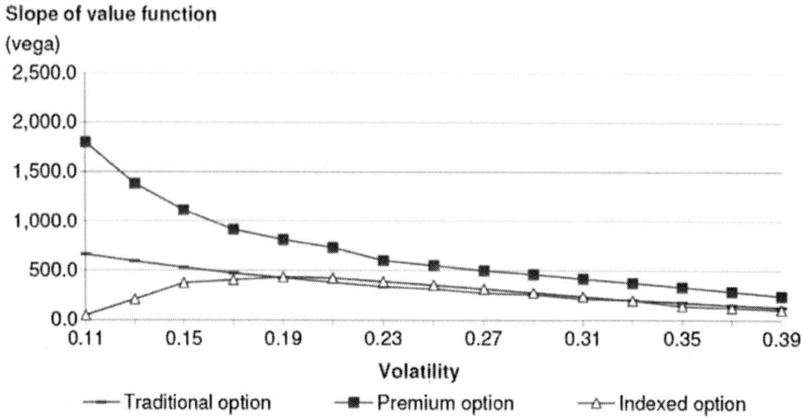

Figure 2 Influence of caps on the vegas of different plan structures. This figure shows the slopes of the plans' value functions with respect to volatility for the case with additional caps. Plan values are standardized at €100. For the traditional option and the premium option, the cap equals two times the current stock price. For the indexed option, the maximum outperformance of the stock equals .72 times the current stock price. Further parameter values are a dividend yield of 2% p.a. for the stock, a dividend yield of 2% p.a. for the reference index, an index volatility of 15% p.a., a risk-free rate of 3% p.a., and a correlation between stock and index returns of 0.7. The time to expiration equals 8 years and the vesting period equals 2 years

As the examples show, the opportunity of early exercise can make caps ineffective to improve robustness. Therefore, we should consider the relation between caps and another feature of stock option plans, the length of the vesting period. As an illustration, let us consider the traditional option. For our base case parameters and a cap that equals two times the stock price at grant date, a vesting period of 2, 4, 6, and 8 years is considered. Vega decreases from 314 to 178 if we lengthen the vesting period from 2 to 4 years and reaches a level of 103 for a vesting period of 6 years. If we lengthen the vesting period to 8 years, which means that the option is a European one, the option is almost vega-neutral (vega equals 13). In conclusion, this example shows that the robustness effects of the two design elements caps and vesting period cannot be judged in isolation. Moreover, it shows that caps have the potential to improve robustness considerably, if the vesting period is long enough.

3.4 Example plans revisited

After the analysis of the different basic plan structures and some additional design features, let us finally go back to our existing stock option plans in Table 3. We have now gained a better understanding of the factors that drive the robustness of these plans. These factors can be summarized as follows: (i) Exchange options have no interest rate sensitivity, which explains the high robustness with respect to interest rates of the plans by Deutsche Boerse and BASF. (ii) The dividend protection of Right B of the plan by BASF makes this plan less sensitive to dividend yield changes. (iii) For the medium volatility level considered, options with absolute exercise condition have a higher robustness with respect to volatility than premium options and exchange options. Therefore, the plans by Allianz (option with absolute exercise condition) and BASF (partly option with absolute exercise condition) are more robust than the plans by Linde (premium option) and Deutsche Boerse (exchange option). (iv) The caps included in the plans by Allianz and BASF can only to a small part explain the relatively high robustness of these plans. (v) The joint-exercise feature of the plan by BASF can't explain robustness.

4 ROBUST PLANS: AN ENGINEERING APPROACH

4.1 Construction of robust plans

In this section we tackle the problem of how different design elements of stock option plans can be combined to improve robustness. As the literature and our results in the previous section have shown, stock option plans are rather non-robust with respect to volatility, dividend yield, and risk-free interest rate.

The granting of American or Bermudan options that allow for early exercise after the end of the vesting period is related to different robustness issues. If the exercise strategies of corporate executives are unknown, the effective time to maturity of the options, which has a strong impact on their value, becomes also unknown. Moreover, as we have seen in the previous section, early exercise reduces the potential of caps to improve robustness with respect to volatility. The vesting period is the design element of stock option plans which is most directly connected with the options' effective time to maturity and can be used to resolve the related problems. In the extreme, early exercise can be completely eliminated by equating the end of the vesting period with the expiration date of the option, i.e., by granting European options.

We furthermore have to acknowledge that there are some potential problems with the granting of European options. A possible objection might be that European options may give rise to particular incentive problems because the executive would concentrate on the firm's success at a single point in time. We believe, however, that these problems are not so special to European options and can be mitigated. Firstly, even with an American option, an executive might also concentrate her efforts on a single point in time, which is, however, unknown to others. Moreover, it might not be in the shareholders' interest that the executive just exploits a lucky situation when stocks are temporarily overvalued. Secondly, one could easily avoid the focus on a single time horizon by granting a portfolio of options with different times to maturity. The current practice of many firms to grant new executive stock options on a regular basis is effectively nothing else than giving employees a portfolio of options with different maturities.

Another issue is the practical problems that arise if an executive leaves the firm before the options can be exercised. Such problems already exist for American options if the leave happens within the vesting period. In this case, the options forfeit worthless. Although the granting of European options doesn't create a completely new problem, one has to admit that the problem is aggravated. When thinking about practical problems that arise from granting European options, one should also bear in mind that European options have already been granted. The stock option plans of the major German telecommunications firm, Deutsche Telecom AG and Metro AG, Germany's largest wholesaler, are examples of such practices. Moreover, at least in Germany, legislation has moved towards European options by requiring a minimum vesting period of 4 years since July 2009.[21] Under previous law, the minimum vesting period was only 2 years.

The use of European options has clear advantages in terms of numerical tractability. The traditional option, the premium option, the option with an absolute performance target and the indexed option have straightforward analytical solutions under standard assumptions, if they are of European style. Even relative performance vested options can be valued analytically, along the lines of Heynen and Kat (1994). Moreover, Johnson and Tian (2000a) and Câmara (2001) have derived analytical valuation formulas for other non-standard options. In addition, caps can also be handled analytically without any further problems. The availability and applicability of analytical valuation formulas give rise to important benefits, especially with regard to accounting valuation. The accuracy of the resulting values is much easier to verify, communicate and audit, if analytical formulas are used. Therefore, the transparency of the valuation process is clearly improved.

The use of European options has also consequences for the robustness issue with respect to the choice of a risk-free interest rate. If the exercise strategy and thus the effective life time of the option is unknown, it is unclear which maturity should be chosen to select the interest rate. This problem can have a serious effect on the stock option plan's value if the term structure of interest rates is steep. However, when European options are granted and the lifetime of the option is known, then the choice of the interest rate is far less problematic, making robustness less important.

A second important input parameter of option pricing models is the dividend yield. Although most executive stock option plans are not protected against dividend effects,[22] we see no good reason for this practice.[23] On the contrary, one could argue that dividends are part of shareholder wealth which is created and management should not be penalized for providing them. Moreover, a missing dividend protection of executive stock option plans should not influence a firm's payout policy.[24] Technically, dividend protection is not difficult to achieve: One either adjusts the strike price at any time a dividend is paid, or one uses a performance index as the underlying of the plan that captures all the benefits for shareholders.[25]

The final remaining problem with respect to parameter robustness is volatility. The main design element that can deal with this problem is the use of caps. In the analysis to follow, we assume that a fully dividend protected European option is granted. In order to achieve robustness, we should set the cap (strike price of the short position in a call) at such a level that makes the whole stock option plan vega neutral.[26] In order to determine the conditions under which this is possible, we have to take a look at the vega of a European call option, as given in Eq. 1,[27] where S_0 denotes the stock price at time zero (grant date), X the strike price, r the risk-free interest rate per year, σ the stock return volatility, and T the time to maturity.

$$\text{Vega} = S_0\sqrt{T}N'(d_1), \quad \text{with}$$

$$N'(d_1) = \frac{1}{\sqrt{2\pi}}e^{\frac{-d_1^2}{2}}, \quad \text{and} \quad d_1 = \frac{\ln(S_0/X) + (r + \sigma^2/2)(T)}{\sigma\sqrt{T}} \tag{1}$$

The vega of a call option depends on the strike price X and can be depicted as a function of the strike price. Figure 3 shows a corresponding graph, using an option with 5 years to maturity,[28] a stock price of 100, a stock return volatility of 25%, and an interest rate of 3%.

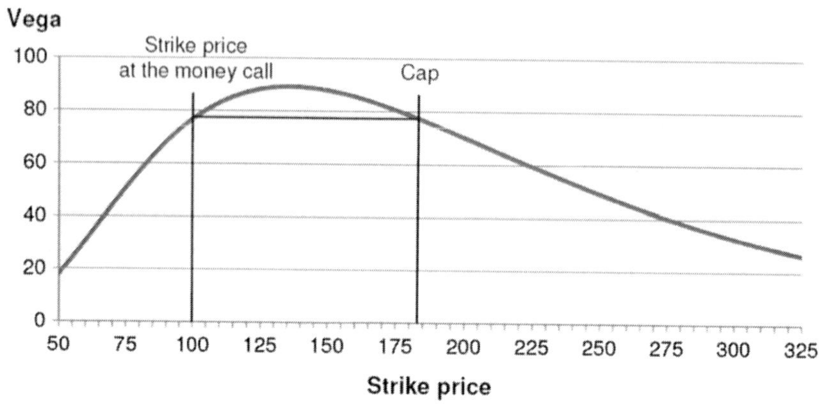

Figure 3 Vega as a function of the strike price. This figure shows the theoretical vega of a single European call option as a function of the strike price. The calculations are based on the theoretical vega, as shown in Eq. 1, which results from the Black and Scholes model. The parameters used for the graph are a current stock price of € 100, a volatility of 25% p.a., a risk-free rate of 3% p.a., and a time to maturity of 5 years

Figure 3 suggests how to construct vega-neutral stock option plans. Vega, as a function of the strike price, is at first increasing and then decreasing. This result explains our previous finding that a premium option has a higher sensitivity with respect to volatility changes than a traditional option. Moreover, it is easy to show that the vega function has a unique maximum at $S_0 e^{T\left(r+\frac{\sigma^2}{2}\right)}$. To obtain a vega-neutral plan, we first choose a strike price *X* for the long call position that lies to the left of the maximum. In Fig. 3, for example, *X* was chosen to be the current stock price of 100. Next we determine the vega of the long call position and find the strike price *Y* to the right of the maximum that leads to the same vega. If this strike price is found, we set the cap accordingly, since the vegas of the long call and the cap (a short call) will offset each other. Figure 3 shows graphically how to find the cap. Analytically, for $X < S_0 e^{T\left(r+\frac{\sigma^2}{2}\right)}$ the strike price *Y* of the short call position is given by

$$Y = \frac{S_0^2}{X} e^{\left(2r+\sigma^2\right)T} \tag{2}$$

This analysis shows under which circumstances we will be able to construct a vega-neutral, non-degenerate plan by means of caps. Firstly, if we set the strike price *X* equal to the current stock price, i.e., if we consider a traditional option, we will always be able to find a cap to the right of $S_0 e^{T\left(r+\frac{\sigma^2}{2}\right)}$ that makes the plan vega neutral. Secondly, with premium options we are more limited. If we choose the risk-free interest rate as a hurdle rate to determine the strike price, once more, we will always be able to find an appropriate cap, although it might already be quite close to the strike price. If we choose a hurdle rate that lies well above the risk-free rate, we could end up with a value of $X > S_0 e^{T\left(r+\frac{\sigma^2}{2}\right)}$ and therefore would not be able to construct a completely vega-neutral plan through a short call with $Y > X$. This means that it might be impossible to obtain a fully robust plan if we require the full risk-adjusted cost of capital as the benchmark that determines the strike price.

Finally, we can also apply our analysis to indexed options. Since the valuation formula for these is structurally identical to the Black–Scholes formula, we simply have to reinterpret some parameters. Thus, for exchange options, X would represent the index level at time zero and σ^2 would equal the sum of the stock return variance and the index return variance minus two times the covariance. Therefore, the relevant variance measure for indexed options would be smaller than the stock return variance if the index return variance is smaller than two times the covariance between stock returns and index returns, which would almost always be true. Since robustness with respect to volatility changes improves with the level of volatility, this smaller volatility explains the relatively low robustness of indexed options in comparison with traditional options that we have seen in Sects. 2 and 3 for medium and high stock volatility levels. Another issue of indexed options is that the interest rate does not enter into the valuation and has to be set to zero in all of the above formulas. Given the correct interpretation and specification of the parameters, we can conclude from our analysis that for at-the-money indexed options ($X = S_0$) we can always find a cap that makes the whole stock option plan vega neutral.

4.2 Properties of robust plans

To get a better understanding of the specific properties of robust plans, let us first look at the cap with strike price Y, as shown in Eq. 2. As we can see, a comparative static increase in either r, σ, or T leads to an increase of Y, i.e., the payoff of a robust stock option plan is capped at a higher level. Such a feature is both reasonable and desirable. Take for example the time to maturity, T. We would naturally think that a plan with a longer time to maturity should have a higher cap than an otherwise identical plan with a shorter time to maturity, since managers should deliver a better total performance if they were given more time. Similar arguments can also be applied to the risk-free interest rate and the volatility.

The comparative static analysis together with the numerical results of Sect. 3.2 (see in particular Fig. 1) show that robustness of stock option plans can be quite different for firms with different firm characteristics. For example, small firms are more likely to have a high return volatility and a low return correlation with the general stock market, which might translate into a low correlation with the benchmark index for indexed options. The first part of Fig. 1 in Sect. 3.2highlights that a higher volatility improves robustness (lower vega) for all basic plan structures. Moreover, for the indexed option, a lower correlation with the reference index has an additional effect to reduce vega. In summary, we see that stock option plans of small firms are likely to be more robust because of higher volatilities and lower correlations with the reference index. The analytical results of Sect. 4.1 complement these findings. In particular, Eq. 2 shows that the cap level increases with growing volatility. Therefore, for small firms with high volatility, a cap is less important to achieve robustness. If volatility grows and return correlation with the benchmark index decreases, the design of a robust plan moves closer and closer towards a standard stock option plan or an indexed option without cap.

If a robust plan comes close to a standard stock option plan or an indexed option for high volatilities, it will also inherit the incentive effects of these basic plan structures. However, it is important to understand more generally whether robustness harms the incentive effects of stock option plans and the extent to which this will actually happen. To obtain some evidence on this issue, we look at the stock price sensitivity of two robust plan structures with either absolute or relative performance target. The first plan structure is a premium option with a strike price equal to the compounded current stock price, where the compounding is done with the risk-free rate over the life of the option. The second plan structure is an at-the-money indexed option. Using cap values according to Eq. 2, "Appendix 3" derives the initial deltas (at grant date) of the plans, which can be interpreted as elasticities of the stock option's value with respect to the stock price. These deltas take the following form:

$$\text{Delta} = \frac{2N\left(\frac{1}{2}\sigma\sqrt{T}\right) - 1}{3N\left(\frac{1}{2}\sigma\sqrt{T}\right) - 2 + e^{\sigma^2 T}N\left(-\frac{3}{2}\sigma\sqrt{T}\right)} \tag{3}$$

For the premium option, σ^2 simply denotes the stock return variance. For the indexed option, σ^2has to be interpreted as the stock return variance plus the index return variance minus two times the covariance.

Equation 3 shows that delta depends only on one parameter, the product $\sigma\sqrt{T}$. This finding allows us to plot delta as a function of σ and *T*, as shown in Fig. 4. According to the figure, delta declines with increasing volatility and increasing time to maturity. In the limit, if either σ or *T* goes to infinity, delta approaches a value of one, which equals the delta of the stock itself. If either σ or *T* goes to zero, delta goes to infinity.[29] Note that such a limiting behavior is not a specific feature of robust stock option plans, because the same results hold for premium options or indexed options without caps. We can draw two main conclusions from Eq. 3 and Fig. 4: Firstly, whatever the parameter values are, the initial incentive effect of a robust stock option plan will always be stronger than the incentive effect of a share of stock with the same value. This finding highlights a trade-off between robustness and incentive effects. An all-stock plan would be perfectly robust, as its value could be calculated from current market prices only. However, as our analysis shows, the incentive effect would be lower. Secondly, if we compare robust premium options with robust indexed options, indexed options always provide a stronger incentive effect if the index return variance is smaller than two times the covariance between stock returns and index returns. This latter condition should almost always be fulfilled in practice.

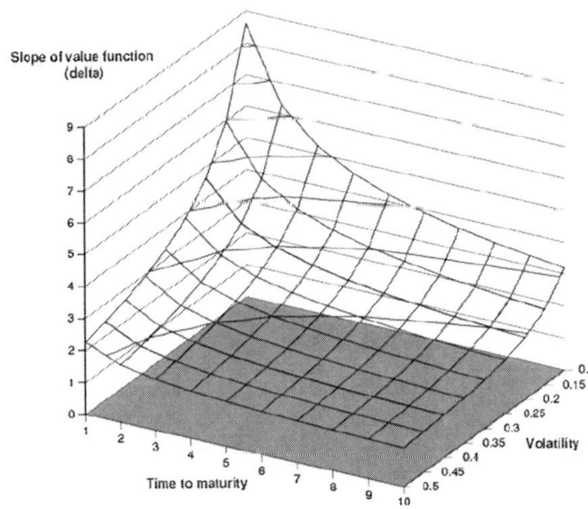

Figure 4 Influence of volatility and time to maturity on the initial delta of a robust premium option (robust indexed option). This figure shows the theoretical initial delta of a robust premium option (robust indexed option) as a function of the volatility and the time to maturity. The calculations are based on the theoretical initial delta, as shown in Eq. 3, which is derived in "Appendix 3". Note that the initial delta depends exclusively on the parameters volatility and time to maturity

Similarly, we can also derive analytical expressions for the deltas of the plans at some later point in time t. Besides t and the maturity T, the delta of the indexed option depends only on the realized stock return normalized to the index level at grant date and on the corresponding volatility. Of course, an extreme change in the stock price significantly affects the plan's delta. However, when focusing, for example, on a move of one standard deviation or below, the resulting deltas are still fairly high. Overall, our findings show that the cap has not destroyed the incentive effects of the options. We still have a delta which is well above one for reasonable scenarios.

5 CONCLUSIONS

In this paper we have addressed the issue of robustness of stock option plans, which is essential for reliable accounting valuations. We have shown that parameter robustness can be very low for a variety of basic plan structures and that none of these plan structures dominate all others in terms of their robustness properties. Furthermore, additional exercise conditions, either absolute or relative, have only a small impact on a plan's robustness. The same applies for the joint exercise of sub-plans in comparison with a separate exercise. The interplay between caps and the length of the vesting period, however, can have strong effects.

Based on our results, we have investigated and shown how to improve robustness by combining certain design elements of stock option plans. An important decision is to lengthen the vesting period and to grant European options instead of American or Bermudan options. As a result, the effective lifetime of the option becomes known, which avoids all complications arising from an unknown exercise behavior of the option holders. The use of European options also improves numerical tractability since usually analytical pricing formulas are available for European style options. Moreover, if the effective lifetime of the option is known, one can use the appropriate interest rate for this period and as a consequence, parameter robustness with respect to interest rates becomes much less important.

Two other important input parameters are the dividend yield and the volatility. As far as the dividend sensitivity is concerned, this can be avoided through dividend protection. Although non-robustness with respect to volatility is relatively more difficult to handle, we have shown that the use of caps can often lead to vega-neutral plans and we have provided solutions for the corresponding strike prices. In conclusion, we believe that almost all of the major issues associated with non-robustness can be resolved by using a proper plan design. Thus, with an appropriate plan design, we can be optimistic about the reliable accounting valuation of stock option plans.

Of course, robustness cannot be the only criterion that a stock option plan should fulfill. There are at least two other essential criteria: incentives to create shareholder value and transparency. Both of these criteria are strongly interrelated with robustness. With respect to the incentive effects of robust stock option plans, we observe that their stock price sensitivity is still very reasonable and at least as high as for an all-stock compensation plan. We have to stress, however, that there can be a trade-off between robustness and incentive effects and that we do not make any attempt in this paper to suggest a plan that leads to a trade-off that is optimal in any sense. In practice, incentive effects of stock option plans can depend on, for example, the competitiveness of the industry, the degree of agency problems, or the extent of operational and financial leverages. Therefore, optimal plan designs might be different for different firms and might also be quite non-robust.

The types of robust plans that we have constructed in this paper are also particularly transparent ones that consist of a few basic building blocks. It is this kind of *transparency* that allows us to achieve robustness, since otherwise we would not be able to solve analytically for a vega neutral plan design. Our building block approach also has advantages with respect to particular requirements for good corporate governance. It is quite easy to include both absolute and relative performance targets in one plan, while maintaining a plan's robustness and transparency. For example, one could grant one sub-plan with a premium option and one sub-plan with an indexed option. Both sub-plans could be designed separately as robust plans.

Finally, we believe that robustness of stock option plans is in itself an element of good corporate governance, as it could make it much harder for managers to understate the values of their compensation packages. It would be an interesting avenue for further research to devise empirical tests of this hypothesis. In this respect, one could analyze, for example, how robustness of compensation packages is related to certain indicators of good corporate governance. Another interesting empirical issue is whether there is a trend in actual compensation packages to be more robust, which could indicate an enforcement of robust plan designs by market discipline. If financial markets see the robustness of a compensation package as an important aspect of good corporate governance, robustness could finally be positively related to a firm's market value. An analysis of robustness properties, as presented in this paper, could provide the basis for this future empirical work.

FOOTNOTES

[1] Compare the accounting standards SFAS 123R of 2004 and IFRS 2 of 2004. Jaggi et al. (2010) provide a perspective on the future of fair value accounting after the financial crisis.

[2] Compare APB Opinion 25 of 1972, whose intrinsic-value method was still permitted under the original SFAS 123.

[3] An interesting exception is the Finnish market, where firms can list their stock option plans on the stock exchange. However, listing occurs only after the options have vested. See Ikäheimo et al. (2006).

[4] Empirical evidence on "too" early exercise is provided e.g. by Hemmer et al. (1996), Huddart and Lang (1996), Bettis et al. (2005), and Brooks et al. (2007)

[5] See e.g. Bebchuk and Fried (2003) and Bebchuk et al. (2002). Sautner and Weber (2011) provide empirical support for the self-dealing hypothesis by showing that when governance structures are weak, stock option plans are designed in a way that is favored by managers.

[6] Corresponding models have been developed and analyzed by Kulatilaka and Markus (1994), Rubinstein (1995), Carpenter (1998), Hull and White (2004), Ammann and Seiz (2004), Cvitanić et al. (2008), and Brisley and Anderson (2008).

[7] Yang and Carleton (2010) analyze the repricing of executive stock options under new accounting rules in an optimal contracting setting.

[8]See Bettis et al. (2005), p. 469.

[9] As the main focus of Johnson's and Tian's (2000a) work is the incentive effect of different kinds of stock option plans, the stock price sensitivity is very important.

[10] See Bettis et al. (2010) for an empirical study on the use of stock option plans with performance contingent vesting conditions.

[11] The DAX 30 comprises 30 companies, 22 of which had installed a stock option plan to compensate executives in 2005. Of the corresponding 22 plans, six belong to the first group, four to the second group, and twelve to the third group.

[12] See the reasoning in the German Corporate Governance Code (Regierugskommission Deutscher Corporate Governance Kodex (2005), p. 7), which recommends caps in stock option plans.

[13] Of course, incentive effects could be quantified within the framework of a particular principal agent model. However, as we make no attempt to

optimize incentive schemes, we employ a simple and general measure, as was used e.g. by Johnson and Tian (2000a). The extensive literature on optimal equity based compensation contracts is surveyed e.g. by Core et al. (2003) and Hall and Murphy (2003).

14 See "Appendix 1" for details.

15 Bakshi et al. (1997) and Chen and Lee (2009) provide empirical evidence on how these characteristics affect the performance of different option pricing models.

16 If stock option plans are settled with cash, as the plans by Allianz, Deutsche Boerse, and BASF, they have to be valued at the balance sheet date. The balance sheet date usually coincides with the end of the calender year.

17 Note that dividends are paid only once a year in Germany.

18 Compare also the pricing formula for European exchange options derived by Margrabe (1978).

19 See "Appendix 2" for details.

20 Table 5 in "Appendix 2" provides these results, showing relative differences between the sensitivity measures ("vega", "dividend yield sensitivity", and "rho") under joint exercise and under separate exercise of sub-plans.

21 See the "Gesetz zur Angemessenheit der Vorstandsvergütung (VorstAG)" of July 2009.

22 Plans also exist that are dividend protected. An example is given by Right B of BASF AG's stock option plan.

23 Historically, a reason has been that dividend protected plans were usually classified as variable plans for accounting purposes in the US, whereas traditional options without dividend protection were classified as fixed plans. Under the new accounting rules, however, this advantage of plans without dividend protection no longer exists.

24 Lambert et al. (1989) and Fenn and Liang (2001) provide empirical evidence that the provision of stock option plans which are not protected for dividends affects corporate payout policy.

25 Compare Arnold and Gillenkirch (2005) for a discussion and analysis of different forms of dividend protection.

26 Although complete robustness calls for a vega of zero, a positive vega could be beneficial in certain situations to mitigate underinvestment in risky projects. See e.g. Smith and Stulz (1985), Hemmer et al. (1999), Feltham and Wu (2001), and Brisley (2006).

[27] See e.g. Hull (2009), p. 373f.

[28] A period of five years to maturity is chosen instead of eight years, which we have used previously as our base case, as five years is a reasonable value for the expected lifetime of an American option with eight years to maturity. Thus, essentially, we aim to leave our base case parameters unchanged.

[29] See "Appendix 3" for a proof of the limiting behavior of delta.

[30] An alternative would be the trinomial model by Chang and Lin (2010).

[31] See also Clewlow and Strickland (2004), p. 44–46.

[32] For such algorithms in the case of barrier options see e.g. Clewlow and Strickland (2004).

APPENDIX 1

All valuations done in Sects. 2 and 3 and of this paper use the same three-dimensional additive tree model with equal up and down factors.[30] Although simpler models exist for some options that we consider, the use of a common framework has the advantage that no price differences can result from different convergence properties and approximation errors of different models.

The concrete model choice was made for three reasons.[31] Firstly, we need a three-dimensional tree to price options with relative performance targets, since their prices depend on two underlyings: the stock price and the value of the reference index. Secondly, the use of an additive tree with equal up and down factors simplifies the valuation of barrier options, like the relative performance vested option. Finally, additive trees (for log prices) have better convergence and approximation properties than comparable multiplicative trees (for prices).

The model is constructed as follows. Assume that the log share price $\ln S$ and the log index level $\ln P$ follow two correlated generalized Wiener processes under the risk-neutral measure, as given in Eqs. 4 and 5, where r is the risk-free interest rate,q_S and q_P denote the dividend yields, σ_S and σ_P the volatilities, and ρ_{SP} the correlation between instantaneous changes in the Wiener processes z_S and z_P.

$$d\ln S = \left(r - q_S - \sigma_S^2/2\right)dt + \sigma_S dz_S \tag{4}$$

$$d\ln P = \left(r - q_P - \sigma_P^2/2\right)dt + \sigma_P dz_P \tag{5}$$

The above processes are approximated by a tree model as follows: We use the up factors $h_S = \sigma_S\sqrt{\Delta t}$ and $h_P = \sigma_P\sqrt{\Delta t}$, respectively, where Δt is the step length. Down factors are just equal to up factors but with opposite signs. These up and down factors are finally combined with risk neutral probabilities for all possible combinations of up and down moves, which are given in Eqs. 6 below

$$\begin{aligned}
p_{uu} &= (h_S h_P + h_P v_S \Delta t + h_S v_P \Delta t + \rho_{SP}\sigma_S\sigma_P\Delta t)/(4h_S h_P),\\
p_{ud} &= (h_S h_P + h_P v_S \Delta t - h_S v_P \Delta t - \rho_{SP}\sigma_S\sigma_P\Delta t)/(4h_S h_P),\\
p_{du} &= (h_S h_P - h_P v_S \Delta t + h_S v_P \Delta t - \rho_{SP}\sigma_S\sigma_P\Delta t)/(4h_S h_P),\\
p_{dd} &= (h_S h_P - h_P v_S \Delta t - h_S v_P \Delta t + \rho_{SP}\sigma_S\sigma_P\Delta t)/(4h_S h_P),
\end{aligned} \tag{6}$$

where v_S and v_P denote the risk-neutral drift rates of ln S and ln P.

The above tree specification needs a slight modification to simplify the valuation of relative performance vested options, which are barrier options with stochastic outside barrier. In fact, the stochastic outside barrier can be transformed into a fixed outside barrier if we consider the process $\ln I = \ln S - \ln P$. Since $\ln I$ provides the outperformance of the stock, a certain level of outperformance can be expressed as a fixed barrier of the $\ln I$ process.

The corresponding stochastic differential equation becomes

$$d\ln I = (-q_S + q_P - \sigma_S^2/2 + \sigma_P^2/2)dt + \sqrt{\sigma_S^2 + \sigma_P^2 = 2\sigma_S\sigma_P\rho_{SP}}dz_I \tag{7}$$

where z_I is a standard Wiener process whose changes have an instantaneous correlation equal to $\rho_{SI} = (\sigma_S - \sigma_P\rho_{SP})\Big/\sqrt{\sigma_S^2 + \sigma_P^2 - 2\sigma_S\sigma_P\rho_{SP}}$ with changes of z_S.

When we use the processes from Eqs. 4 and 7 to construct a three-dimensional tree in the same manner as described above, all kinds of options that we consider can be easily valued by means of recursive algorithms.[32]

APPENDIX 2

This appendix provides some additional information about the similarity of the traditional option, the relative performance vested option, and the absolute performance vested option with respect to their robustness properties. For a broad range of parameter values, we find that the traditional option and the relative performance vested option have almost identical prices, which is due to the fact that the vesting condition of the relative performance vested option is almost always fulfilled. Of course, this feature translates directly into almost identical robustness properties of the two options. When comparing the traditional option with the absolute performance vested option for different parameter combinations, prices can be quite different, however, sensitivities are again very similar. As an example, Table 4 below provides the relative differences in the sensitivity measures "vega", "dividend yield sensitivity", and "rho" for some set of parameter combinations.

Table 4 Differences in robustness between the absolute performance vested option and the traditional option

	int. rate = 2%			int. rate = 3%			int. rate = 4%		
Vol.\Div.	1%	2%	3%	1%	2%	3%	1%	2%	3%
				Vega					
15%	0.2%	0.2%	−0.1%	−1.8%	1.2%	0.2%	−5.5%	−1.2%	1.0%
25%	0.3%	0.2%	0.0%	0.0%	0.4%	0.2%	−0.4%	0.3%	0.4%
40%	0.2%	0.2%	0.1%	0.2%	0.2%	0.2%	0.2%	0.3%	0.2%
				Dividend yield sensitivity					
15%	−0.8%	−4.3%	−2.9%	2.2%	−5.8%	−3.5%	3.5%	1.4%	−5.0%
25%	−1.4%	−1.9%	−1.5%	0.4%	−2.5%	−1.6%	0.8%	−0.9%	−2.0%
40%	−0.8%	−0.8%	−0.7%	−0.3%	−0.9%	−0.7%	0.0%	−0.7%	−0.8%
				Rho					
15%	1.3%	−1.8%	−1.0%	4.4%	−2.7%	−1.1%	6.0%	3.7%	−2.6%
25%	−0.1%	−0.7%	−0.4%	1.5%	−1.2%	−0.5%	1.9%	0.5%	−0.9%
40%	−0.1%	−0.1%	0.0%	0.4%	−0.3%	0.0%	0.7%	0.0%	−0.1%

This table shows the percentage differences between the sensitivity measures ("vega", "dividend yield sensitivity", and "rho") of the absolute performance vested option and the traditional option for different combinations of volatility, dividend yield, and risk-free interest rate ([absolute performance vested − traditional]/absolute performance vested). The absolute performance vested option has a strike price that is 30% above the stock price and both options have a time to maturity of 8 years and a vesting period of 2 years

Table 5 Differences in robustness between two sub-plans that must be exercised jointly and two sub-plans that can be exercised separately

				Vega					
15%	−0.4%	2.7%	4.9%	−0.1%	2.0%	4.6%	−0.1%	3.3%	5.0%
25%	0.1%	1.4%	3.1%	0.0%	1.1%	2.8%	−0.1%	1.3%	2.6%
40%	0.1%	0.5%	1.0%	0.0%	0.4%	0.9%	0.1%	0.4%	0.8%
				Dividend yield sensitivity					
15%	1.9%	7.5%	9.4%	2.0%	6.7%	9.2%	2.6%	6.1%	7.7%
25%	1.5%	3.9%	4.6%	1.5%	3.5%	4.5%	2.1%	2.6%	4.0%
40%	0.9%	1.5%	1.6%	0.9%	1.5%	1.7%	0.9%	1.3%	1.7%
				Rho					
15%	1.8%	8.2%	8.3%	0.1%	1.7%	7.0%	0.0%	−2.3%	2.5%
25%	3.6%	5.0%	5.5%	0.3%	2.3%	4.5%	−0.3%	−1.0%	2.8%
40%	2.6%	2.6%	2.3%	0.6%	1.9%	2.2%	−0.6%	0.8%	1.9%

This table shows the percentage differences between the sensitivity measures ("vega", "dividend yield sensitivity", and "rho") of two sub-plans that must be exercised jointly and two sub-plans that can be exercised separately for different combinations of the stock's return volatility, the stock's dividend yield, and the risk-free interest rate ([joint exercise − separate exercise]/joint exercise). The two sub-plans are a traditional option and an indexed option. The reference index has a volatility of 15% and a dividend yield of 2% p.a.; the correlation between stock returns and index returns equals 0.7. Both options have a time to maturity of 8 years and a vesting period of 2 years

APPENDIX 3

Let us first consider the premium option with strike price $X = S_0 e^{rT}$ Substituting this strike price into Eq. 2, we obtain the following expression for the strike price *Y* of the short call: $Y = S_0 e^{\left(r+\sigma^2\right)T}$. The delta of a portfolio of one call option long with strike price *X* and one call option short with strike price *Y* equals:

$N\left(d_1^x\right) - N\left(d_1^y\right)$, with

$$d_1^x = \frac{\ln(S_0/X) + \left(r + \sigma^2/2\right)T}{\sigma\sqrt{T}} = \frac{\ln(e^{-rT}) + \left(r + \sigma^2/2\right)T}{\sigma\sqrt{T}} = \frac{1}{2}\sigma\sqrt{T}, \quad \text{and}$$

$$d_1^y = \frac{\ln(S_0/Y) + \left(r + \sigma^2/2\right)T}{\sigma\sqrt{T}} = \frac{\ln\left(e^{-\left(r+\sigma^2\right)T}\right) + \left(r + \sigma^2/2\right)T}{\sigma\sqrt{T}} = -\frac{1}{2}\sigma\sqrt{T} \tag{8}$$

Since $N(-\sigma\sqrt{T}/2) = 1 - N(\sigma\sqrt{T}/2)$, we obtain

$$N(d_1^x) - N(d_1^y) = 2N\left(\sigma\sqrt{T}/2\right) - 1 \tag{9}$$

Next consider the value of the portfolio of the two options:

$$C_x - C_y = \left(S_0N(d_1^x) - Xe^{-rT}N(d_2^x)\right) - \left(S_0N(d_1^y) - Ye^{-rT}N(d_2^y)\right), \quad \text{with}$$
$$d_2^x = d_1^x - \sigma\sqrt{T} = -\frac{1}{2}\sigma\sqrt{T}, \quad \text{and}$$
$$d_2^y = d_1^y - \sigma\sqrt{T} = -\frac{3}{2}\sigma\sqrt{T} \tag{10}$$

Using the appropriate values for $N(d_1^x) - N(d_1^y), X, Y, d_2^x$, and d_2^y, the portfolio value in Eq. 10 becomes:

$$S_0\left(2N\left(\sigma\sqrt{T}/2\right) - 1\right) - S_0\left(N(d_2^x) - e^{\sigma^2 T}N(d_2^y)\right)$$
$$= S_0\left(3N\left(\sigma\sqrt{T}/2\right) - 2 + e^{\sigma^2 T}N\left(-3\sigma\sqrt{T}/2\right)\right) \tag{11}$$

In the next step, we have to adjust the number of options in order to obtain a standardized portfolio value equal to the share price in order to facilitate comparisons with a compensation package that consists of one share of stock. To achieve this, we have to divide S_0 by the expression from Eq. 11, which delivers the required number of capped call options in the (standardized) portfolio. Finally, if this number is multiplied by the delta of a single capped call as given in Eq. 9, we obtain the following standardized delta that we require:

$$\text{Delta} = \frac{2N\left(\frac{1}{2}\sigma\sqrt{T}\right) - 1}{3N\left(\frac{1}{2}\sigma\sqrt{T}\right) - 2 + e^{\sigma^2 T}N\left(-\frac{3}{2}\sigma\sqrt{T}\right)} \tag{12}$$

A completely analogous derivation is possible for the indexed option. The only differences are the use of an at-the-money option, i.e., a (normalized) index level that equals the current share price, and the use of the appropriate valuation formula for exchange options. These two modifications exactly offset each other such that the resulting formula is identical to Eq. 12. However, the variance σ^2 does no longer refer to the stock return variance, but to the stock return variance plus the index return variance minus two times the covariance.

Next consider the limiting behavior of delta when either $\sqrt{T}$ or σ go to infinity. Then the numerator of delta in Eq. 12 goes to one. Also the first part of the denominator $\left(3N\left(\frac{1}{2}\sigma\sqrt{T}\right) - 2\right)$ goes to one. Thus, what remains to determine is the limit of the second part of the denominator, which can be written as

$$\frac{N\left(-\frac{3}{2}\sigma\sqrt{T}\right)}{\mathrm{e}^{-\sigma^2 T}} \tag{13}$$

Using de l'Hospital's rule, the above ratio has the same limit as the following ratio:

$$\frac{-\frac{3}{2}N'\left(-\frac{3}{2}\sigma\sqrt{T}\right)}{-2\sigma\sqrt{T}\mathrm{e}^{-\sigma^2 T}} = \frac{-\frac{3}{2}\frac{1}{\sqrt{2\pi}}\mathrm{e}^{-\frac{9}{4}\sigma^2 T}}{-2\sigma\sqrt{T}\mathrm{e}^{-\sigma^2 T}} = \frac{3}{4\sqrt{2\pi}\sigma\sqrt{T}}\mathrm{e}^{-\frac{5}{4}\sigma^2 T} \tag{14}$$

As the last expression in Eq. 14 shows, the ratio goes to zero if either σ or $\sqrt{T}$ go to infinity. This result implies that delta goes to one.

Finally, consider the limit of delta for either σ or $\sqrt{T}$ going to zero. Since both the numerator and the denominator in Eq. 12 go to zero, we have to apply de l'Hospital's rule again. Taking derivatives in the numerator and denominator, we obtain

$$\frac{N'\left(\frac{1}{2}\sigma\sqrt{T}\right)}{\frac{3}{2}N'\left(\frac{1}{2}\sigma\sqrt{T}\right) - \frac{3}{2}N'\left(-\frac{3}{2}\sigma\sqrt{T}\right)\mathrm{e}^{\sigma^2 T} + 2\sigma\sqrt{T}\mathrm{e}^{\sigma^2 T}N\left(-\frac{3}{2}\sigma\sqrt{T}\right)} \tag{15}$$

For either $\sigma \to 0$ or $\sqrt{T} \to 0$ the numerator of the ratio (15) approaches $1/\sqrt{2\pi}$ and the denominator approaches zero. Therefore, the limit of delta is infinity.

REFERENCES

1. Ammann M, Seiz R (2004) Valuing employee stock options: does the model matter? Financial Anal J 60(5):21–34
2. Arnold MC, Gillenkirch RM (2005) Stock options and dividend protection. J Inst Theor Econ 16:453–471
3. Bakshi G, Cao C, Chen Z (1997) Empirical performance of alternative option pricing models. J Finance 52:2003–2049
4. Bebchuk LA, Fried JM (2003) Executive compensation as an agency problem. J Econ Perspect 17:71–92
5. Bebchuk LA, Fried JM, Walker DI (2002) Managerial power and rent extraction in the design of executive compensation. Univ Chicago Law Rev 69:751–846
6. Bettis JC, Bizjak JM, Lemmon ML (2005) Exercise behavior, valuation, and the incentive effects of employee stock options. J Financ Econ 76:445–470

7. Bettis JC, Bizjak JM, Coles J, Kalpathy S (2010) Stock and option grants with performance-based vesting conditions. Rev Financ Stud 23:3849–3888
8. Brenner M, Sundaram RK, Yermack D (2000) Altering the terms of executive stock options. J Financ Econ 57:103–128
9. Brisley N (2006) Executive stock options: early exercise provision and risk-taking incentives. J Finance 61:2487–2509
10. Brisley N, Anderson CK (2008) Employee stock option valuation with an early exercise boundary. Financ Anal J 64(4):88–100
11. Brooks R, Chance D, Cline BN (2007) Private information and the exercise of executive stock options. Working paper
12. Câmara A (2001) The pricing of relative performance based incentives for executive compensation. J Bus Finance Acc 28:1149–1188
13. Carpenter JN (1998) The exercise and valuation of executive stock options. J Financ Econ 48:127–158
14. Chang CC, Lin JB (2010) The valuation of multivariate contingent claims under transformed trinomial approaches. Rev Quant Finance Account 34:23–36
15. Chen RR, Lee CF (2009) Empirical performance of the constant elasticity variance option pricing model. Rev Pac Basin Financ Mark Policies 12:177–217
16. Clewlow L, Strickland C (2004) Implementing derivatives models. Wiley, Chichester
17. Core JE, Guay WR, Larcker DF (2003) Executive equity compensation and incentives: a survey. Econ Policy Rev 9:27–55
18. Cvitanić J, Wiener Z, Zapatero F (2008) Analytic pricing of employee stock options. Rev Financ Stud 21:683–724
19. Dybvig PH, Loewenstein M (2003) Employee reload options: pricing, hedging and optimal exercise. Rev Financ Stud 16:145–171
20. Feltham GE, Wu MGH (2001) Incentive efficiency of stock versus options. Rev Account Stud 6:7–28
21. Fenn GW, Liang N (2001) Corporate payout policy and managerial stock incentives. J Financ Econ 60:45–72
22. Hall BJ, Murphy KJ (2003) The trouble with stock options. J Econ Perspect 17:49–70
23. Hemmer T, Matsunaga S, Shevlin T (1996) The influence of risk diversification on early exercise of employee stock options by executive officers. J Account Econ 21:45–68
24. Hemmer T, Matsunaga S, Shevlin T (1998) Optimal exercise and value of reload options. J Account Res 36:231–255
25. Hemmer T, Kim O, Verrecchia RE (1999) Introducing convexity into optimal compensation contracts. J Account Econ 28:307–327
26. Heynen R, Kat H (1994) Crossing barriers. Risk 7(6):46–51
27. Huddart S, Lang M (1996) Employee stock option exercises: an empirical analysis. J Account Econ 21:5–43
28. Hull JC (2009) Options, futures and other derivatives. Pearson, Prentice Hall, Upper Saddle River
29. Hull JC, White A (2004) How to value employee stock options. Financ Anal J 60(1):114–119
30. Ikäheimo S, Kuosa N, Puttonen V (2006) The true and fair view of executive stock option valuation. Eur Account Rev 15:351–366
31. Jaggi B, Winder JP, Lee CF (2010) Is there a future for fair value accounting after the 2008–2009 financial crisis? Rev Pac Basin Financ Mark Policies 13:469–493
32. Johnson SA, Tian YS (2000a) The value and incentive effects of nontraditional executive stock option plans. J Financ Econ 57:3–34
33. Johnson SA, Tian YS (2000b) Indexed executive stock options. J Financ Econ 57:35–64
34. Kulatilaka N, Markus AJ (1994) Valuing employee stock options. Financ Anal J 50(6):46–56

35. Lambert RA, Lanen WN, Larcker DF (1989) Executive stock option plans and corporate dividend policy. J Financ Quant Anal 24:409–425
36. Margrabe W (1978) The value of an option to exchange one asset for another. J Finance 33:177–186
37. Regierugskommission Deutscher Corporate Governance Kodex (ed.) (2005) German Corporate Governance Code
38. Rubinstein M (1995) On the accounting valuation of employee stock options. J Deriv 3(1):8–24
39. Sautner Z, Weber M (2011) Corporate governance and the design of stock option contracts. Working paper
40. Smith CW, Stulz RM (1985) The determinants of firms' hedging policies. J Financ Quant Anal 20:391–404
41. Yang JT, Carleton WT (2010) Repricing of executive stock options. Rev Quant Finance Account (Online first™, 27 June 2010)

CHAPTER 7

Relative Equity Market Valuation Conditions and Acquirers' Gains

Dimitris Andriosopoulos[1] • Leonidas G. Barbopoulos[2]

[1] *Department of Accounting and Finance, University of Strathclyde, Glasgow G4 0QU, Scotland, UK*

[2] *School of Economics and Finance, University of St. Andrews, St. Andrews KY16 9AR, Scotland, UK*

ABSTRACT

We examine whether the relative equity market valuation conditions (EMVCs) in the countries of merging firms help acquirers' managers to time the announcements of both domestic and foreign targets. After controlling for several deal- and merging firm-specific features we find that the number of acquisitions and acquirers' gains are higher during periods of high-EMVCs at home, irrespective of the domicile of the target. We also find that the higher gains of foreign target acquisitions realized during periods of high-EMVCs at home stem from acquiring targets based in the RoW (=World-G_7), rather than the G6 (=G_7-UK) group of countries. We argue that this is due to the low correlation of EMVCs between the UK (home) and the RoW group of countries. However, these gains disappear or even reverse during the post-announcement period. Moreover, acquisitions of targets domiciled in the RoW (G_6) countries yield higher (lower) gains than acquisitions of domestic targets during periods of high-EMVCs at home. This suggests that the relative EMVCs between the merging firms' countries allow acquirers' managers to time the market and acquire targets at a discount, particularly in countries in which acquirers' stocks are likely to be more overvalued than the targets' stocks.

Keywords: Acquirers' gains Relative equity market valuation conditions Abnormal returns Post-merger performance

1 INTRODUCTION

Record valuations in 2015 point to a new record of Mergers and Acquisitions (M&As) boom with the expectation of M&As continuing for at least another year due to favorable economic conditions (Fontanella-Khan et al. 2015). Favorable market and economic conditions lead to firms financing a large proportion of their financing deficit with net external equity, particularly when the cost of equity is low (Huang and Ritter 2009). An extant literature has emerged over the last few decades investigating the effect of M&As on firm value (i.e. acquirer abnormal returns) by analysing several wider market conditions at the time the investment is initiated, financed, and implemented. Some of the earlier studies focus on the valuation effects of factors pertinent to deal and merger partners,[1] while others analyze the implications of equity market valuation conditions (EMVCs).[2]

To this date, there are still considerable gaps in our understanding of the determinants of merger value creation, and in particular of the impact of EMVCs on the wealth of shareholders of firms engaged in domestic M&As versus those engaged in cross-border mergers and acquisitions (CBAs). While foreign target M&As occur for similar reasons as domestic ones, national borders add a number of complexities to the estimation of synergies and premium to foreign target M&As, relative to that of domestic ones. For instance, Doidge et al. (2007) show that firms do not differ significantly in corporate governance levels only after controlling for country characteristics. Moreover, legal frameworks determine corporate valuation (La Porta et al. 2002), market efficiency (Morck et al. 2000) and the magnitude of currency crises (Johnson et al. 2000), among others. As CBAs are associated with additional frictions relative to domestic acquisitions, such as, cultural and geographical differences, political, economic, foreign exchange, taxation, accounting standards, and potentially corporate governance differences (Bris and Cabolis 2008), such frictions can impede or facilitate M&As. The theoretical framework of Shleifer and Vishny (2003) argues that more overvalued acquirers tend to use their shares to acquire less overvalued targets. We further argue that the imperfect integration of capital markets across countries can also motivate foreign target M&As in which an overvalued acquirer purchases a relatively inexpensive target following changes in exchange rates or stock market valuations in the local currency. Therefore, we thoroughly investigate the extent to which EMVCs within the merging firms' countries:

(a) play a key role in assisting the acquirers' managers to time the announcement of domestic and foreign target acquisitions and, (b) affect acquirers' value gains both in the short- and long-run.

Whether this consensus appeals in domestic versus foreign deals, especially when EMVCs of the merging firms are not synchronized, remains an important question to be investigated. We fill this void by empirically examining whether acquirers' managers time the announcement of both domestic and foreign target M&As, as captured by the EMVCs at home relative to that in the host countries. We perform this by focusing exclusively on UK acquirers as they are the most active worldwide in engaging in CBAs.[3]

Existing evidence shows that takeovers of domestic targets cluster over time and several theories attempt to explain this phenomenon, such as the neoclassical theory of mergers, the strategic theory and the behavioral approach, among others.[4] First, the neoclassical theory of merger waves, which is mainly associated with the work of Gort (1969), posits that M&A activities spike with technological, economic and/or regulatory industry shocks. Similarly, in an earlier work Nelson (1959) showed that M&A activities are highly associated with business cycles. Second, the strategic theory of merger waves focuses on the acquisition performance, motives, and strategies (Kusewitt 1985; Toxvaerd 2008). In particular, Toxvaerd (2008) develops a theoretical model predicting that competitive pressure interacts with the irreversibility of mergers in an uncertain environment, and argues that acquirers either postpone a bid in order to gain from more favorable future market conditions, or enter the bidding contest. He argues that while in a complete information model all acquirers rush to bid creating merger waves, in an incomplete information environment various factors representing economic fundamentals, including the benefits from mergers and target scarcity, affect the timing of the merger waves. Third, the behavioral explanation of Shleifer and Vishny (2003), argues that the observed clustering in M&A activity is largely driven by stock market miss-valuations. Similarly, Rau and Stouraitis (2011) argue that the waves of corporate activity are driven by a time-varying influence of both the neoclassical and miss-valuation hypotheses.

Timing the announcement of M&As is an important factor influencing the likelihood of M&A success, since, for example, M&As announced during periods of high investor sentiment are likely to be associated with greater mispricing levels (Stambaugh et al. 2012). This mispricing stems from the stronger impact of market-wide sentiment on stocks that are difficult to value (Baker and Wurgler 2006) and the positive (flat) relationship between sentiment and expected returns during low (high) sentiment periods (Yu and Yuan 2011). The evidence on M&As in the US suggests that the announcement period (short-run) gains and the post-announcement (long-run) performance of acquirers are also sensitive to factors pertinent to stock market conditions at the time of the M&A announcement. For example, Bouwman et al. (2009) show that M&As announced during periods of high-EMVCs lead to higher acquirers abnormal returns in the short-run, but such gains are eliminated in the long-run. Rosen (2006) also confirms the same pattern in the performance of acquirers and argues that the observed superior announcement period gains of acquirers at the time of high-EMVCs are driven by investors' sentiment. However, the aforementioned studies analyse the impact of EMVCs for domestic target M&As and ignore the the impact of relative EMVCs between the merging firms countries in CBAs on the gains of domestic versus foreign acquirers, which remains to be investigated.

Due to the globalization of financial markets and the relaxation of restrictions in capital mobility, CBAs have increased substantially. For example, the World Investment Report (UNCTAD) (2009, p. 11, Table I.3.) shows that in 2007 the total value of global CBA deals reached over $1,197bn compared to only $39bn in 1987, which represents a record increase of more than 68 % over the value of 2006 ($711bn). In the same year, net purchases of UK firms reached 21.5 % of the global market. Consequently, the literature has turned to investigate the effects of domestic versus CBAs on acquirers' gains.[5] Some of the earlier studies show that the gains of merger partners in CBAs are affected by the strength of the acquirers' domestic currency (Kang 1993). Froot and Stein (1991) predict that when the acquirers' currency is strong against the US dollar they will have an advantage when purchasing US targets. This is empirically confirmed by Harris and Ravenscraft (1991) who study US target firms only and find that a stronger currency—against the US dollar—of foreign acquirers leads to higher wealth effects for US targets at the time of the M&A announcement. di Giovanni (2005) further shows that CBA activities are affected by financial and other institutional factors, with the size of financial markets being one of the most influential factors.

Overall, these findings suggest that M&A activities (domestic and cross-border) are affected by financial and economic conditions in the acquirers' home market. However, little is known about the effect of relative EMVCs on the gains of acquirers targeting domestic versus foreign target firms. We address this issue and identify whether managers of acquiring firms can extract any benefits by timing the market, namely, whether the relative EMVCs in merging firms' countries allow them to time the announcement of both domestic and foreign target M&As.

Our results show that acquirers enjoy higher short-run abnormal returns from deals announced during periods of high-EMVCs at home, regardless of the target being a domestic or a foreign firm, consistent with Bouwman et al. (2009). We further find that foreign target acquirers enjoy greater gains from deals made during high-EMVCs at home only when the EMVCs between the merging firms countries are less likely to be correlated, which falls in the category of the target operating in the RoW (=World-G_7) rather than in the G6 (=G_7-UK) group of countries. However, this is disappearing or even reversed in the post-merger period. We argue that the EMVCs at home, relative to host countries, allow managers to time the market and acquire targets at a discount, particularly in markets in which their stocks are likely to be more overvalued than their targets. UK acquirers also reap superior gains from M&As announced during relatively strong trade-weighted effective exchange rate of the domestic currency. In the context of domestic versus foreign target M&As we find that M&As of foreign targets from the RoW (G_6) group of countries yield higher (lower) acquirer gains compared to acquiring domestic targets when occurring during periods of high-EMVCs at home. We argue that a possible explanation for this finding is that the EMVCs in the RoW (G_6) are lower (similar) to the ones at home, hence, acquirers are more (less) likely to acquire targets at a discount. Overall, our results suggest that acquirers aiming to maximize shareholder wealth can extract signals from EMVCs at home relative to those in host countries, and carefully pick the timing of their M&As announcements.

We contribute to the related literature by demonstrating the role of EMVCs at home relative to EMVCs in the host countries, in the case of CBAs, on the the short- and long-run gains of acquirers. Our findings suggest that firms willing to engage in acquisitions should consider not only the merging firms- and deal-specific features, but also the possible implications of factors pertinent to the wider economy such as EMVCs at home, relative to EMVCs in the host country in the case of CBAs. Therefore, a simultaneous analysis of EMVCs in the merging firms countries can assist managers in timing the announcement of both domestic and foreign target M&As and maximize shareholder wealth.

The rest of the paper is organized as follows. Section 2 develops our hypotheses. Section 3discusses the data and methodology employed. The empirical results are reported in Section 4. Section 5 concludes.

2 RELATED LITERATURE AND HYPOTHESES

2.1 EMVCs and gains from M&A

A growing body of literature (e.g. Shleifer and Vishny 2003) suggests that during high-EMVC periods, managers of firms with overvalued stocks are motivated to use their overvalued stocks to bid for targets that they perceive to be less overvalued than themselves. In the long-run, however, when the market corrects for overvaluation, the acquirers suffer a drop in share price indicating long-term under-performance. This theoretical proposition has been supported by the empirical findings of, among others, Rhodes-Kropf et al. (2005). Although the theoretical proposition of Shleifer and Vishny (2003) is limited to all-share deals, a similar effect of high-EMVCs can occur on the performance of acquirers in other types of deals, such as cash and mixed, due to managerial overconfidence and investors' optimism caused by booming markets. The manager of a company whose market value is increasing is likely to be overconfident about one's managerial ability and start bidding for targets, driven by the maximization of their private benefits and prestige or the enhancement of shareholder wealth (Draper and Paudyal 2006). When share prices increase investors are likely to be more confident in their managers and react positively to acquisition decisions. Consequently, acquirers are likely to experience higher announcement period abnormal returns from bids announced at the time of high-EMVCs irrespective of the payment method. In contrast, at the time of low-EMVCs, investors are likely to be skeptical about the value of the deals and react cautiously. This leads to the first testable hypothesis (*H1a*): *M&A announcements during high-EMVC-periods yield higher short-run abnormal returns than deals announced during low-EMVC-periods.* In the long-run, however, there should be a market correction and M&As announced during during high-EMVC-periods should exhibit a stock performance reversal that wipes out the announcement gains or cause significant losses. This leads to our next hypothesis (*H1b*): *M&As announced during periods of high-EMVCs break-even or experience losses in the long-run.*

The short-run performance of bidding firms is a function of investors' optimism regarding the acquisition's future prospect (Rosen 2006). The implications of investors' sentiment, namely confidence, on M&A announcements are likely to differ by the domicile of the target, given that foreign investments generally add to the level of uncertainty regarding the expected value to be generated from the deal, mainly due to its additional exchange rate, economic and political risks.

During periods of low-EMVCs, this added uncertainty exacerbates the already low investor sentiment. Therefore, during low-EMVC periods, foreign acquisitions are expected to deliver lower yields compared to domestic deals. However, during high-EMVC periods acquirers of foreign targets are expected to yield a more favorable market reaction than domestic deals, driven by higher investor optimism. These arguments lead to the second testable hypothesis (*H2*): *CBAs announced during high-EMVC periods at home outperform domestic deals announced during similar periods.*

2.2 Relative equity market valuation conditions and the gains from CBA

Shleifer and Vishny (2003) argue that during high-EMVC periods managers of firms with overvalued shares are motivated to use shares for bidding to acquire targets they perceive to be less overvalued than their own firm. This is particularly interesting in the CBA context as different markets have different levels of EMVCs at any one time. Therefore, at a given level of EMVCs at home, whether acquirers time the market and bid for targets abroad is limited to their ability to identify less overvalued (foreign) targets than themselves. We argue that in CBAs the relative EMVCs (i.e. the deviation of EMVCs between an acquirer's and a target's country) provide an additional insight to the acquiring firms' managers aiming to maximize the wealth of their shareholders.

At a high level of EMVCs at the acquirer's home market a lower correlation with the EMVCs of the target firm's country implies a low (or lower) EMVCs for the target firm. Therefore, acquirers that are overvalued (or even fairly valued) and have higher EMVCs compared to a foreign target's country, they can exploit this EMVC deviation and yield a more favorable market reaction to the M&A announcement. As discussed earlier, while this effect is more pronounce to share deals, it can also apply to other types of deals, such as cash and mixed (i.e. cash and shares), due to managerial overconfidence and investors' optimism caused by the level of EMVCs at home. This leads to the third testable hypothesis (*H3*): *Acquirers gain significantly more from CBAs announced during periods of high-EMVCs at home and low- or lower-EMVCs in the host country.*

2.3 Strength of domestic currency and gains from M&A

A number of studies on foreign direct investment (for example, di Giovanni 2005) show that the strength of the domestic currency plays a significant role in determining the flow of foreign direct investment. On a theoretical framework Froot and Stein (1991) argue that when the US dollar is weaker compared to foreign acquirers' local currency they will have a purchasing advantage for US targets. In line with this theoretical prediction, Harris and Ravenscraft (1991) find that a weaker US dollar leads to a higher market reaction for US firms being targeted by acquirers with a stronger currency due to their purchasing advantage. When the domestic currency is stronger than its equilibrium value, it is cheaper for acquirers to acquire foreign targets. This should not only enhance M&A activity but also generate higher gains for acquirers. This is because when the domestic currency is strong, the effective price (inclusive of premium) paid to foreign targets is more likely to be less than the equilibrium value of the target. Consequently, such deals are more likely to transfer wealth from target company's shareholders to the shareholders of the acquiring company. This leads to the fourth hypothesis (*H4*): *M&As announced at times of stronger domestic currency generate higher gains to acquirers.*

3 DATA AND METHODOLOGY

3.1 Data

The information on the deal announcements is extracted from Securities Data Corporation (SDC). The sample comprises of 67,229 M&A bids announced by UK firms between 01/01/1986 and 31/12/2010. The sample of M&As analyzed by the paper includes 5880 (4142 domestic and 1738 foreign) deals that meet the following criteria: (1) The acquirer is a UK company traded in the London Stock Exchange; (2) the target is a private, a public, or a subsidiary firm, both domestic and foreign; (3) to avoid the effects of very small transactions, only deals equal to, or greater than, £1 million are included; (4) the market value of the acquirer is greater than £1 million at four weeks prior to the announcement of the deal; (5) acquiring firms are not involved in multiple bids during the 5-day announcement window of $t-2$ to $t+2$ days around the announcement day, where $t=0$ is the announcement day; (vi) the daily return to index, market value, and market-to-book-value ratio of the acquirer are obtained from DataStream. Finally, we collect the monthly P/E ratio for the FTSE All Share market index and the monthly P/E ratio for the 9 industry groups from DataStream in order to measure the EMVC.

3.2 Sample statistics

Table 1 reports the annual distribution of M&As announced by UK acquirers between 01/01/1986 and 31/12/2010. The table shows that almost one in three M&A deals (30 %) involve foreign targets. The proportion of CBAs in the UK increased steadily over time, exceeding 40 % of all deals in 2010. In addition, the proportion of CBA in G6 (=G_7-UK) countries (20 %) is well over the proportion of CBA in the Rest of the World (RoW=World-G_7) group of countries (10 %). We follow this classification due to the evidence that the economic activity and stock market performance of the G7 countries (UK, US, Canada, Germany, France, Italy, and Japan) are more integrated than the rest-of-the-world (Jansen and Stokman 2004). The statistics also show that in approximately 51 % of the deals the merging firms are based in the same-industry (based on the same 2-SIC code) while the remaining 49 % are diversified deals. Approximately half of our sample of M&As are settled in cash (44 %), while share only deals constitute 7 % of the total number of deals in the sample. The proportion of cash only deals is much larger in CBA, while only about one in twenty CBA deals are settled in shares. The largest deals in our sample were announced during the dot com bubble (£479.3 m) followed by deals announced before, and immediately after, the 2007 financial crisis. Moreover, the highest takeover premia are offered during 1997 and 1998, whereas, acquirers enjoyed the higher announcement period abnormal returns during the years 1993, 1999 and 2005.

Table 1 Annual M&A activity by target firm's domicile, industry, and method of payment

Year	ALL	DOM	CBA	G6	ROW	FOC	DIV	CASH	STOCK	UNL	LISTED	DV	MV	PREM	CAR
1988	42	34	8	8	0	12	30	17	4	35	7	38.9	492.0	36.2 %	−0.38 %
1989	336	247	89	67	22	138	198	163	30	294	42	33.6	402.3	45.5 %	−0.18 %
1990	204	149	55	39	16	86	118	105	14	184	20	28.1	420.9	35.4 %	0.11 %
1991	137	113	24	16	8	55	82	54	15	115	22	38.9	768.9	37.2 %	0.43 %
1992	138	107	31	21	10	49	89	59	11	133	5	13.1	534.6	17.0 %	0.42 %
1993	210	162	48	33	15	85	125	90	17	194	16	43.8	439.3	40.1 %	2.49 %
1994	262	201	61	41	20	127	135	118	24	242	20	28.8	438.5	34.5 %	1.06 %
1995	278	200	78	61	17	107	171	106	18	259	19	43.3	556.0	20.9 %	0.90 %
1996	315	233	82	58	24	142	173	138	22	288	27	47.4	520.2	30.1 %	1.47 %
1997	383	262	121	81	40	165	218	152	23	352	31	66.0	626.8	46.8 %	0.96 %
1998	411	282	129	86	43	222	189	212	20	368	43	56.3	674.8	48.5 %	1.35 %
1999	433	298	135	97	38	252	181	197	26	366	67	479.3	1154.2	42.7 %	3.47 %
2000	434	294	140	91	49	244	190	160	43	388	46	113.2	1179.7	40.8 %	0.83 %
2001	302	208	94	69	25	166	136	104	19	275	27	38.4	757.2	44.5 %	0.84 %
2002	217	163	54	32	22	115	102	110	14	203	14	70.9	730.6	43.5 %	1.88 %
2003	193	127	66	50	16	123	70	92	10	175	18	43.6	875.7	38.2 %	2.74 %
2004	214	151	63	45	18	124	90	81	11	205	9	42.0	719.6	26.4 %	1.30 %
2005	292	210	82	52	30	173	119	128	16	266	26	71.3	1036.8	16.4 %	2.56 %
2006	298	202	96	54	42	180	118	127	9	280	18	52.8	674.5	29.9 %	1.45 %
2007	336	226	110	59	51	201	135	129	13	310	26	63.2	1198.5	41.8 %	1.71 %
2008	183	111	72	47	25	107	76	86	8	174	9	67.7	957.8	39.5 %	0.47 %
2009	96	63	33	19	14	56	40	44	12	79	17	48.3	1224.8	45.6 %	1.12 %
2010	166	99	67	39	28	89	77	93	9	153	13	83.1	1705.4	26.0 %	1.49 %
Total	5880	4142	1738	1165	573	3018	2862	2565	388	5338	542	–	–	–	–
% of all	–	70 %	30 %	20 %	10 %	51 %	49 %	44 %	7 %	91 %	9 %	–	–	–	–
Average	–	–	–	–	–	–	–	–	–	–	–	70.1	786.5	35.9 %	1.24 %

The table presents the annual distribution of M&As announced by UK acquirers between 01/01/1986 and 31/12/2010. The distribution of M&As is presented according to the target firm's domicile (domestic, foreign, and foreign deals' subgroups including the G6 (=G7–UK) and the RoW (=World–G7)), merging firms' industry classification (focused and diversifying), cash and stock M&A payment methods, listing status of the target firm (unlisted and listed), annual average deal value (DV), annual average acquirer size measured by its market capitalization (MV) 20 days prior to the M&A announcement, annual average takeover premium 1 week prior to the M&A announcement (offered by SDC Thomson-ONE), and annual average announcement period Cumulative Abnormal Return (CAR as estimated in Eq. 1) earned by acquirers' shareholders. Appendix 1 refers to the definition of each variable

Table 2 (Panel A) shows that unlisted targets account for about 91 % of all deals, while only one in ten deals involve listed targets. The dominance of unlisted targets in UK takeover deals is consistent with Draper and Paudyal (2006). The remaining statistics of either transaction- or firm-specific characteristics reported in Panel A show that the number of deals occurring during periods of high- versus low-EMVCs, is higher (1598 vs. 1545). Noticeably, the sum of deal value in each group is substantially higher for deals occurring during periods of booming than depressed markets across all classifications.

Table 2 Summary statistics

	ALL	DOM	CBA			Focused (FOC)	Diversifying (DIV)	CASH	STOCK	Target listing status	
			ALL	G6	RoW					Unlisted	Public
Panel A											
ALL											
N	5880	4142	1738	1165	573	3018	2862	2565	388	5338	542
%	–	70	30	20	10	51	49	44	7	91	9
Sum of DV (in £ bn)	508	177	331	276	55	363	145	123	54	177	331
Low-EMVCs											
N	1545	1136	409	287	122	742	803	682	117	1406	139
%	26	19	7	5	2	13	14	12	2	24	2
Sum of DV (in £ bn)	71	41	30	24	7	34	37	26	8	43	28
Neutral-EMVCs											
N	2737	1912	825	538	287	1435	1302	1181	165	2498	239
%	47	33	14	9	5	24	22	20	3	42	4
Sum of DV (in £ bn)	149	80	69	40	28	98	51	62	26	79	70
High-EMVCs											
N	1598	1094	504	340	164	841	757	702	106	1434	164
%	27	19	9	6	3	14	13	12	2	24	3
Sum of DV (in £ bn)	288	56	232	212	20	231	57	35	21	55	233

	MV (in £ m)		DV (in £ m)		Deal relative size		MTBV		Age (in years)		Liquidity		Leverage		Premium	
	Mean	Median	Mean	Median	Mean	Median	Mean	Median	Mean	Median	Mean	Median	Mean	Median	Mean	Median
Panel B																
ALL	793.91	121.72	86.33	7.20	0.34	0.07	2.87	2.07	16.02	12.46	0.13	0.08	22.03	17.25	38.71	35.25
Low-EMVCs	620.83	114.14	46.10	6.50	0.28	0.07	3.79	2.05	16.14	13.71	0.13	0.08	18.83	16.68	37.92	36.13
Neutral-EMVCs	805.16	115.23	54.39	7.00	0.32	0.07	2.70	2.05	16.05	12.14	0.14	0.08	19.32	16.81	36.70	31.89

	MV (in £ m)		DV (in £ m)		Deal relative size		MTBV		Age (in years)		Liquidity		Leverage		Premium	
	Mean	Median	Mean	Median	Mean	Median	Mean	Median	Mean	Median	Mean	Median	Mean	Median	Mean	Median
High-EMVCs	941.99	135.83	179.92	8.48	0.43	0.07	2.31	2.16	15.83	11.77	0.14	0.09	29.55	18.36	41.77	38.48
Domestic (DOM)	441.84	81.66	42.71	6.00	0.36	0.08	2.64	1.88	14.65	10.64	0.13	0.07	22.88	16.10	37.29	34.97
CBA (ALL)	1632.95	342.91	190.28	11.63	0.30	0.05	3.42	2.47	19.26	20.00	0.15	0.11	20.03	18.91	42.24	35.65
CBA (G6)	1616.18	353.32	236.95	13.35	0.25	0.05	3.02	2.54	19.78	20.93	0.15	0.11	20.42	19.75	41.42	36.82
CBA (RoW)	1667.03	296.35	95.38	6.70	0.30	0.03	4.22	2.40	18.21	16.37	0.17	0.11	19.24	17.44	44.85	24.48
Focused	780.45	116.20	120.25	7.65	0.28	0.08	2.90	2.05	14.08	9.89	0.14	0.08	24.80	17.23	39.97	34.77
Diversifying	808.10	124.85	50.56	6.80	0.40	0.06	2.84	2.10	18.06	18.31	0.13	0.08	19.06	17.28	37.34	36.13
Unlisted	679.06	114.40	33.16	6.28	0.32	0.06	2.98	2.09	15.85	12.15	0.13	0.08	22.14	17.09	86.94	63.52
Public	1924.99	256.00	609.95	50.23	0.54	0.24	1.83	1.89	17.66	15.69	0.14	0.09	20.84	18.42	38.35	34.97
Cash	1146.28	202.75	47.97	7.58	0.16	0.04	2.28	2.06	18.89	19.88	0.12	0.08	26.60	19.34	39.11	34.57
Stock	574.04	50.93	138.54	10.15	1.48	0.27	2.49	1.87	11.70	7.76	0.19	0.11	17.79	14.27	38.39	31.49

In Panel A the sample is classified by target firm's domicile (domestic, CBA, and CBA sub-groups including the G6 (=G7-UK) and the RoW (=World-G7)), SIC 2-digit industry classification (focused and diversifying), method of payment (cash and stock), and target firm's listing status (unlisted and listed). The sample comprises M&As announced by UK acquirers between 01/01/1986 and 31/12/2010 and recorded by the SDC. Acquirers are firms listed in the London Stock Exchange (LSE). In Panel A, N represents the number of deals; % is number of deals in a group as a proportion of All deals in each column rounded off to the nearest integer; Sum of DV in billions pounds is the sum of deal values of all deals in each group; the sum of DV is rounded off to the nearest billion pounds. In Panel B the mean and median are the group mean and median. The process of de-trending is discussed in Sect. 3.3. All variables are defined in the Appendix 1

The total deal value of CBA deals is almost double than domestic deals (£331bn vs. £177bn), while CBA deals are larger in the G6 countries compared to the RoW countries (£276bn vs. £55bn). Table 2 (Panel B) also shows that the acquirers of foreign targets are more mature (19.26 years) than their counterparts of domestic targets (14.65 years). Furthermore, the average market capitalization of acquirers engaged in domestic M&As is less than the average market capitalization of acquirers engaged in CBAs (£441.84 vs. £1632.95 m). Acquirers of targets in the RoW countries (£1667.03 m) are larger than acquirers in the G6 countries (£1616.18 m) albeit the deals being larger for the latter (£236.95 m) than the former (£95.38 m) group. Furthermore, the average acquirer of a domestic target is less liquid than the corresponding acquirers of a foreign target (0.13 vs. 0.15), has higher leverage (22.88 vs. 20.03), and pays lower premia (37.29 vs. 42.24 %).

3.3 Classification of high-, neutral-, and low-EMVCs

The EMVC is measured by market P/E ratio, using the FT All share index as market proxy and since the measure of EMVCs possesses an up-ward trend, it is de-trended using the method outlined in Bouwman et al. (2009). First, the best straight line fit value of the variable is removed from the observed value in each month and in five years preceding the announcement. Second, each month is categorized as above (below) the average if the detrended value of the variable for the month is above (below) the average of the previous five years. Finally, the upper half of the above-average months (i.e., the 25 % top months) are classified as high value periods and the lower half of the below-average months (i.e., the 25 % bottom months) are classified as low value periods. All other months in the sample (i.e., 50 % of the middle months) are classified as neutral. As implied by the stock market valuation argument of Shleifer and Vishny (2003), merger activity increases during high-EMVC periods (Table 2, Panel A). Moreover, this pattern holds for both domestic and CBAs, focused and diversified deals.

3.4 Methodology

The performance of acquirers is analyzed by indicators of EMVCs using both univariate and multivariate frameworks. For the univariate analysis, first the excess returns of acquirers announcing bids under various financial conditions are tested, followed by the comparison of the gains of such acquirers. Finally, the gains of acquirers by the domicile of their targets (domestic targets, foreign targets, acquirers of targets in G6 countries, and in the RoW) and EMVCs at the time of the bid announcement are also compared. Under multivariate analysis, similar tests on the gains of acquirers are performed after controlling for the effects of other factors using cross-sectional regressions that are known to affect the gains to acquirers.

Because of multiple bids by the same acquirers occurring within a short period, sufficiently long time series of stock returns that are free from the effects of takeover bids are not available to apply conventional time-series based excess-return estimation models. Therefore, we follow Draper and Paudyal (2008) Fuller et al. (2002) and Faccio et al. (2006) among others and use the market adjusted returns to estimate the announcement period abnormal returns.,[67] As a result, for measuring the announcement period abnormal returns of acquirers the market-adjusted model is employed as shown in Eq. 1:

$$AR_{it} = R_{it} - R_{mt} \tag{1}$$

where, R_{it} is the return of firm i on day t, and R_{mt} is the market return measured by the changes in FT-All Share Index (inclusive of dividends). The abnormal returns (AR_{it}) are cumulated for the 5-days ($t-2, t+2$) surrounding the M&A bid announcements, as shown in Eq. 2:

$$CAR_i = \sum_{t-2}^{t+2} AR_{it} \tag{2}$$

3.5 Long-run performance of acquirers

To examine the post-merger performance of acquirers, the standard buy-and-hold-abnormal return (BHARs) approach (Barber and Lyon 1997) is implemented. The monthly abnormal return is calculated as the acquiring firm's BHARs less the BHARs of the market portfolio over the period T (1, 3, or 5 years following the acquisition announcement month), as shown in Eq. (3):

$$BHAR_{it} = \prod_{t=1}^{t=T}(1 + R_{it}) - \prod_{t=1}^{t=T}(1 + R_{mt}) \tag{3}$$

where R_{it} is the monthly return for firm i; R_{it} is the monthly return of the market index; $t$represents the 12, 36, 60 month after the acquisition's month. The monthly BHAR of each bidder is regressed against a set of explanatory variables, similar to the control factors included in the analysis of announcement period returns, which are presented in the Sect. 3.6.

3.6 Multivariate analysis

To further investigate the implications of relative EMVCs on announcement period gains of the UK acquirers, we examine the impact of relative EMVCs after controlling for other factors that are known to affect acquirers' gains simultaneously. We do so by employing Eq. (4) estimated in a nested regression form with various combinations of explanatory variables discussed below

$$alpha_i = \beta_1 + \sum_{j=2}^{k} \beta_j X_{ij} + \varepsilon_i \quad i = 1 \ldots N \tag{4}$$

When Eq. (4) is applied to analyzing the announcement period gains, the dependent variable (alphaialphai) is measured by the 5-days cumulative excess return of acquirers as explained in Eq. (1). Alternatively, when Eq. (4) is applied to analyzing the post-merger acquirer performance, the dependent variable (alphaialphai) is measured by the 1-, 3- and 5-years BHARs of acquirers as explained in Eq. (3). The intercept β_1 in Eq. (4) reflects the excess returns of acquirers after controlling for the effects of all explanatory variables, incorporated in the vector X_{ij}, simultaneously, the impact of which is recorded in the vector βjβj. The vector of explanatory variables $X_{i,j}$ includes a set of explanatory variables discussed below.

Earlier literature, based on the experience of listed targets, shows that cash-only acquirers experience the highest gains, while share-only acquirers suffer a loss. However, Travlos (1987) shows that the method of payment interacts with the listing status of targets in shaping the gains of acquirers. Therefore, to control for the *methods of payment*, deals are classified into cash only, shares only, and mixed. Moreover, on the announcement of M&A deals acquirers of unlisted targets gain more than acquirers of listed targets (Draper and Paudyal 2006). Therefore, we control for the *listing status of targets*, namely listed, private and subsidiary.

Larger acquirers gain less than smaller acquirers on the announcements of bids (Moeller et al. 2004). Therefore, to allow for this effect we control for the *size of the acquirer*, measured by their market capitalization (MV) four weeks before the announcement of the deal. Sudarsanam and Mahate (2003), among others, show that value acquirers (firms with low market-to-book value ratio) outperform glamour acquirers (firms with high market-to-book value ratio) in the short-run. Thus, we control for the *growth opportunity of bidding firms*, measured by the market-to-book value (MTBV) ratio one month prior to the announcement of the deal.

Draper and Paudyal (2008) suggest that acquirers' announcement period gains are dependent on the level of public information. Barry and Brown (1985) show that more information is available in the public domain in relation to firms with a long trading history. Since mature firms are likely to have a longer trading history, consequently more publicly available information, we control for *acquirer's maturity (age)* measured by the number of days from the date a firm is first recorded on DataStream[8] and the date of the announcement. Several studies (Asquith et al. 1993) have shown that the acquirers' announcement period returns increase with the size of the target relative to the value of the acquirer. This may be due to the stronger impact on the structure of the organization of the acquiring firm that relatively larger targets may have. It is also possible that the observed positive relation between the relative size of the deal and gains from acquisitions is simply a statistical effect as a relatively larger deal is likely to generate higher synergy gain, which could be substantial relative to the size of acquirer. *The relative size of the deal* is measured by the ratio of the acquirer's market capitalization to the value of the deal.

We also control for the *Effective Exchange Rate* (Effect. Ex. Rate) of Pound Sterling obtained from the Bank of England's statistical database in order to account for the impact of the strength of the domestic currency on acquirers' gains.[9] In addition, we add a set of dummy variables that is assigned the value of 1 (0 otherwise) to control for: CBA deals, diversifying deals (defined as a deal when target and acquirer do not share the same 2-digit SIC code),[10]private targets, cash-only deals, stock-only deals, high EMVC period, period of strong domestic currency, and frequent acquirers. Finally, variables representing interactions between various measures of EMVCs are also introduced in the model.

4 EMPIRICAL RESULTS

4.1 Acquirers' announcement period gains

This section presents and discusses the findings from the univariate and multivariate analysis of announcement period and long-run gains of acquirers against financial conditions (i.e. EMVCs) at the time of M&As announcements. The gains are also analyzed by the targets' domicile. The results reported in Table 3 (Panel A) show that the average acquirer earns statistically significant gains (1.36 %) on the announcement of a takeover bid. Table 3 (Panels B and C) show similar findings when the 3- and 4-factor models are used to estimate the acquirer gains. This is consistent with UK studies that include M&As of both listed and unlisted targets (Draper and Paudyal 2006). The table further shows that firms acquiring domestic targets earn no significantly different abnormal returns to firms acquiring foreign targets (diff = 0.06 %), consistent with Lowinski et al. (2004). However, further analysis (Panel A) shows that domestic target acquirers: (1) outperform acquirers of foreign targets in the G6 group of countries by 0.42 %; and (2) outperform acquirers of foreign targets in the RoW group of countries by 0.63 %. In Table 3 (Panels B and C) we estimate the CARs based on the 3-factor Fama–French (1993) and 4-factor Carhart (1997) models, respectively, where we report similar findings. Moreover, among the CBAs, acquirers enjoy 1.05 % higher abnormal returns when targets are based in the RoW rather than in the G6 group of countries (Panel A). Whether these gains to acquirers in the framework of domestic deals versus CBAs are driven by the impact of EMVCs at home, relative to those in the host countries, is empirically investigated in this paper and the results are recorded in the following section.

Table 3 Acquirers' announcement period returns by EMVCs and target firm's domicile

	Equity market valuations conditions				
	ALL	**Low-EMVCs**	**Neutral-EMVCs**	**High-EMVCs**	**HML-EMVCs**
Panel A: Market-adjusted model					
ALL					
Mean	1.36***	1.04***	1.18***	1.99***	0.95***
Median	0.52***	0.51***	0.46***	0.66***	0.15*
N	5880	1545	2737	1598	
Domestic					
Mean	1.38***	1.08***	1.21***	2.01***	0.93**
Median	0.51***	0.50***	0.48***	0.60***	0.10
N	4142	1136	1912	1094	
CBA (ALL)					
Mean	1.32***	0.93***	1.12***	1.95***	1.02*
Median	0.52***	0.53**	0.42***	0.79***	0.26
N	1738	409	825	504	
CBA (G6)					
Mean	0.97***	0.75**	0.89***	1.28**	0.54
Median	0.39***	0.37	0.38**	0.57*	0.20
N	1165	287	538	340	
CBA (RoW)					
Mean	2.02***	1.36***	1.55***	3.33***	1.98*
Median	0.84***	1.14**	0.49***	1.29***	0.15
N	573	122	287	164	
Differentials					
Domestic versus CBA (ALL)					
Mean	0.06	0.15	0.09	0.06	
Median	−0.01	−0.02	0.07	−0.19	
Domestic versus CBA (G6)					
Mean	0.42*	0.34	0.32	0.73	

Median	0.13*	0.13	0.10	0.03	
Domestic versus CBA (RoW)					
Mean	−0.63*	−0.27	−0.34	−1.32	
Median	−0.33**	−0.64	0.00	−0.69*	
G6 versus RoW					
Mean	−1.05***	−0.61	0.66	−2.05**	
Median	−0.45***	−0.77**	−0.10***	−0.72**	
Panel B: Carhart (1997) 4-factor model					
ALL					
Mean	1.56***	1.09***	1.36***	2.34***	1.25***
Median	0.62***	0.49***	0.67***	0.62***	0.13
N	5880	1545	2737	1598	
Domestic					
Mean	1.62***	1.16***	1.30***	2.52***	1.36***
Median	0.62***	0.52***	0.71***	0.50***	−0.02
N	4142	1136	1912	1094	
CBA (ALL)					
Mean	1.40***	0.89***	1.33***	1.94***	1.04**
Median	0.62***	0.44***	0.62***	0.78***	0.34
N	1738	409	825	504	
CBA (G6)					
Mean	0.99***	0.88**	1.02***	1.05**	0.17
Median	0.50***	0.46*	0.53***	0.51*	0.05
N	1165	287	538	340	
CBA (RoW)					
Mean	2.25***	0.94**	1.93***	3.78***	2.85***
Median	0.80***	0.41***	0.80***	1.48***	1.07**
N	573	122	287	164	
Differentials					
Domestic versus CBA (ALL)					
Mean	0.22	0.27	0.05	0.59	
Median	0.00	0.08	0.09	−0.28	
Domestic versus CBA					

(G6)					
Mean	0.63**	0.29	0.36	1.48**	
Median	0.12*	0.06	0.18	-0.01	
Domestic versus CBA (RoW)					
Mean	-0.62*	0.23	-0.55	-1.26	
Median	-0.18*	0.11	-0.09	-0.98**	
G6 versus RoW					
Mean	-1.26***	-0.06	-0.91*	-2.74***	
Median	-0.30***	0.05	-0.27	-0.97***	
Panel C: Fama and French (1993) 3-factor model					
ALL					
Mean	1.34***	1.01***	1.29***	1.73***	0.71**
Median	0.46***	0.45***	0.51***	0.40***	-0.05
N	5880	1545	2737	1598	
Domestic					
Mean	1.39***	1.10***	1.31***	1.82***	0.72*
Median	0.48***	0.46***	0.57***	0.30***	-0.15
N	4142	1136	1912	1094	
CBA (ALL)					
Mean	1.21***	0.78***	1.24***	1.53***	0.75
Median	0.41***	0.32**	0.35***	0.62***	0.30
N	1738	409	825	504	
CBA (G6)					
Mean	0.85***	0.78**	0.88***	0.85*	0.07
Median	0.31***	0.46*	0.26*	0.38	-0.07
N	1165	287	538	340	
CBA (RoW)					
Mean	1.96***	0.77*	1.90***	2.94***	2.17**
Median	0.59***	0.23**	0.47***	0.77***	0.54
N	573	122	287	164	
Differentials					
Domestic versus CBA (ALL)					
Mean	0.17	0.32	0.08	0.29	

Median	0.07	0.14	0.22	−0.32	
Domestic versus CBA (G6)					
Mean	0.54**	0.32	0.43	0.97*	
Median	0.17	0.00	0.31	−0.08	
Domestic versus CBA (RoW)					
Mean	−0.57	0.33	−0.59	−1.12	
Median	−0.11	0.23	0.10	−0.47**	
G6 versus RoW					
Mean	−1.11***	0.01	−1.02*	−2.09**	
Median	−0.28*	0.23	−0.21	−0.38**	
Panel D: 1 year (12 months) post-merger BHARs					
ALL					
Mean	−1.99***	−3.74***	−5.10***	5.04***	8.78***
Median	−6.31***	−6.63***	−7.80***	−3.46***	3.17***
N	5876	1543	2735	1598	
Domestic					
Mean	−2.90***	−4.08***	−6.99***	5.47***	9.56***
Median	−7.23***	−8.17***	−9.41***	−2.91**	5.26***
N	4138	1134	1910	1094	
CBA (ALL)					
Mean	0.19	−2.80	−0.71	4.10	6.90*
Median	−4.04***	−3.75**	−4.02***	−4.28**	−0.53
N	1738	409	825	504	
CBA (G6)					
Mean	0.60	−4.23**	−2.49	9.58**	13.82***
Median	−3.22***	−3.58*	−4.17***	−1.18	2.40
N	1165	287	538	340	
CBA (RoW)					
Mean	−0.64	0.58	2.62	−7.25**	−7.83
Median	−5.82***	−4.98*	−3.94	−9.20***	−4.23
N	573	122	287	164	
Differentials					
Domestic versus CBA					

(ALL)					
Mean	−3.09**	−1.29	−6.28***	1.37	
Median	−3.19**	−4.41	−5.39***	1.37	
Domestic versus CBA (G6)					
Mean	−3.50**	0.15	−4.50**	−4.11	
Median	−4.01**	−4.59	−5.24**	−1.73	
Domestic versus CBA (RoW)					
Mean	−2.26	−4.66	−9.61**	12.72***	
Median	−1.41	−3.19	−5.47**	6.29**	
G6 versus RoW					
Mean	1.24	−4.81	−5.11	16.84***	
Median	2.60	1.40	−0.23	8.02**	
Panel E: 5 years (60 months) post-merger BHARs					
ALL					
Mean	−5.99***	−15.00***	−7.10***	4.66	19.66***
Median	−30.35***	−43.75***	−30.35***	−16.86***	26.90***
N	5845	1539	2719	1587	
Domestic					
Mean	−7.55***	−15.70***	−10.71***	6.40	22.10***
Median	−35.42***	−47.25***	−36.44***	−21.58***	25.66***
N	4118	1130	1899	1089	
CBA (ALL)					
Mean	−2.26	−13.06**	1.24	0.86	13.92*
Median	−18.37***	−32.79***	−13.96***	−11.85***	20.94***
N	1727	409	820	498	
CBA (G6)					
Mean	−1.41	−15.60**	4.39	1.45	17.05*
Median	−18.64***	−40.01***	−11.28***	−12.20***	27.81***
N	1159	287	536	336	
CBA (RoW)					
Mean	−3.98	−7.10	−4.70	−0.37	6.73

Median	−17.55***	−21.32***	−19.33***	−10.47	10.85*
N	568	122	284	162	
Differentials					
Domestic versus CBA (ALL)					
Mean	−5.30	−2.64	−11.95**	5.54	
Median	−17.05***	−14.46**	−22.48***	−9.73	
Domestic versus CBA (G6)					
Mean	−6.14	−0.10	−15.10**	4.95	
Median	−16.78***	−7.24	−25.16***	−9.39	
Domestic versus CBA (RoW)					
Mean	3.57	−8.60	−6.00	6.77	
Median	−17.87***	−25.93**	−17.11**	−11.11	
G6 versus RoW					
Mean	2.57	−8.50	9.10	1.83	
Median	−1.09	−18.69	8.05	−1.73	

The table presents 5-day ($t - 2$, $t + 2$) announcement period's cumulative abnormal returns (CAR) in percent measured as in Eq. 2, as well as the 1- and 5-year post-merger buy-and-hold abnormal returns (BHAR) in percent measured as in Eq. 3. Both the CARs and BHARs are reported by the domicile of targets (domestic, CBA, and CBA sub-groups) and equity market valuations conditions (EMVCs). The financial conditions are classified by the de-trended market P/E ratio of FTSE All-Share Market Index. EMVCs is depicted as announcement periods of depressed (Low-EMVCs), neutral (Neutral-EMVCs), and booming (High-EMVCs) markets). The table further reports acquirers' CAR by the state of EMVCs (HML EMVCs stands for High Minus Low EMVCs) and the differentials between acquirers' gains from domestic and foreign (CBA) deals. In Panel A the acquirers' CAR is estimated based on the adjusted market model (as shown in Eq. 1). In Panel B the acquirers' CAR is estimated based on the Carhart (1997) 4-factor model. In Panel C the acquirers' CAR is estimated based on the Fama and French (1993) 3-factor model. In Panel D and E the acquirers' BHAR for 1- and 5-years, respectively. The process of de-trending is discussed in Sect. 3.3. All variables are defined in the Appendix 1. The significance of the difference the means of two groups of acquirers is tested by using the t test of equality of means. The significance of the median is tested by using the 'Sign' test. The significance of the difference between the medians of two groups of acquirers is tested by using the Wilcoxon two-sample test. The number of deals (N) for each group is reported below the estimates of excess returns

***, **, * Significance level at 1, 5, and 10 %, respectively

4.2 EMVCs and acquirers' short- and long-run gains

Analysis of the acquirers' gains by the EMVCs at the time of the M&A announcement reveal that the gains increase monotonically with EMVCs (Table 3, Panels A, B and C). The average gains to all acquirers increase from 1.04 % for M&As announced during low-EMVC periods to 1.99 % during high-EMVC periods (Panel A), or from 1.09 % for M&As announced during low-EMVC periods to 2.34 % during high-EMVC periods (Panel B). These findings are consistent with Bouwman et al. (2009) who show that acquirers enjoy higher short-term gains from announcing bids in high- rather low-EMVC periods. This pattern holds for the analysis based on the de-trended industry P/E ratio, confirming that the results are robust to the methods of assessment of market conditions. This evidence is not sensitive to the choice of the method of payment and confirms the prediction that acquirers should gain more from bids announced at the time of high-EMVCs due to favorable investors' sentiment (optimism). Overall, the results support part of the first hypothesis (*H1a*) and provide additional evidence in support of the neoclassical theory of mergers.

Further analysis shows that EMVCs play a significant role in explaining the difference in acquirers' gains by their targets' domicile (differentials recorded in Panels A, B and C). In particular, acquirers experience similar abnormal from domestic versus foreign M&As (as a whole) regardless the level of EMVC in the UK at the time of M&A announcements. However, we find that the groupings of the countries of foreign targets according to the degree of correlation of the EMVCs between home and host markets plays a significant role. Specifically, acquirers' shareholders earn 1.48 % higher abnormal returns when they acquire targets in the domestic rather than G6 group of countries (Panel B) or 0.97 % higher abnormal returns (Panel C), when the deal is made during periods of high-EMVCs. As the EMVCs between the home and host (G6) countries are highly correlated (see Appendix 2), the impact of EMVCs between the merging firms markets remains trivial. However, M&As of foreign targets in the RoW countries: (a) earn similar abnormal returns to domestic target acquirers, and (b) enjoy significantly higher abnormal returns to acquirers in the G6 countries—especially during high-EMVC periods.[11] We argue that the EMVCs at home relative to those in host countries allow acquirers' managers to time the market and acquire targets at a discount, particularly in markets in which their stocks are likely to be more overvalued than their targets. Therefore, a possible reason for these findings is that the EMVCs in the RoW (G6) are lower (similar) to the home ones and hence British acquirers are able (unable) to acquire targets at a discount. Accordingly, this suggests that the null difference between domestic and foreign targets is balanced out by CBAs in the RoW countries, as CBAs in the G6 group are experiencing losses. These findings support the second and third hypotheses (*H2* and *H3*) relating to the level of EMVCs at home, relative to those in host countries, and the gains from domestic versus CBAs. Overall, the findings reported in Table 3 (Panels A–C) show that EMVCs interact with the domicile of targets in shaping the gains to acquirers.

Consistent with our first hypotheses (*H1a and H1b*) and previous studies (Bouwman et al. 2009), our results show that M&As announced during periods of high-EMVCs, while they yield significant gains in the short-run, in the long-run they either break-even or experience significant losses (1-year and 5-years post-merger BHARs, Panels D and E, respectively). For example, while M&As of foreign targets in the RoW group of countries made during periods of high-EMVCs in the UK yield 3.33 % abnormal returns in the short-run (Panel A), the same deals experience significant losses (−7.25 % monthly BHARs) during 1-year post-merger period (Panel D) or they break-even during the 5-years post-merger period (Panel D). Similarly acquirers of domestic targets (all acquirers) that are announced during period of high-EMVCs gain 2.01 % (1.99 %) significant gains in the short-run, in the long-run they break-even. Therefore, while periods of high-EMVCs are associated with high gains in the short-run due to investors' high sentiment or (over)optimism regarding the high expected returns of their investments, in the long-run there is a market correction and acquirers break-even or even suffer significant losses.

In summary, the results of the univariate analysis reported in Table 3 confirm that acquirers' short- and long-run gains are affected by the level of home-EMVCs at the time of M&A announcements, as well as the domicile of the target firm. This suggests that in making an M&A decision the managers of acquiring firms should consider the levels of EMVCs in the home market.

4.3 Gains of acquirers: a cross-sectional analysis

We further explore in a multivariate framework the impact of EMVCs at home, relative to EMVCs in host countries, along with several other known determinants that influence merger success, on the short- and long-run acquirers' abnormal returns. The results are reported in Table 4.

Table 4 Determinants of announcement period gains to acquirers: cross sectional analysis

Model: sample	(1) All	(2) All	(3) All	(4) CBA	(5) CBA	(6) CBA	(7) CBA	(8) CBA	(9) CBA	(10) CBA
High-EMVCs	0.008***	0.008***	0.003	0.009**	0.014**	0.002	0.008	0.134**	−0.037	−0.048
	(2.84)	(2.84)	(1.18)	(2.10)	(2.12)	(0.37)	(1.07)	(2.35)	(−0.62)	(−0.57)
Stock	0.009**	0.009**	−0.006	0.011	0.021*	0.007	0.021*	0.100	−0.177*	−0.159
	(2.00)	(2.02)	(−0.86)	(0.92)	(1.80)	(0.65)	(1.78)	(1.09)	(−1.72)	(−0.94)
CBA	0.001		0.002							
	(0.51)		(0.57)							
G6		−0.002								
		(−0.58)								
RoW		0.008*		0.011**	0.010*	0.008*	0.006	0.129***	0.134**	0.085
		(1.77)		(2.26)	(1.69)	(1.82)	(1.00)	(2.63)	(1.98)	(0.87)
High-EMVCs × stock			0.053***					−0.008	−0.150	−0.255
			(5.55)					(−0.004)	(−0.79)	(−1.01)
High-EMVCs × RoW						0.014*	0.018*	−0.319***	0.074	0.130
						(1.80)	(1.88)	(−2.98)	(0.53)	(0.75)
Effect. ex. rate	0.008***	0.008***	0.007***	0.001	−0.001	0.001	−0.001	0.006	0.039	0.029
	(2.58)	(2.97)	(2.84)	(0.28)	(−0.02)	(0.20)	(−0.02)	(0.18)	(0.72)	(0.41)
Diversifying	0.001	0.001	0.001	−0.001	0.002	0.002	0.002	−0.004	−0.116**	−0.171**
	(0.22)	(0.24)	(0.17)	(−0.26)	(0.35)	(0.37)	(0.34)	(−0.08)	(−2.09)	(−2.32)
Relative size	0.006***	0.006***	0.006***	0.005***	0.009***	0.006***	0.009***	0.016	0.031**	0.017
	(5.12)	(5.14)	(5.13)	(4.17)	(2.62)	(4.57)	(5.41)	(1.29)	(1.96)	(0.77)
Acquirer age	0.001	0.001	0.001	−0.001	−0.002	−0.001	−0.002	−0.001	0.060**	0.079***
	(0.54)	(0.58)	(0.53)	(−0.61)	(−0.69)	(−0.39)	(−0.69)	(−0.06)	(2.45)	(2.70)
Acquirer MTBV	−0.001	−0.001	0.001	−0.001**	−0.001**	−0.001*	−0.001**	−0.003***	0.001	−0.001
	(−0.08)	(−0.11)	(0.29)	(−2.03)	(−2.25)	(−1.93)	(−2.17)	(−4.98)	(0.39)	(−0.67)
Unlisted target	0.033***	0.033***	0.033***	0.015**	0.018**	0.013*	0.019**	0.029	0.043	0.075
	(6.28)	(6.27)	(6.32)	(2.01)	(1.96)	(1.71)	(2.05)	(0.49)	(0.60)	(0.84)
Liquidity	0.023***	0.022***	0.022***	0.002	−0.039**	0.003	−0.038**	0.276**	0.023	−0.119
	(2.82)	(2.77)	(2.78)	(0.09)	(−2.25)	(0.19)	(−2.24)	(2.24)	(0.14)	(−0.61)
Political stability					0.009		0.009	0.002	0.004	−0.094
					(1.00)		(0.98)	(0.03)	(0.05)	(−0.72)
Rule of law					−0.021**		−0.021**	−0.056	−0.053	0.061
					(−2.00)		(−2.07)	(−0.51)	(−0.38)	(0.35)
Crisis	−0.017**	−0.017**	−0.017**	0.015	0.012	0.012	0.011	−0.006	0.119	0.274
	(−2.10)	(−2.12)	(−2.17)	(1.40)	(1.03)	(1.17)	(0.96)	(−0.09)	(1.04)	(1.48)
Crisis × CBA	0.018*	0.017*	0.017*							
	(1.88)	(1.72)	(1.83)							
Crisis × stock	0.188***	0.188***	0.195***							
	(17.14)	(17.14)	(13.93)							
Crisis × diversified	−0.001	−0.001	−0.001	−0.022	−0.011	−0.015	−0.011	−0.023	0.052	−0.014
	(−0.07)	(−0.02)	(−0.01)	(−1.40)	(−0.74)	(−1.13)	(−0.74)	(−0.30)	(0.38)	(−0.07)
Crisis × RoW				−0.009	−0.020	−0.006	−0.016	−0.262***	−0.300**	−0.359*
				(−0.66)	(−1.48)	(−0.45)	(−1.22)	(−3.05)	(−2.01)	(−1.80)
Intercept	−0.013	−0.013	−0.011	0.019	0.057**	0.020	0.057**	0.072	−0.388*	−0.513*
	(−0.95)	(−0.96)	(−0.84)	(1.09)	(2.16)	(1.08)	(2.20)	(0.36)	(−1.78)	(−1.76)
F-stat	10.19***	9.84***	11.45***	2.70***	3.74***	2.81***	3.62***	3.10***	1.51*	1.29
Adj. R^2 (%)	2.71	2.78	3.21	2.30	5.45	2.40	5.61	5.14	2.57	2.21
Mean VIF	1.57	1.54	1.60	1.65	2.12	1.49	2.16	2.13	2.12	2.11

As indicated by the intercept 'β_1' of models 5 and 7 (models 1–4 and 6), the average acquirer earns significant gains (breaks-even) in the short-run after controlling for the impact of several merging firms- and deal-specific factors, as well as the impact of EMVCs. Models 1–2 show that acquirers enjoy significant gains from deals announced during high-EMVC, despite the domicile of the target firm. This finding supports our first hypothesis (*H1a*). Model 3 further shows that M&As made during periods of high-EMVCs at home and financed with stock are associated with the highest gains to acquirers' shareholders. This provides great further support to our first hypothesis (*H1a*), the empirical findings of Bouwman et al. (2009) and the theoretical findings of Shleifer and Vishny (2003). Among the variables of interest, the strength of the domestic currency (models 1–3) exerts a positive effect on acquirers' gains, which supports our fourth hypothesis (H_4). The favorable effect of the strength of the domestic currency is not limited to acquirers engaged in foreign targets, but it is also important for acquirers of domestic targets. Higher gains from domestic target M&As are possible because the strength of domestic currency reflects expected growth in the economy and investors are willing to pay a higher price for stocks of growing/expanding firms. Hence, the markets react favorably to announcements of (also) domestic M&As.

Model 2 shows that UK acquirers of foreign targets in the RoW countries enjoy significant gains in the short-run, which partly supports our second hypothesis (H_2). Moreover, models 4–6 show that UK foreign acquirers of targets in the RoW rather than G6 countries enjoy significant gains in the announcement period, which is more pronounced if M&As are announced during high-EMVCs periods at home. This supports our third hypothesis (H_3), which predicts that M&As with targets in countries with EMVCs being lower than those at home yield higher gains to acquirers' shareholders. In contrast, foreign target acquirers in G6 countries experience significant losses or break even in the announcement period relative to domestic target acquirers. We argue that this is the outcome of the high correlation of EMVCs between the UK and G6 countries (Appendix 2), which ultimately diminishes the impact of EMVCs in such deals. Hence our findings on domestic versus G6-only CBAs further confirm the findings reported by Moeller and Schlingemann (2005) whereas our results on domestic vs all CBAs confirm Barbopoulos et al. (2012).[12]

The findings also show the importance of firm and deal- and merging firms-specific features in determining the gains to acquirers' shareholders in the domestic versus foreign target acquisitions. Specifically, the relative size of the deal (across models 1–7), stock payments (models 1, 2, 5 and 7), target listing status (i.e. unlisted, across models 1–7), liquidly of the acquiring firm (models 1–3) have a significant and positive effect on acquirers' gains. The positive effect of the relative deal's size is consistent with previous studies, which conclude that the acquiring firm's abnormal returns increase with the target's size relative to the acquirer's size (Asquith et al. 1983). The case of the evidence of positive gains from stock deals is similar to the findings of Draper and Paudyal (2006). In particular, provided that the UK market is overpopulated by unlisted target deals (Draper and Paudyal 2006), stock financing of such deals leads to higher acquirers' abnormal returns in the short-run (Chang 1998). The estimates also suggest that foreign acquirers with higher market-to-book-value gain more from acquisitions (models 4–7). Finally, our estimates show (models 1–3) that while the market reaction to M&A announcements is lower during the financial crisis, all else equal, acquirers engaged in CBAs or financing M&As with stock, during the 2007 financial crisis, enjoy significant short-run abnormal returns.

As indicated by the intercept 'β1β1' of models 9 and 10 (model 8) the average acquirer experiences significant losses (breaks-even) in the long-run from M&As after controlling for the impact of several merging firms- and deal-specific factors, as well as the impact of EMVCs.[13]Moreover, models 9 and 10 highlight that acquirers engaging in M&As during periods of high-EMVCs do not enjoy abnormal returns within 3- and 5-years following the M&A announcement. This is also evident when stock is involved in the M&A financing. Our results are consistent with our hypothesis (*H1b)* and provide some support to the empirical findings of Bouwman et al. (2009). Other results also indicate that diversified acquirers and acquirers bidding for foreign targets from the RoW group of countries during the crisis, experience losses in the post-merger period.

5 CONCLUSION

We examine whether the short- and long-run abnormal returns of the shareholders of UK firms that engaged in domestic and foreign acquisitions is affected by the home EMVCs at the time of M&A announcements, as well as the relative EMVCs between the merging firms' countries in CBAs. We find that acquirers enjoy significantly higher announcement period gains from deals announced during high- rather than low-EMVC at home, regardless of the domicile of the target firm. However, in the post-merger period, M&As announced during periods of high-EMVC at home break-even or experience significant losses. We further show that foreign target acquirers enjoy greater gains from deals made during high-EMVCs at home only when the EMVCs between the merging firms' countries are less likely to be correlated, which falls in the category of the target being in the RoW rather than in the G6 group of countries. Our findings also show that during high-EMVCs at home, domestic deals yield higher gains than foreign deals in the G6 countries. Provided that EMVCs in G6 countries are highly correlated to the ones at home (Appendix 2), UK acquirers are less able to extract any gains from CBAs due to the high-EMVCs in the host countries. Hence, in the absence of the impact of EMVCs, the gains of domestic deals versus CBAs confirm those reported by Moeller and Schlingemann (2005) and Barbopoulos et al. (2012).

However, we find that acquirers of domestic targets, and also foreign targets in the RoW countries, realize similar announcement period gains regardless of the level of EMVCs at home. Moreover, within CBAs, deals of targets in the RoW countries outperform those in the G6 countries when announced during periods of high-EMVCs at home. However, this is reversed in the post-announcement period. Provided that the correlation of EMVCs between the UK and the RoW (G6) countries is low (high), UK acquirers are able (unable) to extract significant gains from foreign deals in the RoW (G6) countries that made during high-EMVCs at home, which is reflected in the short-run acquirers abnormal returns. This is possible due to the low EMVCs in the RoW group of countries (as the correlation of EMVCs between the UK and RoW is low). Hence, managers are able to time the market and acquire targets in the RoW countries at a discount, particularly in markets in which their stocks are likely to be more overvalued than their targets.

Overall, our results suggest that acquirers' managers aiming to maximize their shareholders' wealth can extract signals from EMVCs of the home, relative to host countries, and carefully pick the timing of their M&As announcements in both the domestic and foreign market.

FOOTNOTES

1 These studies are mainly concerned with the sensitivity of acquirers' gains to deal and firm-specific characteristics, such as the relative size of the deal (Asquith et al. 1983), the method of payment (Travlos 1987), the listing status of targets (Draper and Paudyal 2006), the size of the acquirer (Moeller et al. 2004), and the growth opportunities of the acquirer (Sudarsanam and Mahate 2003).

2 See for example, Shleifer and Vishny (2003) and Rhodes-Kropf and Viswanathan (2004).

3 Healy and Palepu (1993) portray the UK as a leader in CBA deals accounting for roughly 30 % of global activity in the late 1980 s. Similarly, data available from the United Nations (UNCTAD 2000) portray the UK as holding the same proportion of CBA activity by the late 1990s.

4 See for example, Rhodes-Kropf et al. (2005), and Bouwman et al. (2009).

5 See for example, Doukas and Travlos (1988), Lowinski et al. (2004), Gregory and McCorriston (2005), Moeller and Schlingemann (2005), and Barbopoulos et al. (2012).

6 A large number of bidder firms are typically engaged in multiple acquisitions and takeover bids which can reduce significantly the number

of sample firms having "sterile" risk parameters, potentially biasing our findings. Never the less, for short event windows, the use of traditional and well established methods of estimating abnormal returns does not improve significantly the accuracy of abnormal returns (Brown and Warner 1980).

7 We employ alternative estimation methods for calculating the acquiring firm's abnormal returns based on the estimation of the model parameters of the 1-factor CAPM, the Fama and French (1993) 3-factor model, and the Carhart (1997) 4-factor model. Our results remain unaltered.

8 This date corresponds to the first trading day of a listed firm. Since the disclosure requirements of listed firms are much more stringent and systematic, the listing period is a good proxy measure of the level of information in the public domain.

9 See www.bankofengland.co.uk. Moreover, the first difference of the effective exchange rate index (i.e. changes is the value) of Pound Sterling does not have any trend and hence its de-trending is not required.

10 Anderson et al. (2011) report mixed evidence on the impact of corporate diversification on firm risk, though, they suggest that on average diversification does not decrease firm risk.

11 This pattern holds when the 9 industry P/E ratios are employed to assess the market valuation conditions.

12 Moeller and Schlingemann (2005) show that US acquires enjoy higher gains from domestic deals rather than foreign target ones. However, Meng and Sutton (2016) show that the listing status of the target drives the cross-border effect in two opposite directions: acquirers of private targets fare worse in CBAs, while acquirers of public targets experience significantly higher gains in acquisitions of foreign targets. Barbopoulos et al. (2012) show that UK bidders' gains from domestic vs. foreign target deals are not significantly different. Similarly Lowinski et al. (2004) show that Swiss bidders' gains from domestic vs. foreign target deals are not significantly different.

13 Allen and Soongswang (2006) show that Thai takeovers result in significant negative abnormal returns in the long-run. Similarly, Du et al. (2016) report negative long-term abnormal returns for Chinese bidders targeting firms across Europe, North America, Australia and Asia.

ACKNOWLEDGMENTS

We thank Sandra Betton, Krishna Paudyal, Sudi Sudarsanam and the participants of the 2015 FEBS conference (Nantes, France) for helpful comments and suggestions.

APPENDIX 1

Table 5 Variables definitions

Variable name	Description	Data source
All deals (ALL)	Refers to the entire sample analyzed in this paper	SDC
Acquirer age (AGE)	Number of days between the day when the acquirer is first recorded on DataStream and bid's announcement day	DataStream
Periods of high equity market valuation conditions (high-EMVCs)	Dummy = 1if the M&A deal is announced during period of high equity market valuations conditions (EMVCs) and = 0 otherwise. The financial conditions are classified by the de-trended market P/E ratio of FTSE All-Share Market Index. See the text for the process of de-trending (Sect. 3.3)	DataStream & SDC
Cash payment (CASH)	Dummy = 1 when the deal is financed with 100 % cash	SDC
Cross-border acquisitions (CBAs)	Dummy = 1 with a UK acquirer and non-UK target and = 0 with both acquirer and target are UK firms (=Domestic [DOM])	SDC
Crisis	Crisis = 1 1if the M&A deal is announced between 01 July 2007 and 30 September 2008 and = 0 otherwise.	SDC
Periods of low equity market valuation conditions	Dummy = 1 if M&A deal is announced during period of low equity market valuations conditions (EMVCs) and = 0 otherwise. The financial conditions are classified by the de-trended market P/E ratio of FTSE All-Share	DataStream & SDC

(low-EMVCs)	Market Index. See the text for the process of de-trending (Sect. 3.3)	
Deal value (DV)	Bid transaction value, in millions pounds	SDC
Diversifying (DIV)	Dummy = 1 when acquirer and target are based in different 2-digit SIC industries and = 0 when both share the same 2-digit industry (=Focused [FOC])	SDC
G6 Group of Countries (G6)	G6 = 1 when a UK acquirer acquires a target that based in the rest of G7 countries (=G7-UK) and = 0 when a UK acquirer acquires a target that based in the rest of the work (RoW=World-G7)	SDC
Cash ratio (liquidity)	Measured by the ratio of total cash and equivalent to the total assets. Since these are annual ratios M&A deals announced before (after) June are matched with the ratio of the previous (same) year	DataStream
Mixed payments (MIXED)	Dummy = 1 when the financing process of the deal includes a mixture of cash, stock, and other methods of payment	SDC
Market-to-book value (MTBV)	Market-to-Book Value of acquirer equity at four weeks and book value of equity from the most recent accounting statement prior to bid announcement	DataStream
Market value (MV)	Acquirer's market value of equity at four weeks prior to bid announcement, in million pounds	DataStream
Periods of neutral equity market valuation conditions (neutral-EMVCs)	Dummy = 1 when the M&A deal is announced during period of neutral equity market valuations conditions (EMVCs) and = 0 otherwise. The financial conditions are classified by the de-trended market P/E ratio of FTSE All-Share Market Index. See the text for the process of de-trending (Sect. 3.3)	DataStream & SDC
Acquirer premium (premium)	Takeover premium from Thomson Financial SDC computed as the difference between the offer price and the target's stock price four weeks before the acquisition announcement	SDC

	divided by the latter; values beyond the range of [0, 2] are winsorized following Officer (2003)	
Private (PRV)	Dummy = 1 if target is private and = 0 otherwise	SDC
Public (PUB)	Dummy = 1 if target is public/listed and = 0 otherwise	SDC
RoW group of countries (RoW)	RoW = 1 when a UK acquirer acquires a target that based outside the G6 countries (=World-G7) and = 0 when a UK acquirer acquires a target that based in the rest of G7 countries (G6=G7-UK)	SDC
Relative size (RS)	Ratio of DV to MV	DataStream & SDC
Strong effective exchange rate (SEER)	Dummy = 1 when M&A deal is announced during period of strong Effective Exchange Rate (EER), or strong domestic currency, and = 0 otherwise	Bank of England
Stock payments (STOCK)	Dummy = 1 when the deal is financed with 100 % stock exchange	SDC
Subsidiary (SUB)	Dummy = 1 if target is subsidiary firm and = 0 otherwise	SDC
Unlisted (UNL)	Dummy = 1 if target is unlisted i.e. private or subsidiary, and = 0 otherwise	SDC
Rule of law (RL)	Rule of law is an index measuring the quality and level of law and order in a country obtained from World Bank's International Country Risk Guide (ICRG)	World Bank-ICRG
Political stability (PS)	Rule of law is an index measuring the quality and level of political stability and tensions in a country obtained from World Bank's International Country Risk Guide (ICRG)	World Bank - ICRG

The table describes and defines the variables used in the paper, as well as it indicates the data source. SDC is Thomson-Reuters SDC M&A database. With a dummy variable, a sample observation without the value of 1 has a value of 0

APPENDIX 2

Table 6 Correlations among equity market valuations conditions (EMVCs) in G7 countries

	US	Germany	France	Australia	Canada	Japan	UK
US	1.00						
Germany	0.62	1.00					
France	0.68	0.79	1.00				
Australia	0.72	0.62	0.60	1.00			
Canada	0.68	0.60	0.54	0.79	1.00		
Japan	0.17	0.26	0.42	0.32	0.17	1.00	
UK	0.86	0.72	0.77	0.80	0.71	0.41	1.00

The table shows the correlation coefficients among the G7 countries' equity market indices. The P/E ratio is used to classify each of the equity market indices as low (depressed), neutral and high (booming), as in Bouwman et al. (2009)

REFERENCES

1. Allen DE, Soongswang A (2006) Post-takeover effects on thai bidding firms: are takeovers in the bidder's interests? Rev Pac Basin Financ Mark Polic 9:509–531
2. Anderson RI, Stowe JD, Xing X (2011) Does corporate diversification reduce firm risk? Evidence from diversifying acquisitions. Rev Pac Basin Financ Mark Polic 14:485–504
3. Asquith P, Bruner RF, Mullins DW (1983) The gains to bidding firms from merger. J Financ Econ 11:121–139
4. Baker M, Wurgler J (2006) Investor sentiment and the cross-section of stock returns. J Financ 61:1645–1680
5. Barber MB, Lyon DJ (1997) Detecting long-run abnormal stock returns: the empirical power and specification of test statistics. J Financ Econ 43:341–372
6. Barbopoulos LG, Paudyal K, Pescetto G (2012) Legal systems and gains from cross-border acquisitions. J Bus Res 65:1301–1312
7. Barry C, Brown S (1985) Differential information and security market equilibrium. J Financ Quant Anal 1:407–422
8. Bouwman C, Fuller K, Nain A (2009) Market valuations and acquisition quality: empirical evidence. Rev Financ Stud 22:633–679
9. Bris A, Cabolis C (2008) The value of investor protection: firm evidence from cross-border mergers. Rev Financ Stud 21:605–648
10. Brown S, Warner J (1980) Measuring security price performance. J Financ Econ 8:205–258

11. Carhart MM (1997) On persistence in mutual fund performance. J Financ 52:57–82
12. Chang S (1998) Takeovers of privately held targets, methods of payment, and bidder returns. J Financ 53:773–784
13. di Giovanni J (2005) What drives capital flows? The case of cross-border M&A activity and financial deepening. J Int Econ 65:127–149
14. Doidge C, Karolyi GA, Stulz RM (2007) Why do countries matter so much for corporate governance? J Financ Econ 86:1–39
15. Doukas J, Travlos NK (1988) The effect of corporate multinationalism on shareholder wealth: evidence from international acquisitions. J Financ 43:1161–1175
16. Draper P, Paudyal K (2006) Acquisitions: private versus Public. Eur Financ Manag 12:57–80
17. Draper P, Paudyal K (2008) Information asymmetry and bidders gains. J Bus Financ Account 35:376–405
18. Du M, Boateng A, Newton D (2016) The impact of state ownership, formal institutions and resource seeking on acquirers' returns of Chinese M&A. Rev Quant Financ Acc 47:159–178
19. Faccio M, McConnell JJ, Stolin D (2006) Returns to acquirers of listed and unlisted targets. J Financ Quant Anal 41:197–220
20. Fama EF, French KR (1993) Common risk factors in the returns on stocks and bonds. J Financ Econ 33:3–56
21. Fontanella-Khan J, Massoudi A, Rennison J (2015) Record valuations drive 2015 M&A boom. Financial Times June, 29
22. Froot KA, Stein JC (1991) Exchange rates and foreign direct investment: an imperfect capital markets approach. Q J Econ 106:1191–1217
23. Fuller K, Netter J, Stegemoller M (2002) What do returns to acquiring firms tell us? Evidence from firms that make many acquisitions. J Financ 57:1763–1793
24. Gort M (1969) An economic disturbance theory of mergers. Q J Econ 83:624–642
25. Gregory A, McCorriston S (2005) Foreign acquisitions by UK limited companies: short- and long-run performance. J Empir Financ 12:99–125
26. Harris RS, Ravenscraft D (1991) The role of acquisitions in foreign direct investment: evidence from the US Stock Market. J Financ 46:825–844
27. Healy PM, Palepu KG (1993) International corporate equity acquisitions: who, where and why? Foreign direct investment. University of Chicago Press, Chicago, pp 231–250
28. Huang R, Ritter JR (2009) Testing theories of capital structure and estimating the speed of adjustment. J Financ Quant Anal 44:237–271
29. Jansen CJ, Stokman CJ (2004) Foreign direct investment and international business cycle comovement. ECB working paper series, no. 401
30. Johnson S, Boone P, Breach A, Friedman E (2000) Corporate governance in the Asian financial crisis. J Financ Econ 58:141–186
31. Kang J (1993) The international market for corporate control: mergers and acquisitions of US firms by Japanese firms. J Financ Econ 34:345–371
32. Kusewitt JB (1985) An explanatory study of strategic acquisition factors relating to performance. Strateg Manage J 6:151–169
33. La Porta R, Lopez-de-Silanes F, Shleifer A, Vishny R (2002) Investor protection and corporate valuation. J Financ 57:1147–1170
34. Lowinski F, Schiereck D, Thomas TW (2004) The effect of cross-border acquisitions on shareholder wealth—evidence from Switzerland. Rev Quant Financ Account 22:315–330
35. Meng Y, Sutton NK (2016) Is the grass on the other side greener? Testing the cross-border effect for U.S. acquirers. Rev Quant Financ Account. doi:10.1007/s11156-016-0573-1
36. Moeller BS, Schlingemann PF (2005) Global diversification and bidder gains: a comparison between cross-border and domestic acquisitions. J Bank Financ 29:533–564

37. Moeller SB, Schlingemann FP, Stulz RM (2004) Firm size and the gains from acquisitions. J Financ Econ 73:201–228
38. Morck R, Yeung B, Yu W (2000) The information content of stock markets: why do emerging markets have synchronous price movements? J Financ Econ 58:215–260
39. Nelson RL (1959) Merger movements in American industry. Princeton University Press, Princeton
40. Officer M (2003) Termination fees in mergers and acquisitions. J Financ Econ 69:431–467
41. Rau RP, Stouraitis A (2011) Patterns in the timing of corporate event waves. J Financ Quant Anal 46:209–246
42. Rhodes-Kropf M, Viswanathan S (2004) Market valuation and merger waves. J Financ 59:2685–2718
43. Rhodes-Kropf M, Robinson DT, Viswanathan S (2005) Valuation waves and merger activity: the empirical evidence. J Financ Econ 77:561–603
44. Rosen R (2006) Merger momentum and investor sentiment: the stock market reaction to merger announcements. J Bus 79:987–1017
45. Shleifer A, Vishny RW (2003) Stock market driven acquisitions. J Financ Econ 70:295–311C
46. Stambaugh RF, Yu J, Yuan Y (2012) The short of it: investor sentiment and anomalies. J Financ Econ 104:288–302
47. Sudarsanam PS, Mahate AA (2003) Glamour acquirers, method of payment and post-acquisition performance: the UK evidence. J Bus Finan Account 30:299–341
48. Toxvaerd F (2008) Strategic merger waves: a theory of musical chairs. J Econ Theory 140:1–26
49. Travlos GN (1987) Corporate takeover bids, methods of payment, and bidding firms stock returns. J Financ 42:943–963
50. UNCTAD (2000) Transnational corporations, agricultural production and development. UNCTAD, New York
51. White H (1980) A heteroskedasticity-consistent covariance matrix estimator and a direct test for heteroskedasticity. Econometrica 48:817–838
52. World Investment Report (2009) Transnational corporations, agricultural production and development. UNCTAD, New York
53. Yu J, Yuan Y (2011) Investor sentiment and the mean–variance relation. J Financ Econ 100:367–381

CHAPTER 8

Option Pricing Applications of Quadratic Volatility Models

Srimantoorao S. Appadoo[1*], Aerambamoorthy Thavaneswaran[2], Saman Muthukumarana[2]

[1]*Department of Supply Chain Management, University of Manitoba, Winnipeg, Canada*

[2]*Department of Statistics, University of Manitoba, Winnipeg, Canada*

ABSTRACT

Recently there has been a surge of interest in higher order moment properties of time varying volatility models. Various GARCH-type models have been developed and successfully applied in empirical finance. Moment properties are important because the existence of moments permit verification of how well theoretical models match stylized facts such as fat tails in most financial data. In this paper, we consider various types of random coefficient autoregressive (RCA) models with quadratic generalized autoregressive conditional heteroscedasticity (GARCH) errors and study the moments, mean, variance and kurtosis. We also consider the Black-Scholes model with RCA GARCH volatility and show that these moments can be used to evaluate the call price for European options.

Keywords: Garch Processes; RCA Models; Garch Models; Time varying volatility; Kurtosis

1. INTRODUCTION

It is well-known that many financial time series such as stock returns exhibit leptokurtosis and time-varying volatility [1]. The generalized autoregressive conditional heteroscedasticity (GARCH) and the random coefficient autoregressive (RCA) models have been extensively used to capture the time-varying behaviour of the volatility. Studies using GARCH models commonly assume that the time series is conditionally normally distributed; however, the kurtosis implied by the normal GARCH tends to be lower than the sample kurtosis observed in many time series Bollerslev [1]. Thavaneswaran et al. [2] use an ARMA representation to derive the kurtosis of various classes of GARCH models such as power GARCH, nonGaussian GARCH, non-stationary and random coefficient GARCH. Recently, Thavaneswaran et al. [3], Appadoo et al. [4] have extended the results to stationary RCA processes with GARCH errors and Paseka et al. [5] further extended the results to RCA processes with stochastic volatility (SV) errors.

Leptokurtosis is commonly observed in financial time series, as well as in currency and commodity markets. The opening and closure of the markets, time-of-the-day and day-of-the-week effects, weekends and vacation periods cause changes in the trading volume that translates into regular changes in price variability. Financial, currency, and commodity data also respond to new information entering into the market, which usually have large kurtosis. Recently, there has been growing interest in using volatility models [3,4]. Most of the studies use GARCH models with dummy variables in the volatility equation, and a few of them have been extended to a more flexible form such as the RCA GARCH. However, even though much research has been performed on volatility models applied to market data such as stock returns, more general specifications accounting for RCA with GARCH errors have been little explored. First we derive the kurtosis of a simple time series model with behaviour in the mean. Then we introduce various classes of RCA GARCH models and study the moments and discuss applications in option pricing. We extend the results for RCA GARCH volatility models to RCA quadratic GARCH models. The RCA GARCH model is appropriate for time series where significant autocorrelation exists. Option pricing with RCA model with quadratic GARCH errors is also discussed in some detail. The moments derived for the RCA GARCH volatility models provide more accurate estimates of market data behaviour and help investors, decision makers, and other market participants develop improved trading strategies. The rest of the paper is organized as follows. In rest of Section 1, we present results on standard GARCH models. These results are interesting for their own sake. In Section 2, we derive the higher order moments of some RCA models with GARCH errors, and in Section 3 we discuss some option pricing applications with RCA models with GARCH errors.

GARCH Models

Consider the general class of GARCH (P,Q) model for the time series y_t, where

$$\varepsilon_t = \sqrt{h_t} Z_t, \quad (1.1)$$

$$h_t = \omega + \sum_{i=1}^{P} \alpha_i \varepsilon_{t-i}^2 + \sum_{j=1}^{Q} \beta_j h_{t-j} \quad (1.2)$$

where Z_t is a sequence of independent, normally distributed random variables with zero mean, unit variance. Let ut = ?^2_t - h_t be the martingale difference and let σ_u^2 be the variance of u_t, (2.12) and (1.2) could be written as:

$$\varepsilon_t^2 - u_t = \omega + \sum_{i=1}^{P} \alpha_i \varepsilon_{t-i}^2 + \sum_{j=1}^{Q} \beta_j h_{t-j} \quad (1.3)$$

$$\left[1 - \sum_{i=1}^{P} \alpha_i B^i - \sum_{j=1}^{Q} \beta_j B^j \right] \varepsilon_t^2 = \omega - \sum_{j=1}^{Q} \beta_j B^j u_t, \quad (1.4)$$

$$\Phi(B)\varepsilon_t^2 = \omega + \beta(B) u_t \quad (1.5)$$

where

$$\Phi(B) = 1 - \sum_{i=1}^{R} \Phi_i B^i,\ \Phi_i = (\alpha_i + \beta_i),\ \beta(B) = 1 - \sum_{j=1}^{Q} \beta_j B^j$$

and R = *max*(*P*,*Q*).We shall make the following stationarity assumptions for ε_t^2 which has an ARMA(R,*Q*) representation.All the zeroes of the polynomial Φ(B) lie outside of the unit circle. $\sum_{i=0}^{\infty} \Psi_i^2 < \infty$ where the $\Psi_i' s$

are obtained from the relation $\psi(B)\Phi(B)$ with

$\Psi(B)=1+\sum_{i=1}^{\infty}\Psi_i B^i$. The assumption sensure that the $u_t's$ are uncorrelated with zero mean and finite variance and that the ε_t^2 process is weakly stationary. In this case, the autocorrelation function of ε_t^2 will be exactly the same as that for astationary ARMA(R, Q) model. For any random variable ε with finite fourth moments, the kurtosis defined by $\frac{E(\varepsilon-\mu)^4}{\left[Var(\varepsilon)\right]^2}$ and if the process $\{Z_t\}$

is normal then the process $\{\varepsilon_t\}$ defined by equations (1.3) and (1.4) is called a normal GARCH (p, q) process. The kurtosis ofthe GARCHprocess is denoted by $k^{(\varepsilon)}$ when it exists.

2. RANDOM COEFFICIENT VOLATILITY MODELS

Consider the class of random coefficient autoregressive (RCA) models defined by allowing random additive perturbations of the autoregressive (AR) coefficients of ordinary AR models. That is, we assume that the process y_t is given by,

$$y_t-\sum_{i=1}^{p}\left(\phi_i+b_i(t)\right)y_{t-i}=e_t \tag{2.1}$$

where the parameters θ_i, i = 2, ⋯, p, are assumed to be known, e^t and $b_i(t)$ are zero mean square integrable independent processes and the variances are denoted by σ_e^2 and σ_b^2. $b_i(t)\ (i=1,2,\cdots)'s$ are independent of e_t and y_{t-1} and may be thought of as incorporating structural changes. In order to motivate nonlinear forecasts for nonlinear models, we consider a class of estimating functions of the form $g_n=\sum_{i=2}^{n}b_{t-i}h_t$ [6], where $h_t=y_t-E\left[y_t\middle|F_{t-1}^y\right]=y_t-\sum_{i=1}^{p}\phi y_{t-i}$ and b_{t-1} is a function of $y_1, y_2, .., y_{t-1}$ and possibly the known parameters $\phi_1,\cdots,\phi_p$ (i.e. We assume that the fitted model is available). If we restrict ourselves to a class of estimating functions of the above form then we can forecast the future value of y_{n+1} based on the observed values

$y_1, y_2, \cdots, y_n$ as $\hat{y}_n(1) = E\left[y_{n+1} \mid y_n, y_{n-1}, \cdots\right]$. That iswhether we have an AR(p) model or RCA(p) model we will get the same linear predictor of y_{n+1}. However, for the RCA model under consideration, we have

$E\left[y_t \mid F_{t-1}^y\right] = \sum_{i=1}^{p} \phi y_{t-i}$ and $\operatorname{var}\left[y_t \mid F_{t-1}^y\right] = \sigma_\varepsilon^2 + \sum_{i=1}^{p} y_{t-i}^2 \sigma_b^2$.

Thus, the conditional variance is a nonlinear function and hence the RCA model may be viewed as a non-linear time series model. Nicholls and Quinn [7] studied linear as well as some nonlinear (proposed) forecast by fitting a nonlinear (RCA) model for the classical lynx cycle data. Using heuristic reasoning they proposed a nonlinear forecast and $\hat{y}_{n+1} = \operatorname{sgn}(\phi_1 \hat{y}_n)\left[\phi_1^2 \hat{y}_n^2 + \sigma_\varepsilon^2\right]^{\frac{1}{2}}$ they showed empirically that the forecast y_{n+1} is a better predictor (having smalller forecast errors when compared with the actual observations) than the linear forecast for the lynx data. It is of interest to note that by defining

$h_t = y_t^2 - E\left[y_t^2 \mid F_{t-1}^y\right]$, the optimal forecast for y_{n+1} can be obtained as

$$y_n^*(1) = \left[E\left[y_t^2 \mid F_{t-1}^y\right]\right]^{\frac{1}{2}} = \operatorname{sgn}\left(\phi y_n^2\right)\left[\psi_1^2 + \sigma_n^2 + \sigma_\varepsilon^2\right]^{\frac{1}{2}}.$$

That is, the estimating function method can be used to obtain a nonlinear forecast for a nonlinear models by considering a class of elementary martingale estimating functions generated by nonlinear functions of the observations. Using a similar argument we could also obtain forecasts for various class of GARCH models, see Thavaneswaran and Heyde [6] for details. The main message is RCA models could be used to improve the forecasting performance of stochastic volatility models.

Lemma 2.1. When Z_t is a standard normal random variable such that $Zt \sim N(0,1)$ then,

$E\left|Z_t\right|^\delta = \frac{2^{\frac{\delta}{2}}}{\sqrt{\pi}} \Gamma\left(\frac{\delta+1}{2}\right)$ Now if $Z_t \sim N\left(0, \sigma_Z^2\right)$ then

$$E|Z_t|^\delta = \frac{(2\sigma_Z^2)^{\frac{\delta}{2}}}{\sqrt{\pi}}\Gamma\left(\frac{\delta+1}{2}\right).$$ As a general case for even powers we have $$E(Z^{2\delta}) = \sigma^{2\delta}\frac{2^\delta}{\sqrt{\pi}}\Gamma\left(\delta+\frac{1}{2}\right).$$

2.1. RCA Models

Random coefficient autoregressive time series were introduced by Nicholls and Quinn [10] and some of their properties have been studied recently by Thavaneswaran [3]. RCA models exhibiting long memory properties have been considered in Leipus and Sugailis [8]. A sequence of random variables $\{y_t\}$ is called an RCA (1) time series if it satisfies the equations

$$y_t = (\phi + b_t)y_{t-1} + \varepsilon_t \qquad t \in Z,$$

where Z denotes the set of integers and

1) $$\begin{pmatrix} b_t \\ \varepsilon_t \end{pmatrix} \sim \left(\begin{pmatrix} 0 \\ 0 \end{pmatrix}, \begin{pmatrix} \sigma_b^2 & 0 \\ 0 & \sigma_\varepsilon^2 \end{pmatrix}\right),$$

2) $\phi^2 + \sigma_b^2 < 1$

The sequences $\{b_t\}$ and $\{\varepsilon_t\}$ respectively, are the errors in the model.

Theorem 2.1. Let $\{y_t\}$ be a modified RCA (1) time series with an absolute value random coefficient satisfying conditions (1) and (2). The modified RCA (1) model is given by

$$y_t = (\phi + |b_t|)y_{t-1} + \varepsilon_t \tag{2.2}$$

$\varepsilon_t \sim N(0, \sigma_\varepsilon^2), b_t \sim N(0, \sigma_b^2)$ Then we have the following

1) $$E(y_t) = 0,\ E(y_t^2) = \left[\frac{\sigma_\varepsilon^2}{1-\left(\phi^2 + 2\phi\sigma_b\sqrt{\frac{2}{\pi}} + \sigma_b^2\right)}\right]$$

2)

$$E\left(y_t^4\right)=\frac{3\sigma_\varepsilon^4\left(1+\phi^2+2\phi\sigma_b\sqrt{\frac{2}{\pi}}+\sigma_b^2\right)}{\left(1-\frac{8\sqrt{2}\sigma_b^3}{\sqrt{\pi}}\phi-\phi^4-3\sigma_b^4-4\phi^3\sigma_b\sqrt{\frac{2}{\pi}}-6\phi^2\sigma_b^2\right)\left(1-\left(\phi^2+2\phi\sigma_b\sqrt{\frac{2}{\pi}}+\sigma_b^2\right)\right)},$$

3)

$$K^{(y)}=\left[\frac{3\left(1-\left(\phi^2+2\phi\sigma_b\sqrt{\frac{2}{\pi}}+\sigma_b^2\right)^2\right)}{\left(1-\frac{8\sqrt{2}\sigma_b^3}{\sqrt{\pi}}\phi-\phi^4-3\sigma_b^4-4\phi^3\sigma_b\sqrt{\frac{2}{\pi}}-6\phi^2\sigma_b^2\right)}\right].$$

The autocovariance and the autocorrelation functions are given by

$$\gamma_k=\left(\phi+\frac{\left(2\sigma_b^2\right)^{\frac{\delta}{2}}}{\sqrt{\pi}}\Gamma\left(\frac{\delta+1}{2}\right)\right)\gamma_{k-1}$$ and $$\rho_k=\left(\phi+\frac{\left(2\sigma_b^2\right)^{\frac{\delta}{2}}}{\sqrt{\pi}}\Gamma\left(\frac{\delta+1}{2}\right)\right)^k.$$

Proof:

$$E\left(y_t\right)=0,\ E\left(y_t^2\right)=\phi^2E\left(y_{t-1}^2\right)+2\phi E\left(y_{t-1}^2\right)E\left(\left|b_t\right|\right)+E\left(y_{t-1}^2\right)E\left(\left|b_t\right|^2\right)+E\left(\varepsilon_t^2\right)$$

Thus, we have

$$E\left(y_t^2\right)=\left[\frac{\sigma_\varepsilon^2}{1-\left(\phi^2+2\phi\sigma_b\sqrt{\frac{2}{\pi}}+\sigma_b^2\right)}\right] \tag{2.3}$$

and we have

$$E\left(y_t^4\right)=\frac{3\sigma_\varepsilon^4}{\left(1-\frac{8\sqrt{2}\sigma_b^3}{\sqrt{\pi}}\phi-\phi^4-3\sigma_b^4-4\phi^3\sigma_b\sqrt{\frac{2}{\pi}}-6\phi^2\sigma_b^2\right)}$$

$$+\frac{6\sigma_\varepsilon^4\left[\phi^2+2\phi\sqrt{\frac{2}{\pi}}\sigma_b+\sigma_b^2\right]}{\left(1-\frac{8\sqrt{2}\sigma_b^3}{\sqrt{\pi}}\phi-\phi^4-3\sigma_b^4-|4\phi^3\sigma_b\sqrt{\frac{2}{\pi}}-6\phi^2\sigma_b^2\right)\left(1-\left(\phi^2+2\phi\sigma_b\sqrt{\frac{2}{\pi}}+\sigma_b^2\right)\right)}$$

$$=\frac{3\sigma_\varepsilon^4\left(1+\phi^2+2\phi\sigma_b\sqrt{\frac{2}{\pi}}+\sigma_b^2\right)}{\left(1-\frac{8\sqrt{2}\sigma_b^3}{\sqrt{\pi}}\phi-\phi^4-3\sigma_b^4-4\phi^3\sigma_b\sqrt{\frac{2}{\pi}}-6\phi^2\sigma_b^2\right)\left(1-\left(\phi^2+2\phi\sigma_b\sqrt{\frac{2}{\pi}}+\sigma_b^2\right)\right)}$$

$$K^{(y)}=\frac{E\left[y_t^4\right]}{E\left[y_t^2\right]^2}=\left[\frac{3\left(1-\left(\phi^2+2\phi\sigma_b\sqrt{\frac{2}{\pi}}+\sigma_b^2\right)^2\right)}{\left(1-\frac{8\sqrt{2}\sigma_b^3}{\sqrt{\pi}}\phi-\phi^4-3\sigma_b^4-4\phi^3\sigma_b\sqrt{\frac{2}{\pi}}-6\phi^2\sigma_b^2\right)}\right] \tag{2.4}$$

When $\sigma_b^2=0$, the kurtosis of the process y_t converge to $K^{(y)}=3$. Thus, the autocorrelation function is given by

$$\rho_k=\left(\phi+E\left|b_{t-k}\right|\right)\rho_{k-1}=\left(\phi+\frac{\left(2\sigma_b^2\right)^{\frac{\delta}{2}}}{\sqrt{\pi}}\Gamma\left(\frac{\delta+1}{2}\right)\right)^k$$

where we use the fact that $\rho_0=1$.

Theorem 2.2. Suppose y_t is an Random Coefficient Moving average process model of the form

$$y_t=\left(\phi+\left|b_t\right|\right)y_{t-1}+\varepsilon_{t-1}^2\varepsilon_t \tag{2.5}$$

where b_t is an uncorrelated Gaussian process with zero mean and with variance σ_b^2. ε_t is an uncorrelated Gaussian process with zero mean and with variance σ_ε^2. Then, we have the following

Proof:

$$E(y_t) = E\left(\left(\phi + |b_t|\right)\right) y_{t-1} + \varepsilon_{t-1}^2 \varepsilon_t = 0 \tag{2.6}$$

and

$$E(y_t^2) = \phi^2 E(y_{t-1}^2) + 2\phi E(y_{t-1}^2)\sigma_b \sqrt{\frac{2}{\pi}} + E(y_{t-1}^2)\sigma_b^2 + 3\sigma_\varepsilon^6$$

We have

$$E(y_t^2) = \left[\frac{3\sigma_\varepsilon^6}{1 - \left(\phi^2 + 2\phi\sigma_b\sqrt{\frac{2}{\pi}} + \sigma_b^2\right)}\right] \tag{2.7}$$

and

$$E(y_{t-1}^4) = \frac{\left(12\phi\sigma_b\sqrt{\frac{2}{\pi}}3\sigma_\varepsilon^4\sigma_\varepsilon^2 + 6\sigma_b^2 3\sigma_\varepsilon^4\sigma_\varepsilon^2 + 6\phi^2 3\sigma_\varepsilon^4\sigma_\varepsilon^2\right)E(y_{t-1}^2) + 105\sigma_\varepsilon^8 3\sigma_\varepsilon^4}{\left(1 - \phi^4 - 4\phi 2\sqrt{\frac{2}{\pi}}\sigma_b^3 - 6\phi^2\sigma_b^2 - 4\phi^3\sigma_b\sqrt{\frac{2}{\pi}} - 3\sigma_b^4\right)}$$

$$= \frac{\left(36\phi\sigma_b\sqrt{\frac{2}{\pi}}\sigma_\varepsilon^6 + 18\sigma_b^2\sigma_\varepsilon^6 + 18\phi^2\sigma_\varepsilon^6\right)E(y_{t-1}^2) + 315\sigma_\varepsilon^{12}}{\left(1 - \phi^4 - 8\phi\sqrt{\frac{2}{\pi}}\sigma_b^3 - 6\phi^2\sigma_b^2 - 4\phi^3\sigma_b\sqrt{\frac{2}{\pi}} - 3\sigma_b^4\right)}$$

Thus we have

$$E\left(y_{t-1}^4\right) = \frac{\left(36\phi\sigma_b\sqrt{\frac{2}{\pi}}\sigma_\varepsilon^6 + 18\sigma_b^2\sigma_\varepsilon^6 + 18\phi^2\sigma_\varepsilon^6\right)}{\left(1-\phi^4-8\phi\sqrt{\frac{2}{\pi}}\sigma_b^3-6\phi^2\sigma_b^2-4\phi^3\sigma_b\sqrt{\frac{2}{\pi}}-3\sigma_b^4\right)}\left[\frac{3\sigma_\varepsilon^6}{1-\left(\phi^2+2\phi\sigma_b\sqrt{\frac{2}{\pi}}+\sigma_b^2\right)}\right]$$
$$+\frac{315\sigma_\varepsilon^{12}}{\left(1-\phi^4-8\phi\sqrt{\frac{2}{\pi}}\sigma_b^3-6\phi^2\sigma_b^2-4\phi^3\sigma_b\sqrt{\frac{2}{\pi}}-3\sigma_b^4\right)}$$

The kurtosis of the process is given by

$$K^{(y)} = \left[\frac{\left(36\phi\sigma_b\sqrt{\frac{2}{\pi}}\sigma_\varepsilon^6 + 18\sigma_b^2\sigma_\varepsilon^6 + 18\phi^2\sigma_\varepsilon^6\right)\left(1-\left(\phi^2+2\phi\sigma_b\sqrt{\frac{2}{\pi}}+\sigma_b^2\right)\right)}{\left(1-\phi^4-8\phi\sqrt{\frac{2}{\pi}}\sigma_b^3-6\phi^2\sigma_b^2-4\phi^3\sigma_b\sqrt{\frac{2}{\pi}}-3\sigma_b^4\right)3\sigma_\varepsilon^6}\right]$$
$$+\left[\frac{35\left(1-\left(\phi^2+2\phi\sigma_b\sqrt{\frac{2}{\pi}}+\sigma_b^2\right)\right)^2}{\left(1-\phi^4-8\phi\sqrt{\frac{2}{\pi}}\sigma_b^3-6\phi^2\sigma_b^2-4\phi^3\sigma_b\sqrt{\frac{2}{\pi}}-3\sigma_b^4\right)}\right] \tag{2.8}$$

When $\sigma_b^2 = 0$, the kurtosis of the process y_t converge to $K^{(y)} = \frac{35-29\phi^2}{\left(1+\phi^2\right)}$ and when $\sigma_b^2 = 0$, and $\phi = 0$ the kurtosis of the process y_t turns out to be 35.

Theorem 2.3. Suppose y_t is an Random Coefficient Moving average process model of the form

$$y_t = \varepsilon_t + \left(\phi + |b_t|\right)\varepsilon_{t-1} \tag{2.9}$$

where b_t is an uncorrelated noise process with zero mean and with variance σ_b^2. ε_t is an uncorrelated noise process with zero mean and with variance σ_ε^2. Then, we have the following relationships

$$E\left(y_t^2\right) = \sigma_\varepsilon^2\left(1+\phi^2+2\phi\sigma_b\sqrt{\frac{2}{\pi}}+\sigma_b^2\right),$$

$$E\left(y_t^4\right) = \sigma_\varepsilon^4\left[3\left(\phi^4+3\sigma_b^4+1\right)+12\sqrt{\frac{2}{\pi}}\left(1+2\sigma_b^2+\phi^2\right)\phi\sigma_b+6\left(\sigma_b^2+\phi^2+3\phi^2\sigma_b^2\right)\right],$$

$$K^{(y)} = \left[\frac{3\left(\phi^4+3\sigma_b^4+1\right)+12\sqrt{\frac{2}{\pi}}\left(1+2\sigma_b^2+\phi^2\right)\phi\sigma_b+6\left(\sigma_b^2+\phi^2+3\phi^2\sigma_b^2\right)}{\left(1+\phi^2+2\phi\sigma_b\sqrt{\frac{2}{\pi}}+\sigma_b^2\right)}\right].$$

and the autocorrelation functions are given by

$$\rho_k = \begin{cases} 1 & k=0 \\ \left[\dfrac{\phi+\sqrt{\frac{2}{\pi}}\sigma_b}{1+\phi^2+2\phi\sigma_b\sqrt{\frac{2}{\pi}}+\sigma_b^2}\right] & k=1 \\ 0 & k=2,3,\cdots \end{cases} \tag{2.10}$$

Proof:

$$E\left(y_t^2\right) = E\left[\varepsilon_t^2+2\varepsilon_t\varepsilon_{t-1}\phi+2\varepsilon_t\varepsilon_{t-1}\left|b_t\right|+\varepsilon_{t-1}^2\phi^2+2\varepsilon_{t-1}^2\phi\left|b_t\right|+\varepsilon_{t-1}^2\left|b_t\right|^2\right],$$

$$= \sigma_\varepsilon^2+\phi^2\sigma_\varepsilon^2+2\phi\sigma_\varepsilon^2\sigma_b\sqrt{\frac{2}{\pi}}+\sigma_\varepsilon^2\sigma_b^2 = \sigma_\varepsilon^2\left(1+\phi^2+2\phi\sigma_b\sqrt{\frac{2}{\pi}}+\sigma_b^2\right).$$

$$E\left(y_t^4\right) = \sigma_\varepsilon^4\left[3\phi^4+9\sigma_b^4+3+12\sqrt{\frac{2}{\pi}}\phi\sigma_b+6\sigma_b^2+6\phi^2+12\phi^3\sqrt{\frac{2}{\pi}}\sigma_b+18\phi^2\sigma_b^2+24\sqrt{\frac{2}{\pi}}\phi\sigma_b^3\right]$$

$$= \sigma_\varepsilon^4\left[3\left(\phi^4+3\sigma_b^4+1\right)+12\sqrt{\frac{2}{\pi}}\left(1+2\sigma_b^2+\phi^2\right)\phi\sigma_b+6\left(\sigma_b^2+\phi^2+3\phi^2\sigma_b^2\right)\right]$$

Thus,

$$K^{(y)} = \frac{E\left(y_t-\mu\right)^4}{\left(Var\left(y_t\right)\right)^2} = \left[\frac{3\left(\phi^4+3\sigma_b^4+1\right)+12\sqrt{\frac{2}{\pi}}\left(1+2\sigma_b^2+\phi^2\right)\phi\sigma_b+6\left(\sigma_b^2+\phi^2+3\phi^2\sigma_b^2\right)}{\left(1+\phi^2+2\phi\sigma_b\sqrt{\frac{2}{\pi}}+\sigma_b^2\right)^2}\right]$$

The autocorrelation function is given by

$$\rho_0 = 1, \rho_1 = \left[\frac{\phi + \sqrt{\frac{2}{\pi}}\sigma_b}{1+\phi^2 + 2\phi\sigma_b\sqrt{\frac{2}{\pi}} + \sigma_b^2} \right], \ \rho_k = 0 \quad k = 2,3,4,\cdots$$

Theorem 2.4. Let $\{y_t\}$ be a Sign RCA-GARCH (1,1) time series satisfying conditions (i) and (ii) given by

$$y_t = \left(\phi + |b_t| + \Phi s_t\right) y_{t-1} + \varepsilon_t \tag{2.11}$$

where

$\varepsilon_t = \sqrt{h_t} Z_t$ (2.12)

$$h_t = \omega + \sum_{i=1}^{p} \alpha_i \varepsilon_{t-i}^2 + \sum_{j=1}^{q} \beta_j h_{t-j} \tag{2.13}$$

where Z_t and b_t are sequences of independent, identically distributed random variables with zeromean, variance given by σ_Z^2 and σ_b^2 respectively,

$$s_t = \begin{cases} +1 & \text{if} \quad y_t > 0 \\ 0 & \text{if} \quad y_t = 0 \\ -1 & \text{if} \quad y_t < 0 \end{cases}$$

ω, α_1, β_1 and Φ are real parameters, satisfying the following conditions, $\omega > 0$, $\alpha_1 \geq 0$, $\beta_1 \geq 0$. $|\Phi x| \leq \omega$. Note: $E\left(s_t^2\right) = 1$, and in order to calculate the kurtosis, we observe that $E\left(s_t^4\right) = 1$. Then, we have the following moment properties

$$K^{(y)} = \frac{\left(1 - \left(\phi^2 + 2\sqrt{\frac{2}{\pi}}\phi\sigma_b + \sigma_b^2 + \Phi^2\right)\right)^2 \dfrac{E(Z_t^4)}{\left(E(Z_t^4) - \left(E(Z_t^4) - 1\right)\sum_{j=1}^{\infty}\psi_j^2\right)}}{\sigma_Z^4\left(1 - \Phi^4 - \phi^4 - 6\Phi^2\left(\phi^2 + \sigma_b^2\right) - \sqrt{\frac{2}{\pi}}\phi\left(12\Phi^2\sigma_b + 8\sigma_b^3\right) - 3\sigma_b^2\left(2\phi^2 + \sigma_b^2\right)\right)}$$

$$E\left(y_t^2\right)=\left[\frac{\omega\sigma_Z^2}{\left(1-\left(\phi^2+2\sqrt{\frac{2}{\pi}}\phi\sigma_b+\sigma_b^2+\Phi^2\right)\right)\left(1-\left(\alpha\sigma_Z^2+\beta\right)\right)}\right]$$

$$E\left(y_{t-1}^4\right)=\frac{E\left(Z_t^4\right)}{\left(1-\Phi^4-\phi^4-6\Phi^2\left(\phi^2+\sigma_b^2\right)-\sqrt{\frac{2}{\pi}}\phi\left(12\Phi^2\sigma_b+4\phi^2\sigma_b+8\sigma_b^3\right)-3\sigma_b^2\left(2\phi^2+\sigma_b^2\right)\right)}E\left(h_t^2\right)$$
$$+\frac{6\sigma_Z^4\left(2\phi\sigma_b\sqrt{\frac{2}{\pi}}+\Phi^2+\phi^2+\sigma_b^2\right)\left(1-\left(\phi^2+2\sqrt{\frac{2}{\pi}}\phi\sigma_b+\sigma_b^2+\Phi^2\right)\right)^{-1}}{\left(1-\Phi^4-\phi^4-6\Phi^2\left(\phi^2+\sigma_b^2\right)-\sqrt{\frac{2}{\pi}}\phi\left(12\Phi^2\sigma_b+4\phi^2\sigma_b+8\sigma_b^3\right)-3\sigma_b^2\left(2\phi^2+\sigma_b^2\right)\right)}E\left(h_t\right)^2$$

(2.14)

Proof:

$$y_t=\left(\phi+\left|b_t\right|+\Phi s_t\right)y_{t-1}+\varepsilon_t.$$
$$E\left(y_t\right)=E\left(\left(\phi+\left|b_t\right|+\Phi s_t\right)y_{t-1}+\varepsilon_t\right)=0$$

$$E\left(y_t^2\right)=E\left(\left(\phi+b_t+\Phi s_t\right)y_{t-1}+\varepsilon_t\right)^2$$
$$=\frac{E\left[h_t\right]\sigma_Z^2}{1-\left(\phi^2+2\sqrt{\frac{2}{\pi}}\phi\sigma_b+\sigma_b^2+\Phi^2\right)}=\frac{\omega\sigma_Z^2}{\left(1-\left(\phi^2+2\sqrt{\frac{2}{\pi}}\phi\sigma_b+\sigma_b^2+\Phi^2\right)\right)\left(1-\left(\alpha\sigma_Z^2+\beta\right)\right)}$$

$$E\left(y_{t-1}^4\right)=\frac{E\left(Z_t^4\right)}{\left(1-\Phi^4-\phi^4-6\Phi^2\left(\phi^2+\sigma_b^2\right)-\sqrt{\frac{2}{\pi}}\phi\left(12\Phi^2\sigma_b+4\phi^2\sigma_b+8\sigma_b^3\right)-3\sigma_b^2\left(2\phi^2+\sigma_b^2\right)\right)}E\left(h_t^2\right)$$
$$+\frac{6\sigma_Z^4\left(2\phi\sigma_b\sqrt{\frac{2}{\pi}}+\Phi^2+\phi^2+\sigma_b^2\right)\left(1-\left(\phi^2+2\sqrt{\frac{2}{\pi}}\phi\sigma_b+\sigma_b^2+\Phi^2\right)\right)^{-1}}{\left(1-\Phi^4-\phi^4-6\Phi^2\left(\phi^2+\sigma_b^2\right)\ \sqrt{\frac{2}{\pi}}\phi\left(12\Phi^2\sigma_b+4\phi^2\sigma_b+8\sigma_b^3\right)-3\sigma_b^2\left(2\phi^2+\sigma_b^2\right)\right)}E\left(h_t\right)^2$$

$$E\left(y_{t-1}^4\right)=AE\left(h_t^2\right)+BE\left(h_t\right)^2$$

where

$$A=\frac{E\left(Z_t^4\right)}{\left(1-6\Phi^2\left(\phi^2+\sigma_b^2\right)-12\phi\Phi^2\sigma_b\sqrt{\frac{2}{\pi}}-4\phi^3\sigma_b\sqrt{\frac{2}{\pi}}-6\phi^2\sigma_b^2-8\phi\sqrt{\frac{2}{\pi}}\sigma_b^3-3\sigma_b^4-\Phi^4-\phi^4\right)}$$

$$B=\frac{6\sigma_Z^4\left(2\phi\sigma_b\sqrt{\frac{2}{\pi}}+\Phi^2+\phi^2+\sigma_b^2\right)\left(1-\left(\phi^2+2\sqrt{\frac{2}{\pi}}\phi\sigma_b+\sigma_b^2+\Phi^2\right)\right)^{-1}}{\left(1-\Phi^4-\phi^4-6\Phi^2\left(\phi^2+\sigma_b^2\right)-\sqrt{\frac{2}{\pi}}\phi\left(12\Phi^2\sigma_b+4\phi^2\sigma_b+8\sigma_b^3\right)-3\sigma_b^2\left(2\phi^2+\sigma_b^2\right)\right)}$$

$$K^{(y)}=\frac{E\left(y_t-\mu\right)^4}{\left(Var\left(y_t\right)\right)^2}=\left(AE\left(h_t^2\right)+BE\left(h_t\right)^2\right)\left(\frac{\left(1-\left(\phi^2+2\sqrt{\frac{2}{\pi}}\phi\sigma_b+\sigma_b^2+\Phi^2\right)\right)^2}{E\left[h_t\right]^2\sigma_Z^4}\right)$$

$$=A\left(\frac{\left(1-\left(\phi^2+2\sqrt{\frac{2}{\pi}}\phi\sigma_b+\sigma_b^2+\Phi^2\right)\right)^2}{\sigma_Z^4}\right)\frac{E\left(h_t^2\right)}{E\left[h_t\right]^2}+B\left(\frac{\left(1-\left(\phi^2+2\sqrt{\frac{2}{\pi}}\phi\sigma_b+\sigma_b^2+\Phi^2\right)\right)^2}{\sigma_Z^4}\right)$$

Using the facts that

$$\frac{E\left(h_t^2\right)}{E\left(h_t\right)^2}=\frac{1}{E\left(Z_t^4\right)-\left(E\left(Z_t^4\right)-1\right)\sum_{j=1}^{\infty}\psi_j^2}$$

$$K^{(y)}=A\left(\frac{\left(1-\left(\phi^2+2\sqrt{\frac{2}{\pi}}\phi\sigma_b+\sigma_b^2+\Phi^2\right)\right)^2}{\sigma_Z^4}\right)\frac{E\left(h_t^2\right)}{E\left[h_t\right]^2}+B\left(\frac{\left(1-\left(\phi^2+2\sqrt{\frac{2}{\pi}}\phi\sigma_b+\sigma_b^2+\Phi^2\right)\right)^2}{\sigma_Z^2}\right)$$

$$=A\left(\frac{\left(1-\left(\phi^2+2\sqrt{\frac{2}{\pi}}\phi\sigma_b+\sigma_b^2+\Phi^2\right)\right)^2}{\sigma_Z^4}\right)\left(\frac{1}{E\left(Z_t^4\right)-\left(E\left(Z_t^4\right)-1\right)\sum_{j=1}^{\infty}\psi_j^2}\right)$$

$$+B\left(\frac{\left(1-\left(\phi^2+2\sqrt{\frac{2}{\pi}}\phi\sigma_b+\sigma_b^2+\Phi^2\right)\right)^2}{\sigma_Z^4}\right)$$

Thus, we have the following expression for the Kurtosis of the process.

$$K^{(y)} = \frac{\left(1-\left(\phi^2+2\sqrt{\frac{2}{\pi}}\phi\sigma_b+\sigma_b^2+\Phi^2\right)\right)^2 \frac{E\left(Z_t^4\right)}{\left(E\left(Z_t^4\right)-\left(E\left(Z_t^4\right)-1\right)\sum_{j=1}^{\infty}\psi_j^2\right)}}{\sigma_Z^4\left(1-\Phi^4-\phi^4-6\Phi^2\left(\phi^2+\sigma_b^2\right)-\sqrt{\frac{2}{\pi}}\phi\left(12\Phi^2\sigma_b+4\phi^2\sigma_b+8\sigma_b^3\right)-3\sigma_b^2\left(2\phi^2+\sigma_b^2\right)\right)} \frac{6\left(2\phi\sigma_b\sqrt{\frac{2}{\pi}}+\Phi^2+\phi^2+\sigma_b^2\right)\left(1-\left(\phi^2+2\sqrt{\frac{2}{\pi}}\phi\sigma_b+\sigma_b^2+\Phi^2\right)\right)}{\sigma_Z^4\left(1-\Phi^4-\phi^4-6\Phi^2\left(\phi^2+\sigma_b^2\right)-\sqrt{\frac{2}{\pi}}\phi\left(12\Phi^2\sigma_b+4\phi^2\sigma_b+8\sigma_b^3\right)-3\sigma_b^2\left(2\phi^2+\sigma_b^2\right)\right)}$$

Special cases, for a Normal GARCH (1,1).

$$K^{(y)} = \frac{3\left(1-\left(\phi^2+2\sqrt{\frac{2}{\pi}}\phi\sigma_b+\sigma_b^2+\Phi^2\right)\right)^2\left(\frac{\left(1\quad(\alpha+\beta)^2\right)}{\left(1-(\alpha+\beta)^2-2\alpha^2\right)}\right)}{\left(1-6\Phi^2\left(\phi^2+\sigma_b^2\right)-\sigma_b\sqrt{\frac{2}{\pi}}\left(12\phi\Phi^2+4\phi^3+6\phi^2\sigma_b+8\phi\sigma_b^2+3\sigma_b^3\right)-\Phi^4-\phi^4\right)} \tag{2.15}$$

Note, that when $\sigma_b^2 = 0, \Phi = 0$, and $Z_t \sim N(0,1)$ in (2.15), the kurtosis of the process converges to

$$K^{(y)} = \left[\frac{6\phi^2}{\left(1+\phi^2\right)}\right] + 3\left(\frac{1-\phi^2}{1+\phi^2}\right)\frac{\left[1-\left(\alpha_1+\beta_1\right)^2\right]}{\left(1-\left(\alpha_1+\beta_1\right)^2\right)-2\alpha_1^2} \tag{2.16}$$

When $\phi = 0$ in (2.16),the kurtosis of the process converge to $K^{(y)} = \frac{3\left(1-(\alpha+\beta)^2\right)}{1-(\alpha+\beta)^2-2\alpha^2} > 3$

Theorem 2.5. Suppose y_t is a modified RCA model with GARCH (p, q) innovations of the form

$$
\begin{aligned}
y_t &= \left(\phi + |b_t|\right) y_{t-1} + \varepsilon_t \\
\varepsilon_t &= \sqrt{h_t}\, Z_t \\
h_t &= \omega + \sum_{i=1}^{p} \alpha_i \varepsilon_{t-i}^2 + \sum_{j=1}^{q} \beta_j h_{t-j}
\end{aligned}
$$

where b_t is an uncorrelated noise process with zero mean and with variance σ_b^2 and Z_t is an uncorrelated noise process with zero mean and with variance σ_Z^2. Then, we have thc following relationship

$$
E\left(y_t^2\right) = \left[\frac{\sigma_Z^2}{\left(1-\phi^2-2\sqrt{\frac{2}{\pi}}\phi\sigma_b-\sigma_b^2\right)}\right] E\left(h_t\right) \tag{2.17}
$$

$$
\begin{aligned}
E\left(y_t^4\right) = &\left[\frac{3\sigma_Z^4}{\left(1-4\phi^3\sigma_b\sqrt{\frac{2}{\pi}}-6\phi^2\sigma_b^2-8\phi\sqrt{\frac{2}{\pi}}\sigma_b^3-\phi^4-3\sigma_b^4\right)}\right] E\left(h_t^2\right) \\
&+\left[\frac{\left(12\phi\sigma_b\sigma_Z^2\sqrt{\frac{2}{\pi}}+6\sigma_b^2\sigma_Z^2+6\phi^2\sigma_Z^2\right)\sigma_Z^2}{\left(1-4\phi^3\sigma_b\sqrt{\frac{2}{\pi}}-6\phi^2\sigma_b^2-8\phi\sqrt{\frac{2}{\pi}}\sigma_b^3-\phi^4-3\sigma_b^4\right)\left(1-\phi^2-2\sqrt{\frac{2}{\pi}}\phi\sigma_b-\sigma_b^2\right)}\right] E\left(h_t\right)^2
\end{aligned} \tag{2.18}
$$

$$
\begin{aligned}
K^{(y)} = &\left(\frac{3\left(1-\phi^2-2\sqrt{\frac{2}{\pi}}\phi\sigma_b-\sigma_b^2\right)^2}{\left(1-4\phi^3\sigma_b\sqrt{\frac{2}{\pi}}-6\phi^2\sigma_b^2-8\phi\sqrt{\frac{2}{\pi}}\sigma_b^3-\phi^4-3\sigma_b^4\right)\left(E\left(Z_t^4\right)-\left(E\left(Z_t^4\right)-1\right)\sum_{j=1}^{\infty}\psi_j^2\right)}\right. \\
&\left.+6\left(\frac{\left(1-\phi^2-2\sqrt{\frac{2}{\pi}}\phi\sigma_b-\sigma_b^2\right)\left(2\phi\sigma_b\sqrt{\frac{2}{\pi}}+\sigma_b^2+\phi^2\right)}{\left(1-4\phi^3\sigma_b\sqrt{\frac{2}{\pi}}-6\phi^2\sigma_b^2-8\phi\sqrt{\frac{2}{\pi}}\sigma_b^3-\phi^4-3\sigma_b^4\right)}\right)\right)
\end{aligned} \tag{2.19}
$$

Proof: Let, $u_t = \varepsilon_t^2 - h_t$

$$\varepsilon_t^2 - u_t = \omega + \sum_{i=1}^{p} \alpha_i \varepsilon_{t-i}^2 + \sum_{j=1}^{q} \beta_j h_{t-j} \left[1 - \left(\sum_{i=1}^{p} \alpha_i B^i + \sum_{j=1}^{q} \beta_j B^j \right) \right] \varepsilon_t^2 = \omega + \left(1 - \sum_{j=1}^{q} \beta_j B^j \right) u_t$$

$$\phi(B)\varepsilon_t^2 = \omega + \beta(B) u_t, \quad \phi(B) = 1 - \sum_{i=1}^{r} \phi_i B^i, \quad \phi_i = (\alpha_i + \beta_i)$$

$$\beta(B) = \left(1 - \sum_{j=1}^{q} \beta_j B^j \right), \quad r = \max(p, q)$$

$$\frac{E(h_t^2)}{E(h_t)^2} = \frac{1}{E(Z_t^4) - \left(E(Z_t^4) - 1\right)\sum_{j=1}^{\infty} \psi_j^2}, \quad K^{(\varepsilon)} = \frac{E(Z_t^4)}{E(Z_t^4) - \left(E(Z_t^4) - 1\right)\sum_{j=1}^{\infty} \psi_j^2}$$

Now we have

$$E(y_t^2) - \phi^2 E(y_{t-1}^2) - 2\sqrt{\frac{2}{\pi}}\phi E(y_{t-1}^2)\sigma_b - E(y_{t-1}^2)\sigma_b^2 = E(h_t)\sigma_Z^2,$$

Thus,

$$E(y_t^2) - \phi^2 E(y_{t-1}^2) - 2\sqrt{\frac{2}{\pi}}\phi E(y_{t-1}^2)\sigma_b - E(y_{t-1}^2)\sigma_b^2 = E(h_t)\sigma_Z^2,$$

and

$$\left(1 - \phi^2 - 2\sqrt{\frac{2}{\pi}}\phi\sigma_b - \sigma_b^2 \right) E(y_t^2) = E(h_t)\sigma_Z^2$$

$$E(y_t^2) = \left[\frac{\sigma_Z^2}{\left(1 - \phi^2 - 2\sqrt{\frac{2}{\pi}}\phi\sigma_b - \sigma_b^2 \right)} \right] E(h_t) \qquad (2.20)$$

$$E\left(y_t^4\right)=\frac{E\left(h_t^2 Z_t^4\right)+12\phi E\left(y_{t-1}^2\left|b_t\right|h_t X_t^2\right)+6E\left(y_{t-1}^2\left|b_t\right|^2 h_t Z_t^2\right)+6\phi^2 E\left(y_{t-1}^2 h_t Z_t^2\right)}{\left(1-4\phi^3 E\left|b_t\right|-6\phi^2 E\left|b_t\right|^2-4\phi E\left|b_t\right|^3-\phi^4-E\left|b_t\right|^4\right)}$$

$$=\frac{3\sigma_Z^4 E\left(h_t^2\right)+12\phi\sigma_b\sigma_Z^2\sqrt{\frac{2}{\pi}}E\left(y_{t-1}^2\right)E\left(h_t\right)+6\sigma_b^2\sigma_Z^2 E\left(y_{t-1}^2\right)E\left(h_t\right)+6\phi^2 E\left(y_{t-1}^2\right)E\left(h_t\right)\sigma_Z^2}{\left(1-4\phi^3\sigma_b\sqrt{\frac{2}{\pi}}-6\phi^2\sigma_b^2-8\phi\sqrt{\frac{2}{\pi}}\sigma_b^3-\phi^4-3\sigma_b^4\right)}$$

$$=\left[\frac{3\sigma_Z^4}{\left(1-4\phi^3\sigma_b\sqrt{\frac{2}{\pi}}-6\phi^2\sigma_b^2-8\phi\sqrt{\frac{2}{\pi}}\sigma_b^3-\phi^4-3\sigma_b^4\right)}\right]E\left(h_t^2\right)$$

$$+\left[\frac{12\phi\sigma_b\sigma_Z^2\sqrt{\frac{2}{\pi}}+6\sigma_b^2\sigma_Z^2+6\phi^2\sigma_Z^2}{\left(1-4\phi^3\sigma_b\sqrt{\frac{2}{\pi}}-6\phi^2\sigma_b^2-8\phi\sqrt{\frac{2}{\pi}}\sigma_b^3-\phi^4-3\sigma_b^4\right)}\right]F\left(y_{t-1}^2\right)E\left(h_t\right)$$

Using the fact that $E\left(y_t^2\right)=\left[\frac{\sigma_Z^2}{\left(1-\phi^2-2\sqrt{\frac{2}{\pi}}\phi\sigma_b-\sigma_b^2\right)}\right]E\left(h_t\right)$, we have

$$E\left(y_t^4\right)=\left[\frac{3\sigma_Z^4}{\left(1-4\phi^3\sigma_b\sqrt{\frac{2}{\pi}}-6\phi^2\sigma_b^2-8\phi\sqrt{\frac{2}{\pi}}\sigma_b^3-\phi^4-3\sigma_b^4\right)}\right]E\left(h_t^2\right)$$

$$+\left[\frac{\left(12\phi\sigma_b\sigma_Z^2\sqrt{\frac{2}{\pi}}+6\sigma_b^2\sigma_Z^2+6\phi^2\sigma_Z^2\right)\sigma_Z^2}{\left(1-4\phi^3\sigma_b\sqrt{\frac{2}{\pi}}-6\phi^2\sigma_b^2-8\phi\sqrt{\frac{2}{\pi}}\sigma_b^3-\phi^4-3\sigma_b^4\right)\left(1-\phi^2-2\sqrt{\frac{2}{\pi}}\phi\sigma_b-\sigma_b^2\right)}\right]E\left(h_t\right)^2$$

For convenience let $E\left(y_t^4\right)=AE\left(h_t^2\right)+BE\left(h_t\right)^2$ where,

$$A = \left[\frac{3\sigma_Z^4}{\left(1 - 4\phi^3\sigma_b\sqrt{\frac{2}{\pi}} - 6\phi^2\sigma_b^2 - 8\phi\sqrt{\frac{2}{\pi}}\sigma_b^3 - \phi^4 - 3\sigma_b^4\right)} \right]$$

$$B = \left[\frac{\left(12\phi\sigma_b\sigma_Z^2\sqrt{\frac{2}{\pi}} + 6\sigma_b^2\sigma_Z^2 + 6\phi^2\sigma_Z^2\right)\sigma_Z^2}{\left(1 - 4\phi^3\sigma_b\sqrt{\frac{2}{\pi}} - 6\phi^2\sigma_b^2 - 8\phi\sqrt{\frac{2}{\pi}}\sigma_b^3 - \phi^4 - 3\sigma_b^4\right)\left(1 - \phi^2 - 2\sqrt{\frac{2}{\pi}}\phi\sigma_b - \sigma_b^2\right)} \right]$$

The Kurtosis of the process is given by

$$K^{(y)} = \frac{E\left[y_t^4\right]}{E\left[y_t^2\right]^2} = \frac{AE\left(h_t^2\right) + BE\left(h_t\right)^2}{E\left[y_t^2\right]^2}$$

$$= \left[AE\left(h_t^2\right) + BE\left(h_t\right)^2\right]\left[\frac{\left(1 - \phi^2 - 2\sqrt{\frac{2}{\pi}}\phi\sigma_b - \sigma_b^2\right)^2}{\sigma_Z^4 E\left(h_t\right)^2}\right]$$

$$= A\left[\frac{\left(1 - \phi^2 - 2\sqrt{\frac{2}{\pi}}\phi\sigma_b - \sigma_b^2\right)^2}{\sigma_Z^4}\right]\frac{E\left(h_t^2\right)}{E\left(h_t\right)^2} + B\left[\frac{\left(1 - \phi^2 - 2\sqrt{\frac{2}{\pi}}\phi\sigma_b - \sigma_b^2\right)^2}{\sigma_Z^4}\right]$$

Using the fact that $\frac{E\left(h_t^2\right)}{E\left(h_t\right)^2} = \frac{1}{E\left(Z_t^4\right) - \left(E\left(Z_t^4\right) - 1\right)\sum_{j=1}^{\infty}\psi_j^2}$ we have

$$K^{(y)} = \left(\frac{3\left(1-\phi^2-2\sqrt{\frac{2}{\pi}}\phi\sigma_b-\sigma_b^2\right)^2}{\left(1-4\phi^3\sigma_b\sqrt{\frac{2}{\pi}}-6\phi^2\sigma_b^2-8\phi\sqrt{\frac{2}{\pi}}\sigma_b^3-\phi^4-3\sigma_b^4\right)\left(E\left(Z_t^4\right)-\left(E\left(Z_t^4\right)-1\right)\sum_{j=1}^{\infty}\psi_j^2\right)} \right.$$
$$\left. +6\left(\frac{\left(1-\phi^2-2\sqrt{\frac{2}{\pi}}\phi\sigma_b-\sigma_b^2\right)\left(2\phi\sigma_b\sqrt{\frac{2}{\pi}}+\sigma_b^2+\phi^2\right)}{\left(1-4\phi^3\sigma_b\sqrt{\frac{2}{\pi}}-6\phi^2\sigma_b^2-8\phi_b^3-\phi^4-3\sigma_b^4\right)} \right) \right) \tag{2.21}$$

2.2. Quadratic GARCH Model

Besides having excess kurtosis market returns may display seriously skewed distributions. Linear GARCH models cannot cope with such skewness, and therefore we can expect forecast of linear GARCH model to be biased for skewed time series. To deal with this problem non-linear GARCH models are introduced, which take into account skewed distributions. The QGARCH model differs from model the classical GARCH model by

$$y_t = \sqrt{h_t} Z_t \tag{2.22}$$

$$\begin{aligned} h_t &= \delta_0 + \delta_1 h_{t-1} + \delta_2 (y_{t-1} + \delta_3)^2 \\ &= \left(\delta_0 + \delta_2 \delta_3^2\right) + \delta_1 h_{t-1} + 2\delta_2\delta_3 y_{t-1} + \delta_2 y_{t-1}^2 \end{aligned} \tag{2.23}$$

This model reduces to the GARCH (1,1) model when the shift parameters $\delta_3 = 0$. The *QGARC* Hmodel can improve upon the standard GARCH since they can cope with positive (or negative) skewness.

Theorem 2.6. Consider the general class of RCA QGARCH (1,1) Volatility Models for the time series y_t, where

$$y_t = \sqrt{h_t} Z_t \tag{2.24}$$

$$h_t = \left(\delta_0 + \delta_2 \delta_3^2\right) + \delta_1 h_{t-1} + 2\delta_2\delta_3 y_{t-1} + \left(\delta_2 + |a_{t-1}|\right) y_{t-1}^2 \tag{2.25}$$

where $Z_t \sim N\left(0, \sigma_Z^2\right)$ and $a_t \sim N\left(0, \sigma_a^2\right)$. Then, we have the following moment properties

$$E(h_t)=\frac{\left(\delta_0+\delta_2\delta_3^2\right)}{\left(1-\left(\delta_1+\delta_2\sigma_Z^2+\sigma_Z^2\sigma_a\sqrt{\frac{2}{\pi}}\right)\right)}$$

$$E\left(h_{t-1}^2\right)=\frac{2\delta_0\delta_2\delta_3^2+\delta_0^2+\delta_2^2\delta_3^4}{\left(1-\left(6\sigma_Z^4\delta_2\sigma_a\sqrt{\frac{2}{\pi}}+\delta_1^2+3\sigma_Z^4\delta_2^2+3\sigma_Z^4\sigma_a^2+2\delta_1\sigma_Z^2\delta_2+2\delta_1\sigma_Z^2\sigma_a\sqrt{\frac{2}{\pi}}\right)\right)}$$
$$+\frac{2\left(3\delta_2^2\delta_3^2\sigma_Z^2+\delta_0\sigma_Z^2\sigma_a\sqrt{\frac{2}{\pi}}+\delta_0\delta_2\sigma_Z^2+\delta_0\delta_1+\delta_2\delta_3^2\delta_1+\delta_2\delta_3^2\sigma_Z^2\sigma_a\sqrt{\frac{2}{\pi}}\right)}{\left(1-\left(6\sigma_Z^4\delta_2\sigma_a\sqrt{\frac{2}{\pi}}+\delta_1^2+3\sigma_Z^4\delta_2^2+3\sigma_Z^4\sigma_a^2+2\delta_1\sigma_Z^2\delta_2+2\delta_1\sigma_Z^2\sigma_a\sqrt{\frac{2}{\pi}}\right)\right)}E(h_{t-1})$$

$$K^{(y)}=\frac{3\left(2\delta_0\delta_2\delta_3^2+\delta_0^2+\delta_2^2\delta_3^4\right)\left(1-\left(\delta_1+\delta_2\sigma_Z^2+\sigma_Z^2\sigma_a\sqrt{\frac{2}{\pi}}\right)\right)^2}{\left(1-\left(6\sigma_Z^4\delta_2\sigma_a\sqrt{\frac{2}{\pi}}+\delta_1^2+3\sigma_Z^4\delta_2^2+3\sigma_Z^4\sigma_a^2+2\delta_1\sigma_Z^2\delta_2+2\delta_1\sigma_Z^2\sigma_a\sqrt{\frac{2}{\pi}}\right)\right)\left(\delta_0+\delta_2\delta_3^2\right)^2}$$
$$+\frac{6\left(\left(3\delta_2^2\delta_3^2+\delta_0\delta_2\right)\sigma_Z^2+\sigma_Z^2\left(\delta_0+\delta_2\delta_3^2\right)\sigma_a\sqrt{\frac{2}{\pi}}+\left(\delta_0+\delta_2\delta_3^2\right)\delta_1\right)\left(1-\left(\delta_1+\sigma_Z^2\left(\delta_2+\sigma_a\sqrt{\frac{2}{\pi}}\right)\right)\right)}{\left(1-\left(6\sigma_Z^4\delta_2\sigma_a\sqrt{\frac{2}{\pi}}+\delta_1^2+3\sigma_Z^4\delta_2^2+3\sigma_Z^4\sigma_a^2+2\delta_1\sigma_Z^2\delta_2+2\delta_1\sigma_Z^2\sigma_a\sqrt{\frac{2}{\pi}}\right)\right)\left(\delta_0+\delta_2\delta_3^2\right)}$$

Proof: is easy and is omitted.

Theorem 2.7. Consider the special class of RCA *QGARCH* (1,1) Sign Volatility Models for the time series y_t, where

$$y_t=\left(\phi+|b_t|\right)y_{t-1}+\varepsilon_t \tag{2.26}$$

$$\varepsilon_t=\sqrt{h_t}Z_t \tag{2.27}$$

$$\begin{aligned}h_t&=\left(\delta_0+\delta_2\delta_3^2\right)+\delta_1h_{t-1}+2\delta_2\delta_3y_{t-1}\\&+\left(\delta_2+|a_{t-1}|+\Phi s_{t-1}\right)y_{t-1}^2\end{aligned} \tag{2.28}$$

where $Z_t\sim N\left(0,\sigma_Z^2\right)$, $a_t\sim N\left(0,\sigma_a^2\right)$, at ~*N*(0,σ2a) and ~*N*(0,σ2*a*)are sequences of independent, identically distributed random variables with zero mean, variance given by σ_Z^2 and σ_b^2 and σ_a^2 respectively, and

$$s_t=\begin{cases}+1 & \text{if} \quad y_t>0\\ 0 & \text{if} \quad y_t=0\\ -1 & \text{if} \quad y_t<0\end{cases}$$

Note: $E\left(s_t^2\right)=1$, and in order to calculate the kurtosis, we observe that $E\left(s_t^4\right)=1$. Then, we have the following moment properties

$$E\left(h_t\right)=\frac{\left(1-\left(\phi^2+2\phi\sigma_b\sqrt{\frac{2}{\pi}}+\sigma_b^2\right)\right)\left(\delta_0+\delta_2\delta_3^2\right)}{\left(1-\delta_1\right)\left(1-\left(\phi^2+2\phi\sigma_b\sqrt{\frac{2}{\pi}}+\sigma_b^2\right)\right)-\sigma_Z^2\left(\delta_2+\sigma_a\sqrt{\frac{2}{\pi}}\right)} \tag{2.29}$$

$$E\left(y_t^2\right)=\frac{\sigma_Z^2\left(1-\left(\phi^2+2\phi\sigma_b\sqrt{\frac{2}{\pi}}+\sigma_b^2\right)\right)\left(\delta_0+\delta_2\delta_3^2\right)}{\left(\left(1-\delta_1\right)\left(1-\left(\phi^2+2\phi\sigma_b\sqrt{\frac{2}{\pi}}+\sigma_b^2\right)\right)-\sigma_Z^2\left(\delta_2+\sigma_a\sqrt{\frac{2}{\pi}}\right)\right)\left(1-\left(\phi^2+2\phi\sigma_b\sqrt{\frac{2}{\pi}}+\sigma_b^2\right)\right)} \tag{2.30}$$

$$E\left(y_t^4\right)=\frac{3\sigma_Z^4}{\left(1-\left(\phi^4+3\sigma_b^4+4\phi^3\sigma_b\sqrt{\frac{2}{\pi}}+6\phi^2\sigma_b^2+8\phi\sqrt{\frac{2}{\pi}}\sigma_b^3\right)\right)}E\left(h_t^2\right)$$
$$+\frac{\sigma_Z^4\left[6\sigma_b^2+6\phi^2+12\phi\sigma_b\sqrt{\frac{2}{\pi}}\right]\left[1-\left(\phi^2+2\phi\sigma_b\sqrt{\frac{2}{\pi}}+\sigma_b^2\right)\right]^{-1}}{\left(1-\left(\phi^4+3\sigma_b^4+4\phi^3\sigma_b\sqrt{\frac{2}{\pi}}+6\phi^2\sigma_b^2+8\phi\sqrt{\frac{2}{\pi}}\sigma_b^3\right)\right)}\left(E\left[h_t\right]\right) \tag{2.31}$$

$$K^{(y)}=\frac{\left[1-\left(\phi^2+2\phi\sigma_b\sqrt{\frac{2}{\pi}}+\sigma_b^2\right)\right]^2}{\sigma_Z^4}\left(A\frac{E\left(h_t^2\right)}{E\left[h_t\right]^2}+B\right) \tag{2.32}$$

where

$$A=\frac{3\sigma_Z^4}{\left(1-\left(\phi^4+3\sigma_b^4+4\phi^3\sigma_b\sqrt{\frac{2}{\pi}}+6\phi^2\sigma_b^2+8\phi\sqrt{\frac{2}{\pi}}\sigma_b^3\right)\right)} \tag{2.33}$$

$$B = \frac{\sigma_Z^4\left[6\sigma_b^2 + 6\phi^2 + 12\phi\sigma_b\sqrt{\frac{2}{\pi}}\right]\left[1-\left(\phi^2 + 2\phi\sigma_b\sqrt{\frac{2}{\pi}} + \sigma_b^2\right)\right]^{-1}}{\left(1-\left(\phi^4 + 3\sigma_b^4 + 4\phi^3\sigma_b\sqrt{\frac{2}{\pi}} + 6\phi^2\sigma_b^2 + 8\phi\sqrt{\frac{2}{\pi}}\sigma_b^3\right)\right)} \tag{2.34}$$

$$E\left(h_t^2\right) = \frac{M_1 + M_2}{\left[\left(1-\delta_1^2\right) - M_9 A\right]} + \frac{\left(M_3 + M_4 + M_5 + M_6 + M_7 + M_8\right)\sigma_Z^2}{\left[1-\left(\phi^2 + 2\phi\sigma_b\sqrt{\frac{2}{\pi}} + \sigma_b^2\right)\right]\left[\left(1-\delta_1^2\right) - M_9 A\right]} E\left[h_t\right] + \frac{M_9 B\left(E\left[h_t\right]\right)^2}{\left[\left(1-\delta_1^2\right) - M_9 A\right]} \tag{2.35}$$

$$M_1 = \left(\delta_0^2 + 2\delta_0\delta_2\delta_3^2 + \delta_2^2\delta_3^4\right)$$

$$M_2 = \frac{\left(2\delta_0\delta_1 + 2\delta_2\delta_3^2\delta_1\right)\left(1-\left(\phi^2 + 2\phi\sigma_b\sqrt{\frac{2}{\pi}} + \sigma_b^2\right)\right)\left(\delta_0 + \delta_2\delta_3^2\right)}{\left(\left(1-\delta_1\right)\left(1-\left(\phi^2 + 2\phi\sigma_b\sqrt{\frac{2}{\pi}} + \sigma_b^2\right)\right) - \sigma_Z^2\left(\delta_2 + \sigma_a\sqrt{\frac{2}{\pi}}\right)\right)}$$

$$M_3 = 2\delta_0\delta_2, \qquad M_4 = 6\delta_2^2\delta_3^2$$

$$M_5 = \frac{2\delta_1\delta_2\left(1-\left(\phi^2 + 2\phi\sigma_b\sqrt{\frac{2}{\pi}} + \sigma_b^2\right)\right)\left(\delta_0 + \delta_2\delta_3^2\right)}{\left(\left(1-\delta_1\right)\left(1-\left(\phi^2 + 2\phi\sigma_b\sqrt{\frac{2}{\pi}} + \sigma_b^2\right)\right) - \sigma_Z^2\left(\delta_2 + \sigma_a\sqrt{\frac{2}{\pi}}\right)\right)}$$

$$M_6 = 2\delta_0\sigma_a\sqrt{\frac{2}{\pi}}$$

$$M_7 = \frac{2\delta_1\sigma_a\sqrt{\frac{2}{\pi}}\left(1-\left(\phi^2 + 2\phi\sigma_b\sqrt{\frac{2}{\pi}} + \sigma_b^2\right)\right)\left(\delta_0 + \delta_2\delta_3^2\right)}{\left(\left(1-\delta_1\right)\left(1-\left(\phi^2 + 2\phi\sigma_b\sqrt{\frac{2}{\pi}} + \sigma_b^2\right)\right) - \sigma_Z^2\left(\delta_2 + \sigma_a\sqrt{\frac{2}{\pi}}\right)\right)}$$

$$M_8 = 2\delta_2\delta_3^2\sigma_a\sqrt{\frac{2}{\pi}}$$

$$M_9 = \left(\Phi^2 + 2\sigma_a\sqrt{\frac{2}{\pi}}\delta_2 + \delta_2^2 + \sigma_a^2\right)$$

Proof: is easy and is omitted.

3. OPTION PRICING WITH VOLATILITY

Option pricing based on the Black-Scholes model is widely used in the financial community. The BlackScholes formula is used for the pricing of European-style options. The model has traditionally assumed that the volatility of returns is constant. However, several studies have shown that assetre turns exhibit variances that change time [9,10,] and others derived closed form option pricing formulas for different models which are assumed to follow a GARCH volatility process. Most recently, Gong et al. [11] derive an expression for the call price as an expectation with respect to random GARCH volatility. The model is then evaluated in terms of the moments of the volatility process. Their results indicate that the suggested model outperforms the classic BlackScholes formula. Here we apply [11] and propose an option pricing model with RCA GARCH volatility as follows:

$$dS_t = rS_t dt + \sigma_t S_t dW_t \tag{3.1}$$

$$y_t = \log\left(\frac{S_t}{S_{t-1}}\right) - E\left[\log\left(\frac{S_t}{S_{t-1}}\right)\right] = \sigma_t Z_t \tag{3.2}$$

$$\Phi(B)\sigma_t^2 = \omega + \beta(B) y_t^2 \tag{3.3}$$

where S_t is the price of the stock, r is the risk-free interest rate, $\{W_t\}$ is a standard Brownianmotion, σ_t is the time-varying RCA *GARCH* volatility process, $\{Z_t\}$ is a sequence of i.i.d. randomvariables with zero mean and unit variance and $\Phi(B)$, and $\beta(B)$ have been defined in (1.5). Theprice of a call option can be calculated using the option pricing formula given in [11]. The call priceis derived as a first conditional moment of a truncated lognormal distribution under the martingalemeasure, and it is based on estimates of the moments of the *GARCH* process. The call price basedon the Black-Scholes model with seasonal GARCH volatility is given by:

$$\begin{aligned} C(S,T) = S&\left(f\left[E(\sigma_t^2)\right] + \frac{1}{2} f''\left[E(\sigma_t^2)\right]\left(\frac{1}{3}\kappa^{(y)} - 1\right) E^2(\sigma_t^2)\right) \\ &- Ke^{-rT}\left(g\left[E(\sigma_t^2)\right] + \frac{1}{2} g''\left[E(\sigma_t^2)\right]\left(\frac{1}{3}\kappa^{(y)} - 1\right) E^2(\sigma_t^2)\right), \end{aligned} \tag{3.4}$$

where f and g are twice differentiable functions, S is the initial value of S_t, K is the strike price, T is the expiry date, σ_t is a stationary process with finite fourth moment and

$$\kappa^{(y)} = \frac{E\left(y_t^4\right)}{\left[E\left(y_t^2\right)\right]^2}.$$

Also,

$$f\left[E\left(\sigma_t^2\right)\right],\ g\left[E\left(\sigma_t^2\right)\right],\ f''\left[E\left(\sigma_t^2\right)\right]$$

and

$$g''\left[E\left(\sigma_t^2\right)\right]$$

are given by:

$$f\left[E\left(\sigma_t^2\right)\right] = \mathrm{N}(d) = \mathrm{N}\left(\frac{\log(S/K) + rT + \frac{1}{2}E\left(\sigma_t^2\right)}{\sqrt{E\left(\sigma_t^2\right)}}\right).$$

$$g\left[E\left(\sigma_t^2\right)\right] = \mathrm{N}\left(d - \sqrt{E\sigma_t^2}\right) = \mathrm{N}\left(\frac{\log(S/K) + rT - \frac{1}{2}E\left(\sigma_t^2\right)}{\sqrt{E\sigma_t^2}}\right).$$

$$f''\left[E\left(\sigma_t^2\right)\right] = \frac{1}{\sqrt{2\pi}}\left[-\left(\frac{E\left(\sigma_t^2\right) - 2\left(\log(S/K) + rT\right)}{4E\left(\sigma_t^2\right)\sqrt{E\left(\sigma_t^2\right)}}\right)\left(\frac{\left[E\left(\sigma_t^2\right)\right]^2 - 4\left(\log(S/K) + rT\right)^2}{8\left[E\left(\sigma_t^2\right)\right]^2}\right)\right.$$
$$\left. + \left(\frac{6\left(\log(S/K) + rT\right) - E\left(\sigma_t^2\right)}{8\left[E\left(\sigma_t^2\right)\right]^2\sqrt{E\left(\sigma_t^2\right)}}\right)\right] \times \exp\left\{-\frac{\left(2\left(\log(S/K) + rT\right) + E\left(\sigma_t^2\right)\right)^2}{8E\left(\sigma_t^2\right)}\right\}.$$

$$g''\left[E\left(\sigma_t^2\right)\right] = \frac{1}{\sqrt{2\pi}}\left[\left(\frac{E\left(\sigma_t^2\right) + 2\left(\log(S/K) + rT\right)}{4E\left(\sigma_t^2\right)\sqrt{E\left(\sigma_t^2\right)}}\right)\left(\frac{\left[E\left(\sigma_t^2\right)^2 - 4\left(\log(S/K) + rT\right)^2\right]}{\left[E\left(\sigma_t^2\right)^2\right]}\right)\right.$$
$$\left. + \left(\frac{6\left(\log(S/K) + rT\right) + E\left(\sigma_t^2\right)}{8\left[E\left(\sigma_t^2\right)\right]^2\sqrt{E\left(\sigma_t^2\right)}}\right)\right] \exp\left\{-\frac{\left(2\left(\log(S/K) + rT\right) - E\left(\sigma_t^2\right)\right)^2}{8E\left(\sigma_t^2\right)}\right\},$$

where N denotes the standard normal CDF, and under the option pricing model with RCA GARCH volatility,

$$E\left(\sigma_t^2\right)=\frac{\omega}{\left(1-\sum_{i=1}^{p}\phi_i\right)\left(1-\sum_{i=1}^{p}\Phi_i\right)}, \kappa^{(y)}=\frac{3}{3-2\sum_{j=1}^{\infty}\Psi_j^2}.$$

4. CONCLUDING REMARKS

Srimantoorao S. Appadoo, Aerambamoorthy Thavaneswaran, Saman Muthukumarana Financial time series exhibit excess kurtosis and in this paper, we propose various classes of RCAGARCH volatility models and derive the kurtosis in terms of model parameters. We consider time series models such as RCA with GARCH errors and quadratic GARCH errors. The models introduced here extend and complement the existing volatility models in the literature to RCA models with quadratic GARCH models by introducing more general structures. The results are primarily oriented to financial time series applications. Financial time series often meet the large data set demands of the volatility models studied here. Also, financial data dynamics and higher order moments are of interest to many market participants. Specifically, we consider the Black-Scholes model with RCA GARCH volatility and show that these moments can be used to evaluate the call price for European options.

REFERENCES

1. T. Bollerslev, "Generalized Autoregressive Conditional Heteroscedasticity," Journal of Econometrics, Vol. 31, No. 3, 1986, pp. 307-327. doi:10.1016/0304-4076(86)90063-1
2. Thavaneswaran, S. S. Appadoo and M. Samanta, "Random Coefficient GARCH Models," Mathematical and Computer Modelling, Vol. 41, No. 6-7, 2005, pp. 723-733. doi:10.1016/j.mcm.2004.02.032
3. Thavaneswaran, S. S. Appadoo and M. Ghahramani, "RCA Models with GARCH Errors," Applied Mathematics Letters, Vol. 22, No. 1, 2009, pp. 110-114.doi:10.1016/j.aml.2008.02.015
4. S. S. Appadoo, A. Thavaneswaran and S. Mandal, "RCA Models with Quadratic GARCH Innovations Distributions," Applied Mathematics Letters, 2012, in Press.
5. Paseka, S. Appadoo and A. Thavaneswaran, "Random Coefficient Autoregressive (RCA) Models with Nonlinear Stochastic Volatility Innovations," Journal of Applied Statistical Science, Vol. 17, No. 3, 2010, pp. 331-349.
6. Thavaneswaran and C. C. Heyde, "Prediction via Estimating Functions," Journal of Statistical Planning and Inference, Vol. 77, No. 1, 1999, pp. 89-101. doi:10.1016/S0378-3758(98)00179-7
7. D. Nicholls and B. Quinn, "Random Coefficient Autoregressive Models: An Introduction," Lecture Notes in Statistics, Vol. 11, Springer, New York, 1982.
8. R. Leipus and D. Surgallis, "Random Coefficient Autoregression,regime Switching and Lond Memory," Advances in Applied Probability, Vol. 35, No. 3, 2003, pp. 737-754.doi:10.1239/aap/1059486826

9. S. Heston and S. Nandi, "A Closed-Form GARCH Option Valuation Model," The Review of Financial Studies, Vol. 13, No. 3, 2000, pp. 585-625. doi:10.1093/rfs/13.3.585
10. R. Elliot, T. Siu and L. Chan, "Option Pricing for GARCH Models with Markov Switching," International Journal of Theoretical and Applied Finance, Vol. 9, No. 6, 2006, pp. 825-841. doi:10.1142/S0219024906003846
11. H. Gong, A. Thavaneswaran and J. Singh, "A BlackScholes Model with GARCH Volatility," The Mathematical Scientist, Vol. 35, No. 1, 2010, pp. 37-42.

CHAPTER 9

Risk Premia: Asymmetric Tail Risks and Excess Returns

Y. Lempérière, C. Deremble, T. T. Nguyen, P. Seager, M. Potters & J. P. Bouchaud*

Capital Fund Management, 23 rue de l'Université, 75007 Paris, France

1.INTRODUCTION

One of the pillars of modern finance theory is the concept of 'risk premium', i.e. that amore risky investment should, on the long run, also be more profitable. The argument leading to such a conclusion looks, at least at first sight, rock solid: given the choice between two assets with the same expected return, any rational investor should prefer the less risky one. The resulting smaller demand for riskier assets should make their prices drop, and therefore their yields go up, until they become attractive enough to lure in investors in spite of their higher risk. This simple idea appears to explain why stock indices on the long run yield higher returns than monetary funds, for example. It is also at the heart of the well-known Capital Asset Pricing Model (CAPM), which predicts that the excess return of a given stock (over the risk free rate) is proportional to its so-called 'β', i.e. to its covariance with the market risk (Sharpe 1964).

The idea has become so pervasive that its logic has actually inverted: efficient market theorists assume that any return in excess of the risk free rate should in fact be associated with a non-diversifiable risk factor of some sort—sometimes argued to be unobservable or 'latent'. It is often invoked to do away with any market anomaly or discrepancies between market prices and theoretical prices, as it allows one to introduce an extra fitting parameter conveniently called the 'market price of risk'.†

Although the argument underlying the existence of risk premia is cogent, estimating and rationalizing their order of magnitude in financial markets is still very much a matter of debate. This is quite surprising in view of the importance of the issue for the asset management industry as a whole, and reveals how primitive is our understanding of financial prices and returns. Empirically, the equity risk premium (ERP) is found to be in the range of 3–9% depending on the period and country (see Fernandez *et al.* 2000, Mehra and Prescott 2003, Damodaran 2012, and our own data in section 2). Mehra and Prescott have pointed out that these values are too high to be explained within a general equilibrium model of the economy and have coined this discrepancy as the 'equity premium puzzle' (Mehra and Prescott 1985).‡ More recently, several authors (Ciliberti *et al.* 2015, Ang *et al.* 2006, 2009, Blitz and van Vliet 2007, Baker *et al.* 2011, Frazzini and Pedersen 2014) have reported an inverted relation between the volatility of a stock (or its β) and its excess return. In other words, contrarily to intuition, less volatile stocks appear to be *more* profitable.

As pointed out in numerous papers—many of them quite recent—the problem may reside in the very definition of risk. Classical theories identify risk with volatility σ. This (partly) comes from the standard assumption of a Gaussian distribution for asset returns, which is entirely characterized by its first two moments: mean μ and variance σ2. But in fact fluctuations are known to be strongly non Gaussian, and investors are arguably not much concerned by small fluctuations around the mean. Rather, they fear large negative drops of their wealth, induced by rare, but plausible crashes. These negative events are not captured by the r.m.s. σ but rather contribute to the negative skewness of the distribution. Therefore, an alternative idea that has progressively emerged in the literature is that a large contribution to the 'risk premium'is in fact a compensation for holding an asset that provides positive average returns but may occasionally erase a large fraction of the accumulated gains. This was in fact proposed as a resolution of the Mehra and Prescott 'equity premium puzzle' by Kraus and Litzenberger (1976), Rietz (1988) and then in many follow-up papers with a similar message, e.g. (Harvey and Siddique 2000, Barro 2006, Santa-Clara andYan 2010, Bollerslev and Todorov 2011), (see also Kelly and Jiang 2013, Kozhan *et al.* 2013, van Oordt and Zhou 2013). Risk premium might be more a question of skewness and negative tail events than related to volatility *per se*.

A more formal way to frame this story is in terms of the utility function supposed to describe investors' preferences. Assuming that the 'happiness' of the investor depends on his/her wealth W as a certain function $U(W)$, the preference for higher profits imposes that $U(W)$ is a monotonically increasing function of W, whereas the condition $U''(W) \leq 0, \forall W$ ensures that a less volatile investment with the same expected gain is always preferred. Assuming that the proposed investment yields an uncertain final wealth W distributed around a mean E[W] with small fluctuations of variance σ^2, and expanding in powers of the difference W–E[W], one finds that the expected utility is given by:†

$$\mathbb{E}[U(W)] = U(\mathbb{E}[W]) - \frac{1}{2}\lambda^2\sigma^2 + \cdots$$
$$\lambda^2 := -U''(\mathbb{E}[W]) \geq 0. \tag{1}$$

Hence, within a small risk expansion, utility reduction is entirely described by the variance $\sigma 2$. The above approximate formula is the starting point for most risk-return portfolio optimization schemes in the classical literature.But as intuition suggests, skewness matters. Lottery experiments (Garrett and Russell 1999, Astebro *et al.* 2009) demonstrate that agents prefer lotteries with very large potential gains—even when the expected gain is negative—to lotteries with small positive expected gain but no possibility of large payoffs. Symmetrically, one expects agents to shy away from investments with a high degree of negative skewness. Clearly, this cannot be captured by equation (1) above where skewness is absent; it may well be the case that the whole concept of utility is anyway insufficient to account for behavioural biases, see (Kahneman and Tversky 1979). Still, a large number of papers have been devoted to generalizing Markowitz's portfolio optimization and the CAPMto account for skewness preferences (Kraus and Litzenberger 1976, Harvey and Siddique 2000, Athayde and Flores 2004, Harvey *et al.* 2010). In terms of utility functions, one can show that the additional condition *U__ (W)* ≥ 0, ∀*W* ensures that more positively skewed investments are preferred (Chiu 2010).‡

Our work is clearly in the wake of the above-mentioned literature on skewness preferences and tail-risk aversion. We present extensive evidence that 'risk premium' is indeed strongly correlated with the skewness of a strategy but very little with its volatility, not only in the equity world—as was emphasised by previous authors—but in other sectors as well. We will investigate in detail many classical so-called 'risk premium'strategies (in equities, bonds, currencies, options and credit) and elicit an approximately linear relation between the Sharpe ratio of these strategies and their negative skewness. We will find however that some well-known strategies, such as trend following and to a lesser extent the Fama–French 'High minus Low' factor and the 'Low Vol' strategy, are clearly not following this rule, suggesting that these strategies are not risk premia but genuine market anomalies.

Compared to the previous abundant literature, the present results are new in different respects:

- First, at variance with most previous investigations (that mostly focuses on stock markets, whereas we analyse a broad spectrum of investment strategies), we do not attempt to frame our empirical analysis within the constraining framework of asset pricing and portfolio theory, but rather let the data speak for itself. This is especially important when studying, as we do here, risk premia across a large universe of assets, where the notion of a global 'risk factor' (generalizing the market factor in the equity space) is far from clear, as we discuss below (see section 8).
- Second, we introduce a simple way to plot the returns of a portfolio that reveals its skewness to the 'naked eye' and suggests an intuitive and robust definition of skewness that is much less sensitive to extreme events.
- Third, our empirical conclusion that for a wide spectrum of 'risk premia'strategies, skewness rather than volatility is a determinant of returns is, to the best of our knowledge, new, as is the finding that some investment strategies— like trend following—seem to behave quite differently.

The paper is organized as follows: we first start in section 2 with the equity market as a whole and revisit the ERP worldwide, and its (negative) correlation with the volatility.We then introduce our new, intuitive definition of skewness that we use throughout the paper and that we justify in the appendix.

We focus on the Fama–French factors in section 3 and study the statistics of market neutral portfolios, including a 'Low Volatility' portfolio. We move on to the fixed income world (Section 4), where we again build neutral portfolios. Section 5 is devoted to an account of risk premia on currencies (the so called 'Carry Trade'), and finally, in Section 6, to the paradigmatic case of selling options. We summarize our findings in Section 7 with a suggestive linear relation between the Sharpe ratio and the skewness of all the Risk Premium strategies investigated in the paper, and discuss some exceptions to the rule—i.e. positive Sharpe strategies with zero or positive skewness—that we define as 'pure α' strategies. We then end the paper in Section 8 by comparing our results to the extended CAPM framework *à la* Kraus and Litzenberger (1976), Harvey and Siddique (2000). We propose instead a simple argument to rationalize for the Sharpe ratio/skewness trade-off and an objective criterion to assess the quality of a risk premium portfolio.

2. BACK TO BASICS: RISK PREMIUM IN GLOBAL EQUITY MARKETS, AND A NEW DEFINITION OF SKEWNESS

2.1. No 'volatility premium' in stock indices

According to the standard framework described in the previous section, investing in the stock market rather than in risk-free instruments only makes sense if stocks provide, in the long run, better returns than the risk-free rate. The existence and the strength of this ERP have been much debated amongst both the academic community and the asset management industry. The general consensus (confirmed below) is that this premium exists and is in the range of 3–9%, with large fluctuations, both across epochs and countries (Fernandez *et al.* 2000, Mehra and Prescott 2003, Damodaran 2012).Within a classical general equilibrium framework, this premium appears to be too large. With such a level of ERP, rational investors should put money more massively in equities, and the demand for bonds should be much weaker than what we observe, which should lead to a smaller risk premium. This is the 'equity premium puzzle' brought to the fore by Mehra and Prescott (1985).

In order to estimate the ERP, it is customary to compute the total return of indices (dividend included), and to subtract the risk-free rate.† We have therefore computed this cumulated difference for various countries, using the 10Y government rate as a risk-free asset (using three-month bills would only give a more flattering result, since short rates are usually lower than 10Y rates). The result is, as expected, overwhelmingly in favour of the existence of a risk premium, as can be seen from table 1 and figure 1. All countries but one display positive excess return on the long run (some time series go back as far as 1870). The global t-stat of a strategy equi-invested in all available indices at any moment in time is 4.2. The average value (weighted by the number of available years for each country) is found to be ≈ 5% with a cross-sectional dispersion of ±3%, compatible with the numbers quoted above (Fernandez *et al.* 2000, Mehra and Prescott 2003, Damodaran 2012).

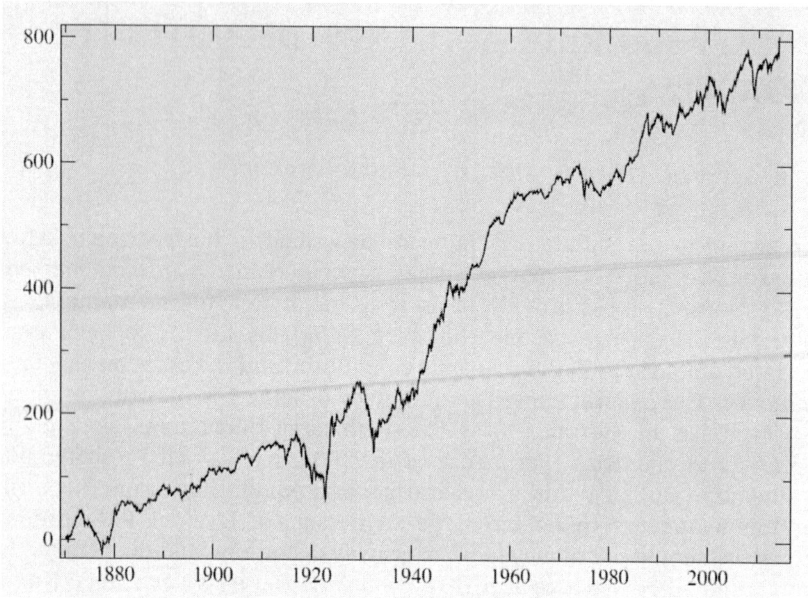

Figure 1. Cumulated ERP since 1870, corresponding to a strategy where one holds a long portfolio of indices, with equal weight on all available contracts at any instant of time, i.e. 1 in 1870 and 27 in 2014. The *t*-stat is 4.2.

In the fourth column of table 1, we give the average annual volatility of the corresponding index. This allows us to compute the Sharpe ratio $S = \mu/\sigma$ of each risk premium. Note indeed that the value of the risk premium itself is not meaningful since it can be arbitrarily leveraged. Therefore here and throughout the paper, we use the a-dimensional Sharpe ratio (SR) meaningfully compare different risk premia. Figure 2 shows a scatter plot of S as a function of the volatility of the index, together with a regression line. We see that the two quantities are in fact *negatively* correlated ($\rho \approx -0.27$). Even the risk premium itself (i.e. without dividing by the volatility σ) is found to be weakly anti-correlated with σ ($\rho \approx -0.1$) but this is not statistically significant. This is at odds with the standard interpretation of the risk premium, but confirms once again the 'low-volatility' anomaly reported by many authors (Ciliberti *et al.* 2015, Ang *et al.* 2006, 2009, Blitz and van Vliet 2007, Baker *et al.* 2011, Frazzini and Pedersen 2014): the risk-adjusted returns of low-vol stocks is found to be higher than for high-vol stocks not only within one market, but across international indices

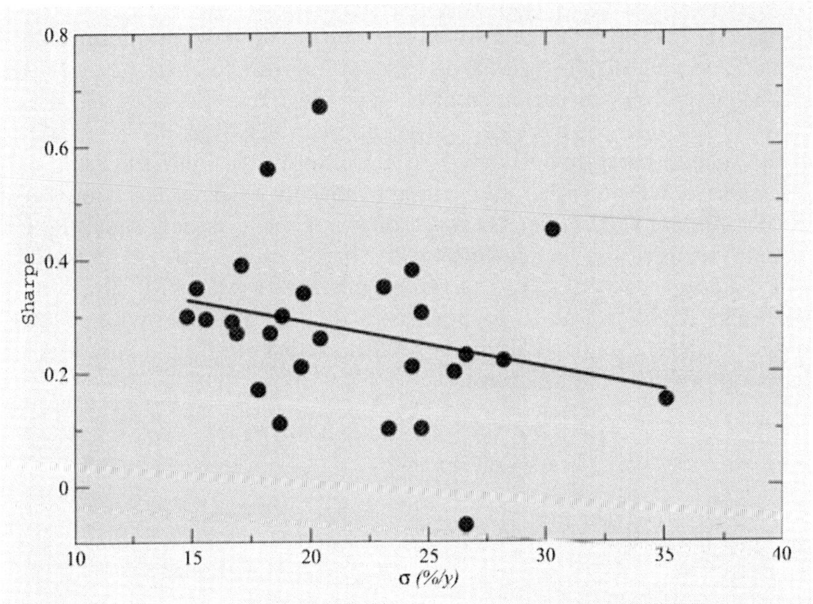

Figure 2. Sharpe ratio of the excess return (over the 10Y bond) for each index listed in table 1 as a function of the volatility of that index. We also plot the regression line, with a *negative* slope and a correlation coefficient $\rho \approx -0.27$. Note that the (flat) average Sharpe ratio over all contracts is $S \approx 0.3$.

2.2. Downside tail risk and skewness premium

However, something interesting occurs when we suggestively re-plot the cumulated ERP(say for the risk managed SPX since 1928†) in the following way. Instead of considering the returns of the strategy in chronological order, we sort these returns in terms of their absolute value and plot the cumulated ERP $F(p)$ as a function of the normalized rank $p = k/N$, starting from the return with the smallest amplitude ($k = 1$) and ending with the largest ($k = N$). The result in the case of the US since 1928 is shown in figure 3. We immediately see that while the small returns contribute positively to the average, the largest returns, contrarily, lead to a violent drop of the P&L. Strikingly, the 5% largest returns wipe out half of the ERP of the 95% small-to-moderate returns.

Table 1. Average annualized excess return ("risk premium") in %, annualized volatility in %, Sharpe ratio, standard skewness $\zeta 3$ and ranked P&L skewness ζ * for various indices over the 10Y bond rates, all based on monthly returns. The average Sharpe ratio is ≈ 0.3. Start dates as indicated in the second column; end date Jan 2014. Note that while the ranked P&L skewness ζ 3 is negative across most markets as expected, the classical third cumulant skewness $\zeta 3$ is very noisy and has a random sign. In the last column, we give the co-skewness with the US index, defined as in Harvey and Siddique (2000). The correlation between the ERP and this co-skewness is also found to be positive ≈ 0.18, at variance with results on single stocks Harvey and Siddique (2000). Data from Global Financial Data, (www.globalfinancialdata.com).

Country	Start date	Premium	Vol.	Sharpe	Skew ζ_3	Skew ζ^*	Co-skew/US
Australia	1983	4.77	16.7	0.29	−2.0	−2.35	−0.29
Canada	1934	4.44	14.8	0.30	−0.6	−1.68	−0.12
Germany	1870	4.23	19.6	0.21	1.7	−0.31	−0.15
France	1898	6.65	19.7	0.34	1.7	1.05	−0.11
Finland	1970	8.09	23.1	0.35	0.2	0.8	−0.10
Italy	1925	6.24	26.6	0.23	0.	1.30	−0.06
Hong Kong	1996	5.24	26.1	0.20	−0.2	−1.87	−0.11
Hungary	1999	2.51	24.7	0.10	0.5	−1.79	−0.20
India	1988	13.6	30.3	0.45	−0.1	0.63	−0.07
Indonesia	2009	13.8	20.4	0.67	0.4	−3.26	0.12
Japan	1980	2.04	18.7	0.11	−0.2	−0.85	−0.21
South Korea	2000	9.33	24.3	0.38	0.1	−0.08	−0.19
Malaysia	1974	6.32	28.2	0.22	−0.5	−0.14	−0.14
Mexico	2001	10.1	18.2	0.56	−0.5	−2.26	−0.32
The Netherlands	1957	4.57	16.9	0.27	−0.1	−1.7	−0.17
New Zealand	1984	3.02	17.8	0.17	0.2	−1.67	−0.12
Norway	1969	5.21	24.3	0.21	−0.5	0.17	−0.11
Philippines	1996	−1.96	26.6	−0.07	0.1	−0.42	−0.11
Poland	1999	2.29	23.3	0.17	−0.2	−0.62	−0.26
Singapore	1973	7.51	24.7	0.31	−0.1	−0.86	−0.15
South Africa	1983	5.26	20.4	0.26	−0.6	−0.14	−0.20
Spain	1948	5.25	18.3	0.27	−0.6	−0.12	−0.08
Sweden	1960	6.69	17.1	0.39	−0.1	−0.42	−0.05
Switzerland	1966	4.58	15.6	0.26	−0.5	−2.05	−0.16
Thailand	1997	5.25	35.1	0.15	0.1	−0.94	−0.12
United Kingdom	1958	5.71	18.8	0.30	1.	−2.31	−0.13
United States	1871	5.33	15.2	0.35	0.3	−0.32	−0.02

In the same graph, we also show the P&L $Fs\ (p)$ that would have been observed if the distribution of returns was exactly symmetrical around the same mean—i.e. such that the final point $Fs\ (p = 1)$ coincides with $F(p = 1)$.‡ In this case, it is easy to show that the P&L is (for large N) a monotonously increasing function of the normalized rank p. One finds that generically $F(p)$ behaves as $p3$ at small p—explaining the strong curvature seen in figure 3. The comparison between real returns and symmetries returns therefore reveals the *strongly skewed* nature of the ERP.

The ranked amplitude P&L representation in fact suggests a new general definition of skewness, as follows: one first normalizes the returns such that their average is zero and their rms is unity. Then the ranked amplitude P&L is plotted, and defines a certain function $F0(p)$, such that $F0(0) = F_0(1) = 0$.

By construction, the symmetric version $F_{0s}(p)$ is now zero (up to small fluctuations). Our definition of the skewness ζ^* is simply (minus) the area underneath the function $F_0(p)$:§

$$\zeta^* := -100 \int_0^1 \mathrm{d}p \, F_0(p) \tag{2}$$

where the arbitrary factor 100 is introduced such that the skewness is of order unity.

We give several properties of $F_0(p)$ in the appendix and show how ζ^* can be related to the classical definition of skewness ζ_3 defined as the third cumulant of the distribution of returns. Although standard, the latter definition is awkward for heavy-tailed distributions (as those describing financial returns) since a few extreme events can completely dominate the empirical determination of the skewness ζ_3. In these cases, it is strongly advised to use low-moment estimates of the skewness, such as the normalized mean-minus-median, for example.We prefer our new definition ζ^* above, not only because of its intuitive interpretation, but also because ζ^* can be defined even when the mean of the distribution diverges. (Note however that all the results reported in this paper are actually robust to the chosen definition of skewness, provided it is based on low moments.) We also show in the appendix that whenever $F_0(p)$ has a *single hump*, as is usually the case in financial applications, then the cumulative distribution of returns crosses its symmetrized version *twice and only twice*. This means, as announced in the introduction, that investors with a utility U such that U⍰⍰⍰ > 0 will shy away from investments with negative ζ^*, i.e. will demand a higher risk premium.

In the case of the US market factor, we find $\zeta^* \approx -1.47$ for daily returns (for a Sharpe ratio of 0.57), and $\zeta^*_{\mathrm{month}} \approx -0.32$ for monthly returns. When analysing the data given in table 1,we now find a *positive* correlation between $-\zeta^*$ and the ERP across different countries, $\rho \approx 0.2$. This is encouraging but not particularly convincing.† We have therefore extended our analysis to different contracts and different risk premia, and indeed find that in most cases, excess returns lead to a humpedshape function $F_0(p)$, meaning that returns of large amplitude have a negative mean. This will lead us to propose a possibly universal definition of 'risk premium'in terms of the skewness of the returns—in a precise sense, the premium compensates for a tail risk that is systematically biased downwards.

3. ZOOMING INTO STOCK MARKETS: THE FAMA–FRENCH FACTORS

We now consider the celebrated extended Fama–French model, which attributes excess stock returns to 4 components, corresponding to 4 orthogonal portfolios: the ERP discussed above (i.e. market minus the risk-free rate, MKT), SMB ('Small'caps minus 'Big' caps), UMD ('Up'—previous winners minus 'Down'—previous losers) and HML (High book to price—'value' stocks minus Low book to price—'growth' stocks), each of which is interpreted as a reward for a specific risk, even if the origin of said risk is not always clear.

The first, long market portfolio is tautologically not market neutral, while the other three are explicitly constructed in a market neutral way, in order to be orthogonal to the market risk factor MKT. SMB consists of being long the small cap stocks and short the large caps. This sounds like an intuitive risk premium strategy, since it seems plausible that small caps have a greater expected future growth but are more 'risky' than large caps. However, if we look at table 2, we see that the volatility of large or small cap indices is practically the same in the US and that, if anything, volatility is actually *smaller* for small caps than for large caps in all other countries. Therefore, volatility alone is unable to explain the existence of a possible SMB risk premium.

But if we now observe the above defined skewness ζ^* of large, mid and small cap indices reported in table 2, we clearly conclude that skewness does indeed become more negative as we move to smaller cap indices. This behaviour is furthermore quite universal: across the 10 cases we have investigated, only the Netherlands and New Zealand show a slight inversion of the skewness of mid- and small cap stocks, while still maintaining the general trend noted above. Let us insist that we have also computed three other (more standard) estimates of the skewness with exactly the same conclusion. To add to the significance of this observation, we have built market neutral portfolios that are long 1$ of the small-cap index, and short 1$ of the large-cap index for each of the 10 countries we consider. We then globally risk manage these portfolios by rescaling the position with a 20-day moving average of their own volatility. In this way, we achieve approximately constant volatility throughout the simulation. For each country in our pool, the skewness of these portfolios is found to be negative, and they all yield positive performance, except one of them (Australia)—see table 2, last column. This is quite significant, since these portfolios are completely decorrelated from one another.

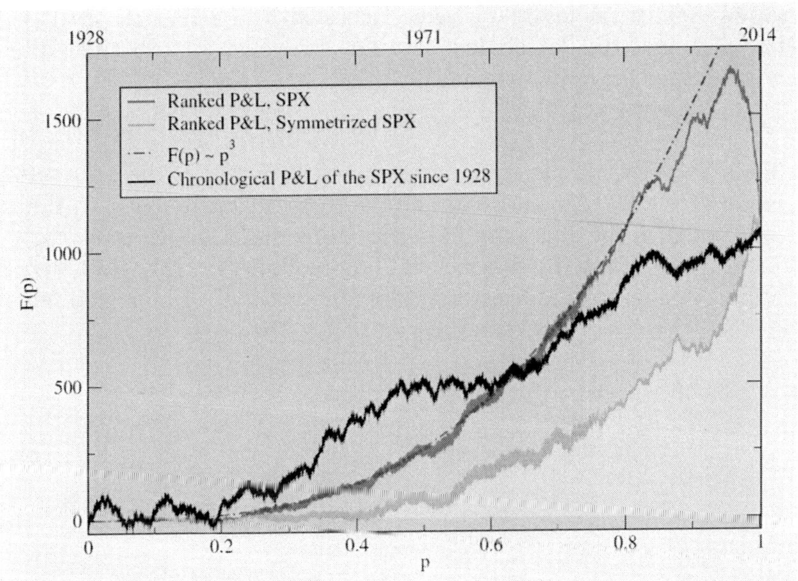

Figure 3. The ranked amplitude P&L representation: plot of the cumulated daily P&L *F(p)* for the SPX (at constant risk) since 1928, as a function of the normalised rank of the *amplitude* of the returns, $p = k/N$ (in red). The standard, chronological P&L corresponding to holding the US equity market index at constant risk is shown in black, that by construction ends at the exact same point. We also show for comparison (in green) *Fs (p)*, corresponding to returns symmetries around the same global mean (i.e. *Fs* $(p = 1) = F(1)$, again by construction). Our definition of the skewness ζ^* is, up to a negligible correction, minus the area enclosed between *F(p)* and *Fs (p)* (shown in yellow).

We can actually dwell further on this issue using the 'decile' portfolios for the three Fama–French factors in the US equity market.‡ We report in table 3 the volatility, skewness and Sharpe ratio for being long each decile portfolio since 1950, for the three factors SMB, HML and UMD. Again, there is little evidence that the volatility plays any role at all for any of the three factors. The skewness, on the other hand, becomes more negative for the portfolios with the highest returns for the SMB and UMD (momentum) factors. The smallest caps have (when traded together) a skewness $\zeta^* = -1.83$, whereas the largest caps have much weaker skewness $\zeta^* = -0.39$, confirming the conclusion above. A long portfolio of winners has a skewness $Z^* = -0.90$ and is prone to large reversal events, whereas a long portfolio of losers are essentially not skewed.† For HML, surprisingly, there is no such effect; if anything, the skewness appears to be slightly less negative for high deciles, i.e. for value stocks with large book to price ratios. Notice also that the correlation between skewness and Sharpe ratios is clearly inverted for HML, see table 3.

Table 2. Volatility (in per cent per day), and skewness of the daily returns of large(B)/mid(M)/small(S) cap indices in various countries. In the last column, we give the Sharpe ratio of portfolios that are long small caps (S) and short large caps (B), and neutral on medium caps (M). Note that the volatility of small cap indices is not larger than that of large cap indices, while their skewness is substantially more negative.

Index	Start date	Cap	σ (%/day)	ζ^*	SR of SMB
		B	1.1	−0.66	
S&P	1990	M	1.2	−1.12	0.32
		S	1.2	−1.16	
Russell	1978	B	1.1	−0.58	0.07
Russell		S	1.2	−1.43	
		B	1.3	0.08	
Japan	1980	M	1.1	−0.63	0.02
		S	1.0	−0.94	
		B	1.4	−0.78	
France	1999	M	1.1	−1.66	0.40
		S	0.9	−2.40	
		B	1.4	−0.92	
The Netherlands	1995	M	1.3	−1.71	0.33
		S	1.0	−1.68	
		B	1.2	−1.68	
Italy	2001	M	1.2	−1.86	0.70
		S	1.0	−2.98	
		B	1.0	−0.78	
Australia	1994	M	1.0	−0.88	−0.27
		S	1.0	−1.41	
		B	0.8	−0.90	
New Zealand	1998	M	0.7	−1.23	0.39
		S	0.5	−1.06	
		B	1.7	0.15	
Poland	1998	M	1.2	−0.92	0.34
		S	1.2	−1.17	
		L	1.3	−0.68	
South Africa	1996	M	0.9	−1.36	0.26
		S	0.7	−1.59	

As a visual confirmation, we plot in figure 4 the ranked amplitude P&L of the three Fama–French factors in the US since 1950. Clearly, SMB and UMD show the typical humped shape observed for the market ERP in the previous section, therefore corresponding to a negative overall skewness equal, respectively, to $\zeta^* = -1.39$ and -1.38. For UMD, the 5% largest events contribute to losses that amount to 20% of the gains accumulated by the 95% small to moderate returns, whereas for SMB nearly all the gains are erased by the 5% largest events! For HML however the situation is inverted, reflecting the known anti-correlation between 'Value' (HML) and 'Momentum'(UMD): large returns contribute positively to the gains, therefore leading to a positive skewness $\zeta^* = +0.25$. We will discuss this feature of HML in section 7.2.

Apart from the special HML case, the above findings give further credence to the idea that risk premium is actually *not* compensating investors for taking a symmetric risk (as measured by the volatility), but, as suggested by Kraus and Litzenberger (1976), Rietz (1988), Barro (2006), Harvey and Siddique (2000) and Santa-Clara and Yan (2010), for carrying the risk of large losses. However, at variance with Rietz (1988), Barro (2006), we do not attempt to define catastrophic macroeconomic events and estimate their probability, but rather directly measure the skewness of these risk premia portfolios through our single humped function *F(p)* and its integral–ζ*.

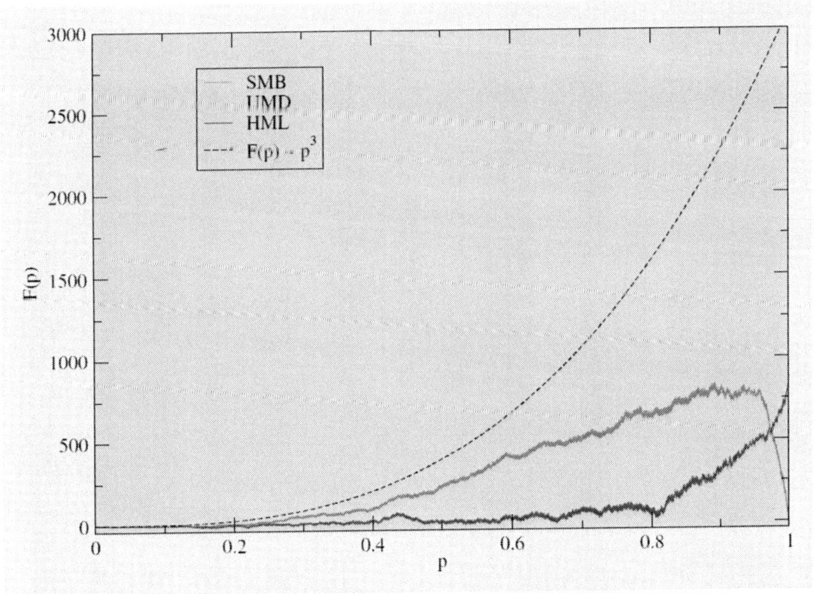

Figure 4. The ranked amplitude P&L *F(p)* of the three Fama–French factors, SMB, UMD and HML. Note the familiar humped shape for SMB and UMD, revealing strong negative skewness (respectively, $\zeta^* = -1.39$ and -1.38), but an inverted behaviour for HML ($\zeta^* = +0.43$). The corresponding Sharpe ratios are, respectively, 0.14, 0.67 and 0.64.We also show for reference the *p*3 behaviour expected at small *p* (for clarity, only shown for UMD).

These quantities immediately reveal the 'tail risk' embedded in these strategies, i.e. that large returns systematically have a negative mean.

For completeness, we have also given in table 3 the volatility, skewness and Sharpe ratios of the volatility deciles, clearly revealing the 'Low Vol' anomaly, in line with Ang *et al.* (2006, 2009), Blitz and van Vliet (2007), Baker *et al.* (2011), Frazzini and Pedersen (2014), Ciliberti *et al.* (2015).

There is however no clear pattern in the skewness; in fact a market neutral 'Low Vol' portfolio is found to have a slightly positive skewness (see figure 7).

Table 3. Volatility, skewness and Sharpe ratio of the daily returns of the decile portfolios corresponding to SMB, UMD and HML in the US equity market since 1950. See also figure 7 for a plot of Sharpe ratios vs. skewness. Decile 1 corresponds, respectively, to Small caps, Down stocks and Low earning to price stocks, whereas decile 10 corresponds to Big caps, Up stocks and High book to price. For completeness, the last three lines correspond to the 'Low Vol' (LoV) deciles (the first decile is the most volatile); see Ciliberti *et al.* (2015) for details.

	1	2	3	4	5	6	7	8	9	10
SMB σ (%/d)	0.83	1.00	1.00	0.98	0.97	0.92	0.93	0.94	0.93	0.96
SMB ζ^*	−1.83	−1.53	−1.49	−1.52	−1.42	−1.45	−1.26	−1.28	−0.93	−0.39
SMB SR	0.56	0.48	0.53	0.50	0.53	0.53	0.53	0.50	0.48	0.39
UMD σ (%/d)	1.48	1.21	1.05	0.99	0.95	0.93	0.92	0.94	1.01	1.23
UMD ζ^*	0.00	−0.14	0.00	−0.24	−0.24	−0.30	−0.58	−0.71	−0.79	−0.90
UMD SR	−0.07	0.17	0.33	0.37	0.39	0.46	0.45	0.62	0.53	0.67
HML σ (%/d)	1.05	0.97	0.93	0.95	0.94	0.92	0.91	0.97	0.98	1.10
HML ζ^*	−0.70	−0.69	−0.52	−0.68	−0.52	−0.66	−0.59	−0.65	−0.55	−0.36
HML SR	0.33	0.41	0.43	0.44	0.50	0.52	0.52	0.59	0.61	0.60
LoV σ (%/d)	1.9	1.5	1.3	1.2	1.1	1.1	1.0	0.9	0.9	0.7
LoV ζ^*	−0.20	−0.40	−0.43	−0.43	−0.43	−0.30	−0.33	−0.35	−0.24	−0.18
LoV SR	0.49	0.59	0.71	0.73	0.79	0.78	0.88	0.86	0.90	1.01

4. RISK PREMIA IN THE FIXED-INCOME WORLD

In the world of fixed income, credit risk is gauged by the investment grade of the corresponding bond, and one should in principle expect higher yields for lower investment grades (i.e. riskier bonds).We have studied corporate bond portfolios, using Global Financial Data time series forUS bond total return data.†We have used theBofACorp. total return index suite that provides daily bond returns since 1997 for grades going from AAA (safe) to CCC and below (risky). We have summarized the average yield, volatility and skewness for each of these indices in table 4. The first striking observation is that although we cover a variety of grades and yields (ranging from 4.7% annual for AAA grade to 15.5% annual for CCC and below), the volatility of these indices is found to be remarkably constant at around 4.8% annualized, except for the lowest grade which has a 7.3%

volatility. Similar to what we found above for stock indices corresponding to different market caps, the skewness is much more revealing than the volatility as it clearly drops between BBB and BB grades (see table 4).

Table 4. Return and volatility (in per cent), and skew of BofA US Corp. indices, using Global Financial Data (www.globalfinancialdata. com). All data since 1997.

Grade	Yield (%/y)	σ (%/y)	ζ^*
AAA	4.7	5.1	−0.87
AA	4.8	4.6	−0.74
A	5.3	4.9	−0.83
BBB	6	4.8	−0.81
BB	7.7	4.0	−1.78
B	9.5	4.8	−1.99
CCC and below	15.5	7.3	−1.94

Another way to elicit this skewness premium is to build a portfolio where we invest 1$ every day in a high-yield index (CCC or BB), and sell 1$ of the AAA index.To achieve constant volatility throughout the simulation, the portfolio is (as above) risk-managed based on its past volatility, as measured by a 20-day moving average of the absolute returns of the strategy itself. The performance of the BB/AAA or CCC/AAA strategies is shown in figure 5, Left. The two strategies are similar and yield positive (but marginally significant) performance. While the explanation in terms of volatility difference is not very convincing, we again find that the ranked amplitude P&L plot has a clear hump shape (see figure 5, Right), indicating again that large returns are systematically biased downwards.

5. RISK PREMIA IN THE FX WORLD: THE 'CARRY TRADE'

Similarly to the previous section where we traded bonds of differing quality, one may want to buy short-term debt in a country whose short-term rate is high using borrowed money from a country whose short-term rate is low. This idea is a textbook strategy in the world of FX finance called the 'Carry Trade'. In the absence of adverse exchange rate moves, this allows the investor to pocket the difference between the high interest rate and the low interest rate. This excess return is again usually thought of as a risk premium: one invests in the economy of a 'risky' country while getting financed in a safe country. The questions we want to ask are, as above: (a) is

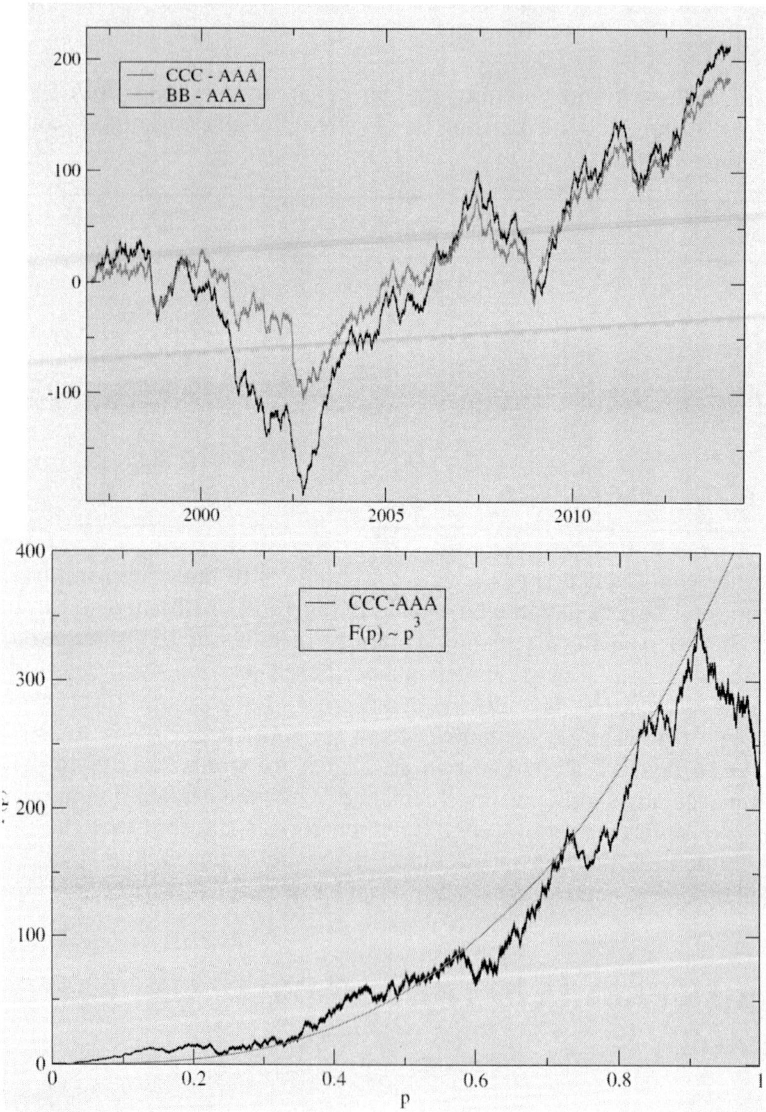

Figure 5. Left: Performance of a risk-managed portfolio long a high yield index (CCC or BB), and short itsAAAcounterpart. SR: 0.54/0.45, *t*-stat: 2.2/1.9, ζ^* : −0.84/ −0.43. Right: Ranked amplitude P&L of the CCC/AAA portfolio, showing the characteristic humped shape of portfolios with downside tail risk.

the excess return compensating for volatility? and (b) is the strategy negatively skewed?

We have simulated this currency pair strategy on a pool of 20 developed countries using data from Global Financial Data (www.globalfinancialdata.com), both for spot currencies and for interest rates, from January 1974 to January 2014.† We use intra-bank rates when they are available, and central bank discount rates otherwise. For each day, we consider all 20 × 19/2 currency ordered pairs, using the difference of interest rate in the two countries to order the pairs in a systematic way. We then make 10 deciles (following the Fama–French construction) corresponding to pairs with the smallest interest rate difference (i.e. the smallest 'carry') to the largest. We then compute, as above, the volatility, skewness and Sharpe ratio corresponding to these 10 portfolios. The data are shown in table 5 and as small pluses in figure 7. We find again that skewness and Sharpe ratio are quite strongly correlated $\rho = -0.76$, with a correlation that is significantly larger in the period 1994–2014 than in the previous period (1974–1993). In this case however ,we also find a substantial positive correlation between volatility and Sharpe ratio (ρ = +0.78). The Sharpe ratio of the fully diversified Carry strategy (i.e. with an equal weight over all ordered FX pairs) is $S \approx 0.85$ for a skewness $\zeta^* \approx -0.94$.

The above results contrast with the recent conclusion of Jurek (2014) who finds that the profitability of the FX carry trade is not substantially degraded when hedged with out of the-money currency options. This is somewhat surprising in view of the general fact that options are overpriced, but this leads Jurek to the conclusion that the carry trade P&L is *not* related to the (implied) tail-risk. However, we have not been able to confirm these findings. We found that when hedging the carry trade strategy with at-the-money options instead of out-of-the-money options (for which our data are less reliable), the carry P&L instead completely vanishes. Since out-of-the-money volatilities are usually more expensive than at-the-money volatilities, we conclude that when hedged with out-of-the-money currency options, the carry trade P&Lwould become negative, in agreement with our general skewness story.

6. A PARADIGMATIC EXAMPLE: RISK PREMIA IN OPTION MARKETS

The textbook example of a risk-premium strategy is selling options, i.e. insurance contracts against large movements of financial assets—for example, stock indices. In this case, there is a built-in asymmetry between the risk taken by the option seller (the insurer) and the option buyer. The option seller bears the risk of arbitrarily large losses, and hence the skewness of the P&L of the insurer is (almost by definition) negatively skewed—even after hedging in a world where perfect replication is not possible (see the discussion in Bouchaud and Potters (2004), in particular Ch.14). Conversely, the skewness of the option buyer is positive. Therefore, one should expect that option sellers require a 'skew premium'as a compensation for the asymmetric risk they agree to cover. In order to confirm this rather basic intuition, we have computed the performance of selling variance swaps, which are a basket of options whose return is directly proportional to the difference between realized and implied variance. This difference is exactly the risk premium we seek to elicit, so these baskets are perfectly suited to our purpose.

We test this risk premium strategy on a variety of products, since the above skew premium argument should be a universal feature valid across the board, from commodity (Crude Oil, Gold) and currency options (EURXUSD, GBPXUSD) to stock index variance swaps (S&P 500, DAX) since Jan 1996. Selling each of these variance swap contracts, and risk-managing using the last 20-day's volatility of the P&L itself (as we did in the previous sections) leads to the global performance shown in figure 6. As we can see, the return is clearly positive, with a *t*-stat over 4, while the strategy is strikingly skewed, as expected for this paradigmatic risk premium strategy .We have summarized the Sharpe ratio and the skewness ζ^* for all six contracts in table 6. Clearly, the skewness is indeed negative in all cases. In fact, the skewness ζ^* is substantially larger than all the examples considered so far in this paper and the performance of the individual short var-swap strategies is also somewhat better than all the risk premia analyzed above.

Table 5. Volatility, skewness and Sharpe ratio of the daily returns of the decile portfolios corresponding to FX Carry on G20 currencies since 1974. See also figure 7 for a plot of Sharpe ratios vs. skewness. Decile 1 corresponds to pairs with small interest rate difference whereas decile 10 corresponds to pairs with small interest rate difference. Note that in this case, there is a positive correlation between the Sharpe ratio and the volatility, on top of the now familiar negative correlation between Sharpe ratio and skewness.

	1	2	3	4	5	6	7	8	9	10
FX Carry σ (%/d)	0.45	0.47	0.45	0.50	0.55	0.58	0.64	0.68	0.74	0.93
FX Carry ζ^*	−0.41	−0.35	−0.60	−0.50	−0.87	−0.97	−0.86	−1.04	−0.81	−1.05
FX Carry SR	−0.28	−0.05	0.56	0.44	0.51	0.40	0.78	0.62	0.91	1.04

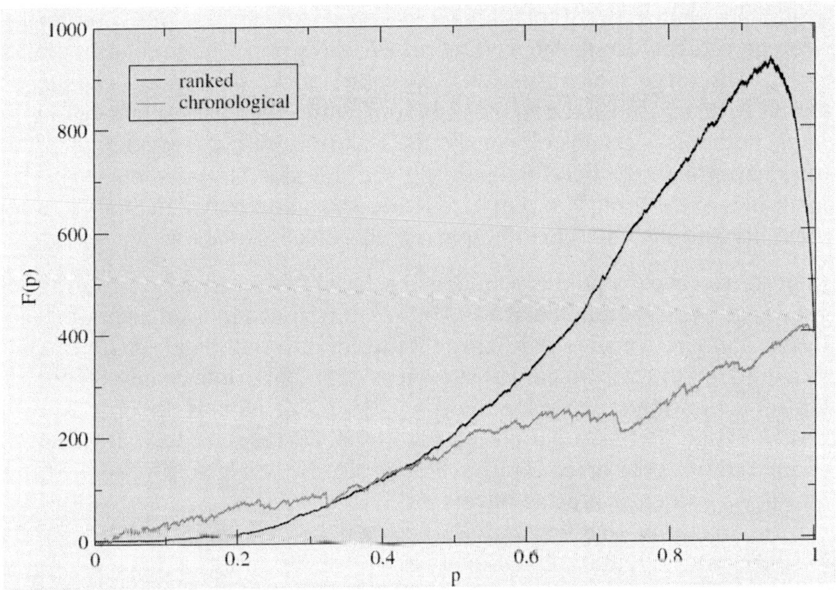

Figure 6. Performance of a risk managed short-option strategy, equi-weighted on S&P 500, DAX, Crude oil, Gold, EURXUSD and GBPXUSD variance swaps from January 1996 until January 2014. The Sharpe ratio is 1.26.We also show the ranked P&Lwhich reveals the strongly skewed nature of selling volatility.

Interestingly, shorting the VIX future instead leads to a P&L that is both less skewed and with a lower Sharpe ratio (see figure 7). This again suggests that there might exist a systematic relation between risk premia and skewness. This is the point we now discuss.

7. RISK PREMIUM IS SKEWNESS PREMIUM

7.1. A remarkable plot

The consistent picture that seems to emerge from all the above empirical results is that 'risk premia' are in fact very weakly (if at all) related to the volatility of an investment, but rather to its skewness, or more precisely to the fact that the largest returns of that investment are strongly biased downwards. This has already been discussed several times in the context of stock markets, from different standpoints (economic, econometric and financial). For example, the 'Low Volatility' puzzle goes against the intuitive argument that more (volatility) risk must be accompanied by higher average returns. Similarly, we find no evidence of a higher volatility in the small caps, large momentum or high value stocks that would explain the excess returns associated to these factors, while skewness seems to play an important role. Perhaps obviously, the risk premium collected by shorting options appears to be compensating for the substantial skewness of an insurance P&L. Table 7 summarizes the correlation between Sharpe ratios, volatility and skewness for all the strategies discussed above.

In order to reveal a possibly universal relation between excess returns and skewness, we have summarized all our results above in a single scatter plot, figure 7, where we show the Sharpe Ratios of different portfolios/strategies as a function of their (negative) skewness $-\zeta^*$. Quite remarkably, all but one scatter around the regression line $S \approx 1/3 - \zeta^* /4$. This is the central result of our paper. The two parallel dashed lines correspond to a 2-σ channel, computed with the errors on the SR and the skewness of the Fama–French strategies (other strategies have even larger error bars), see table 8. The glaring exception, that we discuss in more detail below, is the 50-day trend following strategy on a diversified set of futures contract since 1960 (see Lempérière *et al.* 2014, for details). The other outliers, although less clear-cut, are the Fama–French HML factor and the Low-Vol (LoV) strategy, which also have positive skewnesses.

Interestingly, we have also considered the returns of a portfolio of four credit indices since 2004, and of the HFRX global hedge fund index, which provides us with daily data since 2003. Although this history is relatively short, the Sharpe ratio/ skewness of both credit and of the hedge fund index fall in line with the global behaviour (once fees are taken into account in the HFRX case†).

Let us insist once more that figure 7 is a only suggestive summary of our empirical results. It does not follow from the prediction of a valuation theory (although see the qualitative discussion in section 8.2). Figure 7 is not intended either to establish a convincing linear relation between skewness and Sharpe ratios, only to suggest a substantial correlation. In this respect, note that figure 7 looks qualitatively similar for other low-moment definitions of skewness, but becomes structureless when the skewness is replaced with a measure of coskewness with the equity market, as a naive extension of the CAPM model, along the lines of Kraus and Litzenberger (1976), Harvey and Siddique (2000) would suggest.

Figure 7 not only efficiently captures all our results, but also suggests a rule-of-thumb to classify and rank different investment strategies. It is tempting to *define* a risk premium strategy as one that compensates for the skewness of the returns, in the sense that its Sharpe ratio lies close to the 'skew-rewarding'line $S \approx 1/3 - \zeta^* /4$. Interestingly, the aggregate returns of hedge fund strategies fall on this line, suggesting that a substantial fraction of hedge fund strategies is indeed risk premia.

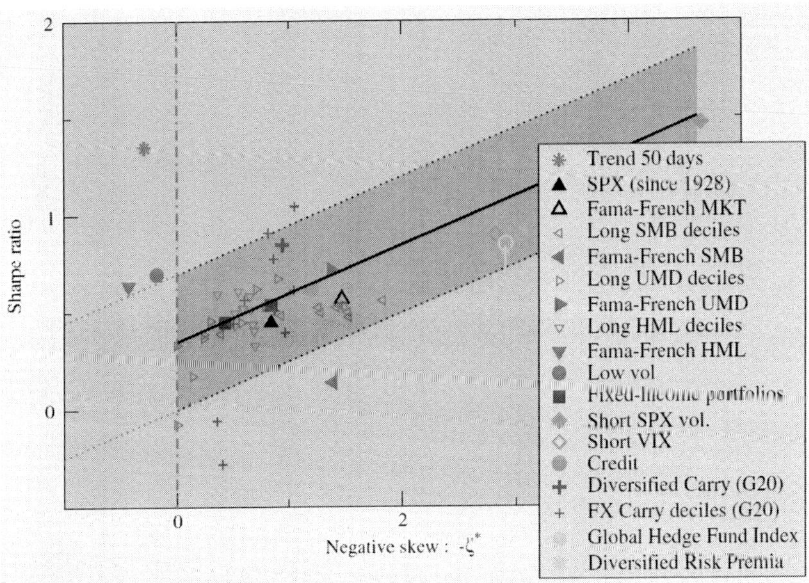

Figure 7. Sharpe ratio vs Skewness $-\zeta *$ of all assets and/or strategies considered in this paper (see legend). The relatively low Sharpe of the HFRX index can be explained by the hedge fund fees (the vertical arrow corresponds to a tentative estimate of these fees). The plain line $S \approx 1/3 - \zeta * /4$ corresponds to the regression line through all positively skewed risk premia marked with filled symbols. The two dotted lines correspond to a 2-σ channel, computed with the errors on the SR and on the skewness of the Fama–French strategies. The small empty triangles correspond to the Fama–French decile portfolios. Trend following is a clear outlier marked as a red point, characterized by a positive skewness *and* a positive SR. The same is true, perhaps to a lesser extent, of HML. Note also the interesting difference between short var-swap and short VIX strategies.

Strategies that lie significantly below this line take toomuch tail risk for the amount of excess return. Contrarily, strategies that lie significantly above this line, in particular those with *positive* skewness such as trend following, seem to get the best of both worlds. But by the same token, these strategies cannot meaningfully be classified as risk premia;† rather, as argued in Lempérière *et al.* (2014), these excess returns must represent genuine market anomalies, or 'pure α'.

Table 6. Sharpe ratio and skewness ζ *of short var-swap, risk managed strategies on different underlyings since January 1996.

Underlying	Sharpe ratio	ζ^*
S&P 500	1.47	−4.64
DAX	0.79	−5.53
Crude oil	0.74	−5.28
Gold	0.71	−4.79
EURXUSD	0.88	−5.32
GBPXUSD	0.38	−5.01
Total	1.26	−4.61

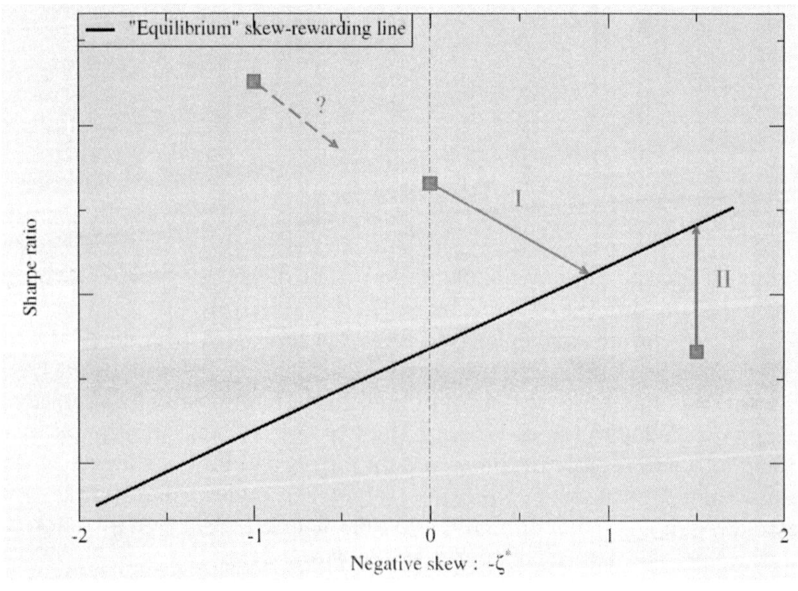

Figure 8. The skew-reward trade-off line $S \approx 1/3 - \zeta* /4$ and different risk premium dynamical paths. A first path (I) starts with a symmetric asset that pays dividends, and therefore attracts investors. The trade becomes crowded and this creates potential deleveraging spirals and downside tail risk. A second path (II) corresponds to an initially skewed payoff (like options) that moves up in Sharpe ratio as a result of skewness aversion and a dearth of option sellers. The path indicated with a questionmark (?) might describe the fate of pure alpha strategies as they become more crowded and prone to deleveraging spirals of the type discussed in Brunnermeier and Pedersen (2009).

Table 7. Correlation coefficient ρ between volatility and Sharpe ratio, and between skewness and Sharpe ratio for all strategies investigated in this paper, and for a 50-day trend following strategy across various futures markets. In most cases, and counter-intuitively, correlation with volatility is found to be *negative* or zero—note in particular the Low Vol anomaly (LoV). Correlation with skew, on the other hand, is chiefly negative, showing that the main determinant of risk premium must be skewness and not volatility. Trend, HML and LoV are clear outliers.

Underlying	Vol/SR corr.	Skewness/SR corr.
Bonds	−0.69	−0.36
Intl. IDX	−0.45	−0.38
SMB	−0.42	−0.89
UMD	−0.63	−0.85
FX Carry	+0.78	−0.76
HML	+0.03	+0.64
LoV	−0.98	+0.23
TREND	+0.23	+0.58

7.2. Three interesting exceptions: HML, LoV and trend following

Let us dwell a little longer on our outliers: trend following, HML and LoV. The daily skewness of these portfolios is significantly positive, respectively $\zeta^* = +0.43$, $\zeta^* = +0.27$ and $\zeta^* = +0.18$, and yet their performance is highly significant and universal across geographical zones.

The concept of risk premium applied to HML and to LoV is indeed problematic. Intuitively, high book-to-price corresponds to 'value' stocks that are usually considered safe and defensive, while low book-to-price corresponds to 'growth' stocks, i.e. a risky bet on future earnings. This is confirmed by the results of table 3, which shows that growth stocks are indeed more negatively skewed than value stocks, in line with our definition of risk, but indeed making HML an outlier in figure 7.‡ For LoV, the paradox is even greater: in what sense can it be risky to invest on low-volatility stocks and short highvolatility stocks? It seems to us quite reasonable that LoV cannot be understood as a risk premium (see Ciliberti *et al.* (2015)).

Table 8. Error bars on the SR and skewness of the strategies appearing in figure 7. The errors on the skewness are estimated using a standard bootstrap method.

Strategy	Error on the skewness	Error on the SR
FXR pairs	0.19	0.18
Fixed income	0.27	0.24
Agg. varswaps	0.22	0.24
Global hedge fund index	0.41	0.30
Trend 50d fut.	0.16	0.14
Fama–French MKT	0.12	0.13
Fama–French SMB	0.13	0.13
Fama–French HML	0.12	0.13
Fama–French UMD	0.12	0.13

Turning now to the nature of trend following, it could in fact have been anticipated that its skewness should be positive. This is because trend following is profitable, by definition, when the long-term realized volatility is larger than the short-term volatility (Fung and Hsieh 1997). In other words, trend following is akin to a 'long gamma' strategy, and is thus expected to have an oppositely signed skewness to that of options. Correspondingly, the skewness of monthly returns is found to be even more positive than that of daily returns (ζ * month = +1.72). Hence, the highly significant excess returns of trend following strategies in the last two centuries (see Lempérière *et al.* 2014, for a recent discussion) seem difficult to ascribe to any reasonably defined risk premium. The profitability of trend following appears to be a genuine market anomaly, plausibly of behavioural origin.† We find our plot figure 7 interesting in the sense that trend following and (albeit to a lesser extent) HML and LoV can be clearly identified as outliers.

8. DISCUSSION AND CONCLUSIONS

8.1. Skewness vs. co-skewness

Our central result, conveyed by figure 7, is that skewness is indeed the main determinant of risk premia, with an approximately linear relation between the Sharpe ratio S of a risk premium strategy and its skewness ζ^*.

From a formal point of view, this finding can be qualitatively interpreted within the classical framework of utility theory. Provided the third derivative of the utility function is positive, skewness-comparable P&Ls can be ordered, and negatively skewed strategies should be compensated by higher returns to remain attractive, as amply confirmed by lotteryexperiments (Garrett and Russell 1999, Astebro *et al.* 2009). However, previous attempts to include formally the effect of skewness in valuation theories have led to an extended formulation of theCAPM(Kraus and Litzenberger 1976, Harvey and Siddique 2000), where the *co-skewness* of a given stock with themarket—rather than the skewness itself—should determine the excess return of that stock. The idea here is that idiosyncratic skewness is diversifiable and should not lead to excess returns; only the exposure to global market shocks should play a role. As shown by Kraus and Litzenberger (1976), Harvey and Siddique (2000), Harvey *et al.* (2010) (and more recently in Kelly and Jiang (2013)), this idea seems to have clear merits in the context of stock markets.

However, this is not the route we have taken in this paper, for several reasons. First, we tend to be very suspicious about equilibrium arguments leading to CAPM-like specifications. As amply demonstrated empirically—including in the present study—market anomalies are numerous and strong, and theoretical predictions based on arbitrage and rational behavior not very compelling. Second, our results do *not* concern the behaviour of individual stocks but rather focus on the profitability of 'factors' in a wide sense—from Fama– French factors to bond, FX and volatility carry trades. Transposing the coskewness idea of Kraus and Litzenberger (1976), Harvey and Siddique (2000) in the present context would first require the definition of a global 'risk factor' that drives all risk premia, much as the market factor drives individual stocks. This is essentially the content of the proposal of Rietz (1988), Barro (2006), Santa-Clara and Y an (2010): all these risk premia should in fact represent exposure to the same 'catastrophic' Black-Swan risk, that would—if realized—spread out over many different asset classes.

We have tried to test this idea directly by studying the correlation matrix of all the risk premia strategies studied above, with the hope of identifying a dominant mode in the PCA, that would define a global risk factor. Perhaps surprisingly, we found that the top and second-largest eigenvalue of this correlation matrix are not clearly separated (at variance with the correlation matrix of single stocks, that lead to a clear separation between the market and other sub-dominant factors). Furthermore, the structure of the top eigenvector is not stable in time, which means that it cannot be used as a benchmark to define a meaningful co-skewness. If we declare the SPX to be the global risk factor, the risk premia co-skewnesses defined à la Kraus and Litzenberger (1976), Harvey and Siddique (2000) were found to be uncorrelated with the corresponding Sharpe ratio, i.e. all the structure suggested in figure 7 vanishes. When testing the co-skewness idea restricted to international equity markets, we found that the Sharpe ratio of the ERP is actually positively (rather than negatively) correlated with the co-skewness with the US market—see table 1.

8.2. Skewness, premia and crowded trades

Instead of trying to develop a formal equilibrium argument to explain our findings, we now propose a hand-waving intuitive picture of the mechanism enforcing the trade-off between skewness and excess returns. Our view is that there may in fact be *two fundamentally different* ways in which risk premia reach the skew-reward trade-off line in figure 7

Positive returns generate skewness: In many situations, one buys an asset because of an expected stream of payments, like dividends for stocks, coupons for bonds, interest rate differences for currency pairs, etc. This intrinsic source of returns attracts a crowd of investors that generate a price increase but simultaneously create the risk of a crash, induced by a self-fulfilling panic or 'bank run' mechanism, due to the crowdedness of the trade. Schematically:

cash flow → buyers → crowdedness→ price increase and downside tail risk.

An obvious example are stock markets themselves: the prospect of dividends attracts investors but the 'madness of crowds'may induce a crash. As illustrated in figure 8, crowdedness decreases returns and increases downside tail risk. Both effects limit the number of additional investors and stabilize the market around an acceptable skewness/excess return trade-off.

• Skewness demands risk premium: In other situations, downside tail risks pre-exist and it is their very existence that leads to excess returns. A perfect example is provided by option markets. Since options are insurance contracts, their pay-off profile is *by construction* skewed—negatively for option sellers and positively for option buyers. In an efficient market, the fair price of options is such that their average pay-off is zero, no risk premium exists and the Sharpe ratio of being long or short options is zero. However, the presence of fat tails and a natural investor preference for positive skewness leads to a rise in the market price of options as investors seek insurance against sudden adverse price movements. The negative skewness from the short option position, arising from the rare large payouts, remains constant and ever present. The Sharpe ratio of the short option position in an inefficient market rises to reflect the extra premium required to act as insurer.We would therefore summarize this scenario as:

downside tail risk → dearth of option sellers→ increased premium and Sharpe,

until the risk premium collected by selling options becomes large enough to lure in enough supply and cap the price of these options.

A crucial assumption for the first mechanism to work is that negative skewness is induced by crowdedness and de-leveraging spirals, as was convincingly argued in Brunnermeier and Pedersen (2009). While it seems to us highly plausible that this is the case, it would be satisfying to find some direct empirical evidence confirming the above 'crowdedness' story. Interestingly, we have checked that the 50-day trend following strategy, that has attracted a large amount of assets in the last 20 years, is characterized by a positive skewness, as shown in figure 7, that has nevertheless been slowly drifting towards zero in recent years. Similarly, the skewness of the carry trade has increased (i.e ζ^* has become more negative) in the period 1994–2014 as compared to the period 1974–1994, perhaps indicating that the carry trade is more popular than it used to be. Unfortunately, even our low-moment skewness is noisy and it is hard to be definitive about the real-time evolution of ζ^*.

8.3. Diversifying skewness?

From a practical point of view, our results provide an objective definition of risk premia strategies and a criterion to assess their quality based on the comparison between Sharpe ratio and skewness, suggested by figure 7. Clearly, not all excess returns can be classified as risk premia—trend following being one glaring counter-example. On the other hand, if the different risk premia are somewhat decorrelated, as suggested by the absence of a global 'risk mode' in the correlation matrix discussed above, then skewness can be diversified away, allowing risk premia portfolios to move *above* the regression line shown in figure 7. That this may indeed be possible is illustrated by the orange 'star' shown in figure 7, corresponding to a synthetic Diversified Risk Premia portfolio, with an equal weight on Long stock indices, Short Vol, FX Carry and CDS indices (but no Trend-following). We believe that this is what good 'alternative beta' managers should strive to achieve.

DISCLOSURE STATEMENT

CFM is an asset management firm specialized in quantitative strategies, including risk premia, trend following and other technical and behavioral anomalies.

REFERENCES

1. Ang, A., Hodrick, R.J., Xing, Y. and Zhang, X., The cross-section of volatility and expected returns. *J. Finance*, 2006, **61**, 259–299
2. Ang, A., Hodrick, R.J., Xing, Y. and Zhang, X., High idiosyncratic volatility and low returns: International and further US evidence. *J. Financ. Econ.*, 2009, **91**, 1–23
3. Astebro, T., Mata, J. and Santos-Pinto, L., Preference for skew in gambling, lotteries and entrepreneurship. Working Paper, DRUID Summer Conference, 2009
4. Athayde,G. and Flores, R., Finding amaximum skewness portfolio:A general solution to three-moments portfolio choice. *J. Econ. Dyn. Control*, 2004, **28**, 1335–1352
5. Baker, M., Bradley, B. and Wurgler, J., Benchmarks as limits to arbitrage: Understanding the low volatility anomaly. *Financ. Anal. J.*, 2011, **67**, 40–54
6. Barro, R.J., Rare disasters and asset markets in the twentieth century. *Q. J. Econ.*, 2006, **121**, 823–866
7. Blitz, D.C. and van Vliet, P., The volatility effect: Lower risk without lower return. *J. Portfolio Manage.*, 2007, **34**, 102–113
8. Bollerslev,T. andTodorov,V.,Tails, fears, and risk premia. *J. Finance*, 2011, **66**, 2165–2211
9. Bouchaud, J.-Ph. and Potters, M., *Theory of Financial Risks and Derivative Pricing*, 2nd ed., 2004 (Cambridge University Press: Cambridge)
10. Brunnermeier,M.K. and Pedersen, L.H., Market liquidity and funding liquidity. *Rev. Financ. Stud.*, 2009, **22**, 2201–2238
11. Chiu, W.H., Skewness preference, risk taking and expected utility maximization. *Geneva Risk Ins. Rev.*, 2010, **35**, 1–22
12. Ciliberti, S., Lempérière, Y., Beveratos, A., Simon, G., Laloux, L., Potters, M. and Bouchaud, J.-Ph., Deconstructing the Low-Vol anomaly, *J. Portfolio Manage.*, 2015, to appear. Available online at: http://ssrn.com/abstract=2670076, and refs. therein
13. Damodaran, A., Equity risk premiums (ERP): Determinants, estimation and implications. Working Paper, Stern School of Business (NYU), 2012
14. Daniel, K. and Moskowitz, T.J., Momentum crashes. Swiss Finance Institute Research Paper Series 13-61, 2013
15. Fernandez, P., Aguirreamalloa, J. and Avendaño, L.C., Market risk premium used in 82 countries in 2012:Asurveywith 7,192 answers, 2000. IESE Business School Working Paper No. WP-1059-E. Available online at SSRN: http://ssrn.com /abstract=2084213
16. Fouque, J.-P., Papanicolaou, G., Sircar, R. and Sølna, K., *Multiscale Stochastic Volatility for Equity, Interest Rate, and Credit Derivatives*, 2011 (Cambridge University Press: Cambridge)
17. Frazzini,A. and Pedersen,L.H., Betting against beta. *J. Financ.Econ.*, 2014, **111**, 1–25
18. Fung,W. and Hsieh, D., Survivorship bias and investment style in the returns of CTAs. *J. Portfolio Manage.*, 1997, **23**, 30–41
19. Garrett, T. and Russell, S., Gamblers favor skewness not risk: Further evidence from United States lottery games. *Econ. Lett.*, 1999, **63**, 85–90
20. Harvey, C.R., Liechty, J.C., Liechty, M.W. and Muller, P., Portfolio selection with higher moments. *Quant. Finance*, 2010, **10**, 469–485
21. Harvey, C. and Siddique, A., Conditional skewness in asset pricing tests. *J. Finance*, 2000, **55**, 1263–1295
22. Jones, M.C. and Faddy, M.J., A skew extension of the *t* distribution, with applications. *J. Roy. Stat. Soc., Ser. B*, 2003, **65**, 159–174
23. Jurek, J.W., Crash-neutral currency carry trades. *J. Financ. Econ.*, 2014, **113**, 325–347
24. Kahneman, D. and Tversky, A., Prospect theory: An analysis of decision under risk. *Econometrica*, 1979, **47**, 263–291
25. Kelly, B.T. and Jiang, H., Tail risk and asset prices, 2013. Chicago Booth Research Paper No. 13-67.Available online at SSRN: http:// ssrn.com/abstract=2321243

26. Kozhan, R., Neuberger,A. and Schneider, P., The skew risk premium in the equity index market. *Rev. Financ. Stud.*, 2013, **26**, 2174– 2203
27. Kraus, A. and Litzenberger, R.H., Skewness preference and the valuation of risk assets. *J. Finance*, 1976, **31**, 1085–1100
28. Lempérière, Y., Deremble, C., Seager, P., Potters, M. and Bouchaud, J.-Ph., Two centuries of trend following. *J. Invest. Strategies*, 2014, **3**, 41–61
29. Mehra, R. and Prescott, E.C., The equity premium: A puzzle. *J. Monetary Econ.*, 1985, **15**, 145–161
30. Mehra, R. and Prescott, E.C., The equity premium in retrospect. In *Handbook of the Economics of Finance*, edited by G. Constantinedes, M. Harris and R.M. Stulz, pp. 889–938, 2003 (Elsevier: Amsterdam)
31. Oja, H., On location, scale, skewness and kurtosis of univariate distributions. *Scand. J. Stat.*, 1981, **8**, 154–168
32. van Oordt,M.R.C. and Zhou, C., Systematic tail risk.Working Paper, De Nederlandsche Bank, Amsterdam, 2013
33. Rietz, T.A., The equity premium puzzle: A solution? *J. Monetary Econ.*, 1988, **22**, 117–131.
34. Santa-Clara, P. and Yan, S., Crashes, volatility, and the equity premium: Lessons from S&P 500 options. *Rev. Econ. Stat.*, 2010, **92**, 435–451.
35. Sharpe, W.F., Capital asset prices: A theory of market equilibrium under conditions of risk. *J. Finance*, 1964, **19**, 425–442.

APPENDIX 1. RANKED AMPLITUDE P&LS AND AN ALTERNATIVE DEFINITION OF SKEWNESS

Let us consider a random variable *r* (the returns) with a certain probability density *P(r)*.We assume that *r* has been standardized (i.e.*r* has zero mean and unit variance). We will denote as $x = |r| \geq 0$ the amplitude of *r* . From the definition of the ranked P&L function *F0(p)*, one has:

$$F_0(p) = \int_0^{x(p)} \mathrm{d}y\, y\,[P(y) - P(-y)] \tag{A1}$$

where *x(p)* is the *p*-quantile of |*r* |, defined as $p = \int_0^{x(p)} \mathrm{d}r\,(P(r) + P(-r))$ We will introduce the symmetric and antisymmetric contributionsto *P* as:

$$P_s(r) = P(r) + P(-r), \qquad P_a(r) = P(r) - P(-r) \tag{A2}$$

Note that for generic distributions, $P(r) \approx_{r \to 0} P(0) - P'(0)r + \cdots$, leading to lowest order (when $p \to 0$) to $x(p) \approx p/2P(0)$ and therefore

$$F_0(p) \sim -P'(0)/12 \times (p/P(0))^3 \tag{A3}$$

i.e. a generic $\propto p3$ behaviour for small *p*.

Now, our definition of skewness is:

$$\zeta^* := -100 \int_0^1 \mathrm{d}p\, F_0(p) = -100 \int_0^\infty \mathrm{d}x\, P_s(x) \int_0^x \mathrm{d}y\, y P_a(y) \tag{A4}$$

It is interesting to give an alternative, intuitive interpretation of this definition. After simple manipulations, one finds:

$$F_0(p) = E[|r||r < -x(p)] - E[|r||r > -x(p)] \tag{A5}$$

F0(p) therefore compares the average amplitude of large negative and large positive returns. *Z** is an average of this difference over all possible quantile choices. It might also be useful to relate ζ * to the standard definition of skewness $\zeta 3$ (defined through the third cumulant of a distribution) in the limit of weakly non-Gaussian distributions:

$$P(r) = \left[1 - \frac{\zeta_3}{3!}\frac{d^3}{\mathrm{d}r^3} + \frac{\kappa}{4!}\frac{d^4}{\mathrm{d}r^4} + \cdots\right] \frac{1}{\sqrt{2\pi}} e^{-r^2/2} \tag{A6}$$

where κ is the kurtosis; finally leading to

$$\zeta^* = \frac{25}{6\pi}\zeta_3\left(1 - \frac{\kappa}{24} + \cdots\right) \approx 1.273\,\zeta_3\left(1 - \frac{\kappa}{24} + \cdots\right) \tag{A7}$$

Note that ζ* does not require the existence of the third moment of the distribution, and remains well defined for all distributions with afinite first moment. In fact, an even weaker condition is sufficient: thedistribution should fall out faster than $|r|-3/2$.

It is interesting to compare ζ * to the classical definition of skewness $\zeta 3$ is a concrete case.We choose an asymmetric Student-*t* distribution, defined as (Jones and Faddy 2003):

$$P(r) = \mathcal{N}\left(1 + \frac{r}{\sqrt{\frac{\nu_+ + \nu_-}{2} + r^2}}\right)^{\frac{\nu_- + 1}{2}} \left(1 - \frac{r}{\sqrt{\frac{\nu_+ + \nu_-}{2} + r^2}}\right)^{\frac{\nu_+ + 1}{2}} \quad \text{(A8)}$$

which behaves asymptotically as:

$$P(r) \sim_{r \to \pm\infty} \text{const.}|r|^{-1-\nu_\pm} \quad \text{(A9)}$$

The classical skewness $\zeta 3$ is finite only when $\nu\pm > 3$, whereas ζ * remains finite as long as $\nu\pm > 1/2$. While the latter condition is always satisfied by financial data, many authors have reported that $\nu\pm$ is actually close to 3 for most markets. As a numerical exercise,

we compute both $\zeta 3$ and ζ^* as a function of $\nu+ > 3$, for a fixed valueof $\nu- = 3.5$. The results are shown in figureA1.Clearly, the skewnessof the distribution is positive when $\nu+ < \nu-$ and negative otherwise.What we see is that, as expected from the above general formula, $\zeta^*$and $\zeta 3$ behave similarly when they are both small. However, as $\nu+$decreases towards 3, $\zeta 3$ diverges whereas ζ^* remains well behaved. In the opposite direction, we see that $\zeta 3$ quickly saturates as $\nu+$ increases,while ζ^* continues to decrease. Therefore, ζ^* is a better discriminant of the asymmetry of the distribution.

Finally, let us assume that *F0(p)* is a humped function of *p* with a single maximum, corresponding to a negatively skewed distribution. This means that necessarily *Pa(y > 0)* is negative for large enough *y > y** and positive for smaller *ys*, and vice-versa for *y* < 0. Now the cumulative function of *P* compared to its symmetrized version is precisely the cumulative of *Pa(y)*:

$$G(y) = \int_{-\infty}^{y} dr[P(r) - P_s(r)] = \int_{-\infty}^{y} dr\, P_a(r) \quad \text{(A10)}$$

Since $P_a(r\ )$ vanishes three times and is positive for large negative *r* ,it is clear that *G(y)* has two symmetric maxima located at ±*y** and a minimum for *y* = 0. Now using 0 = *F0(p* = 1*)* $< y^* \int_0^\infty dr\, P_a(r)$ and *G(0)* = –_∞0 dr *Pa(r*), one immediately finds that *G(0)* < 0for non degenerate distributions. This proves that *G(y)* crosses zero twice and only twice, and therefore that *P(r)* and *Ps (r)* are skewness comparable in the sense of Oja (1981) whenever *F0(p)* has a unique maximum (or minimum).

NOTES

1)†In the context of option pricing, see e.g. Fouque *et al.* (2011). Note, however, that since themarket price of risk itself is usually found to be strongly fluctuating in time, the predictive strength of these theories becomes frustratingly close to zero. ‡However, one should note that the Mehra and Prescott framework fails to explain the high level of the volatility of the stock market, i.e. it does not explain the 'excess volatility puzzle' in the first place!

2)†A way to avoid this expansion is to assume an exponential (CARA) utility function $U(W) \equiv -\exp(-\lambda W)$ and Gaussian fluctuations for W. In this case,

$$\mathbb{E}[U(W)] \equiv -\exp\left[-\lambda\mu + \frac{1}{2}\lambda^2\sigma^2\right]$$

3)‡More precisely, this preference ordering requires in general the distributions to be 'skewness comparable', which means that their cumulative functions must cross twice and only twice (Oja 1981, Chiu 2010). This turns out to be often the case in financial applications, see below and appendix.

4)†This is virtually equivalent to buying a future contract on indices.

5)†Here and below, we will consider 'risk managed' strategies that run at approximately constant risk over time. This is achieved by normalizing the position by a 20-exponential moving average of the volatility of the bare strategy itself. In the case of the SPX, this simply amounts to the recent volatility of the SPX itself. ‡Technically, this amounts to transforming the returns *rt* as: *m + _t (rt −m)* where *m* is the average return and *_t* = ±1 an independent random sign for each *t*.

6)§One could actually directly take the area between *F(p)* and *Fs (p)* as our definition of skewness, i.e. without first removing the average drift *μ*. Up to relative corrections of order *(μ/σ)*2, the result would be identical. For all purposes, this correction is extremely small, at least for daily returns (i.e. less than 10−3).

7)†Note that the correlation between the ERP and the standard skewness *ζ*3 is close to zero, due to the large amount of measurement noise in the latter quantity. In fact, the standard skewness *ζ*3,month of the US market is actually positive≈ +0.3. The correlation between the ERP and the co-skewness is also positive ≈ 0.18, see table 1.

8)†This may look surprising at first sight. In fact, the zero skewness of being long the 'Down' portfolio comes from a compensation between the natural negative skewness of stocks, and a strong positive skewness contribution coming from periods where the market rebounds strongly after prolonged negative trends; see e.g. Daniel and Moskowitz (2013).

9)†In this paper, we only study US corporate debt, but we believe our results have a more universal scope. For example, we have checked that a portfolio of international 10Y bonds also exhibits the same features

10)†The pool of countries is made up of:Australia, Canada, Switzerland, Czech Republic, Germany, Hungary, Israel, Iceland, Japan, Korea, Mexico, Norway, New Zealand, Poland, Sweden, Singapore, Thailand, UK, USA and South Africa.

11)†The Sharpe ratio of HFRX with fees is 0.48, and we have corrected this to yield a Sharpe ratio without fees of 0.85. This assumes that funds run at a volatility of around $\sim$ 10% and have a classic 2 and 20 fee structure.

12)†Barring, of course, the existence of an unidentified risk factor that has never materialized, but that would contribute to a negative skewness if it did.

13)‡However, to add to the confusion, the skewness of the monthly returns of the HML strategy, although more noisy, appears to change sign (ζ * month = −0.56). This suggests that more work is necessary to fully understand the nature of the HML factor, which is actually known to have long periods of draw-downs (like during the Internet bubble) where the effect appears to be more in line with a standard risk premium interpretation. †The attentive reader might be puzzled by the fact that trend following has a positive skewness, while market neutral momentum (i.e. UMD) has a negative skewness, as shown above. Still, the two strategies should hinge upon the same underlying behavioural biases. This paradox will be discussed in a forthcoming publication.

CHAPTER 10

A Geometric Treatment of Time-Varying Volatilities

Chulwoo Han, Frank C. Park, Jangkoo Kang

[1] *Durham University, Durham, UK*

[2] *Seoul National University, Seoul, South Korea*

[3] *KAIST, Seoul, South Korea*

ABSTRACT

In this article, we propose a new framework for addressing multivariate time-varying volatilities. By employing methods of differential geometry, our model respects the geometric structure of the covariance space, i.e., symmetry and positive definiteness, in a way that is independent of any local coordinate parametrization. Its parsimonious specification makes it particularly suitable for large dimensional systems. Simulation studies suggest that our model embraces much of the nonlinear behaviour of the covariance dynamics. Applied to the US and the UK stock markets, the model performs well, especially when applied to risk measurement. In a broad context, our framework presents a new approach treating nonlinear properties observed in the financial market, and numerous areas of application can be further considered.

Keywords: Multivariate volatility model, Differential geometry, Riemmanian metric, GARCH

1 INTRODUCTION

Since the introduction of the ARCH model by Engle (1982), time-varying volatility models have played an important role in finance and have been successfully applied to various financial problems. From a portfolio viewpoint, it is natural to extend the GARCH-type models to a multivariate system, as first done by Bollerslev et al. (1988). Since their work, various types of multivariate time-varying volatility models have flourished. Some of the notable examples are the BEKK model of Engle and Kroner (1995), the DCC model of Engle (2002), and the matrix exponential GARCH model of Kawakatsu (2006). For more comprehensive references, we refer the reader to the survey papers by Bauwens et al. (2006) and Silvennoinen and Teräsvirta (2009).

The areas of application of time-varying volatility models are extensive; potentially all the areas where covariance dynamics comes into play, such as asset pricing and portfolio optimization, can be considered. It was first applied to asset pricing by Bollerslev et al. (1988), and later by Santis and Gerard (1998) and Hafner and Herwartz (1998). Cross relations in volatilities of several markets were studied by Kearney and Patton (2000) and Karolyi (1995), and Lien and Tse (2002) computed time-varying hedge ratios using a multivariate GARCH model. Wang and Chen (2007) explore the dynamics between the spot market and its derivative markets using a multivariate GARCH-M model.

Despite the wide areas of application, development of multivariate GARCH models has been neither as active nor as rapid as univariate GARCH models. Perhaps the most obvious reason for this is the proverbial "curse of dimensionality": because there are $n(n+1)/2$ elements in the covariance matrix of an n-variable system and the number of parameters to be estimated, in the most general case, will be of $O(n^4)$, the computational burden rapidly becomes untenable for even moderately large n.

Another important—and more subtle—difficulty in designing a time series model of a covariance matrix is the need to preserve its geometric structure, i.e., positive definiteness and symmetry. While symmetry can be easily guaranteed, ensuring that a covariance matrix remains positive definite in a mathematically consistent way (we illuminate on this point further below) is not necessarily straightforward. Earlier treatments have often been ad hoc and not taken the underlying geometric structure of the positive definite matrices into account. For example, the *vech*()-type model proposed by Bollerslev et al. (1988) takes the crude and obvious approach of transforming the covariance matrix into a column vector, and specifying the vector dynamics accordingly; in this case positive definiteness is ensured only after imposing the relevant algebraic constraints on the vector parameters. Some other recent models address the positive definiteness issue by imposing other sets of assumptions: see, for example, Tse and Tsui (2002) and Engle (2002). The well-known BEKK model of Engle and Kroner (1995) and its variants ensure positive definiteness by assuming a quadratic form for each term in the dynamics equation.

More recently, in work that is particularly relevant to ours, Kawakatsu (2006) extends the exponential GARCH model of Nelson (1991) to a multivariate system, and assumes a *vech*()-type time series model for the log of the covariance matrix. Since the matrix exponential of a symmetric matrix is always symmetric and positive definite, the geometric structure of the covariance matrix is preserved in this model without additional constraints. The exponential map, it turns out, plays a particularly important role in the geometric characterization of matrix groups, and the work of Kawakatsu (2006) is a promising first step that implies the important fundamental connections between geometry and the positive definite matrices.

Why is taking geometry into account important? This point can perhaps be best illustrated via an example of a point moving on the surface of a sphere. In principle, one could describe the motion of the point using standard three-dimensional Cartesian coordinates $x \in \Re$, with further imposition of the constraint that length of $\boldsymbol{x}$ be one, i.e., $\|\boldsymbol{x}\|=1$. Imposing the constraint is cumbersome, and it is more convenient to use a set of local two-dimensional coordinates for the sphere; spherical coordinates immediately come to mind. However, there are many alternative choices of two-dimensional coordinates that parametrize the sphere, e.g., stereographic projection. In principle, there should be no reason to prefer one choice of local coordinates over another, as long as essential geometric quantities like length of the curve, and the area of a patch, are preserved. In other words, it is important to distinguish geometric properties from those that are coordinate-dependent. Any meaningful analysis must take into account the geometric properties of the underlying space, and preserve them in a coordinate-invariant way.

Likewise, the space of covariance matrices is not a vector space, but a curved space. More precisely, it has the structure of a differentiable manifold that we denote $P(n)$·$P(n)$ is a differentiable Riemannian manifold and this fact motivates us to specify the dynamics of the covariance matrix using an appropriate set of local coordinates, and also taking care to ensure that the geometric structure of $P(n)$ is always respected. In fact, there is well-developed machinery to specify continuous-time dynamics on differentiable manifolds, and we adopt this as the starting point for developing a general multivariate volatility model.

Beyond the aesthetic appeal of describing the dynamics on a curved space in an intrinsic fashion, however, the most critical reason, that also has a profound effect on computational results, is that ignoring the geometric structure more often than not will lead to arbitrary results without physical meaning. To illustrate this with an example pointed out by Fletcher and Joshi (2004), one can define a linear average of a collection of positive definite symmetric matrices by taking their sum and dividing by the number of elements. This procedure ensures that the resulting average is also positive-definite. However, such linear averages fail to interpolate natural properties. The linear average of matrices of the same determinant, for example, can result in a matrix with a larger determinant. This issue is even worse for second-order statistics. The standard principal component analysis as formulated on vector spaces is no longer valid, since the "straight lines" corresponding to the eigenmodes of variation do not remain on the space $P(n)$—standard vector space principal component analysis fails to preserve positive-definiteness.

The aforementioned limitations of the traditional approaches can be overcome using methods of differential geometry. With a proper Riemannian metric, a "straight line" (geodesic) connecting two covariance matrices that remains on $P(n)$ can be defined, and the distance can be measured. The notion of geodesic can then be utilized to define statistics on $P(n)$, such as mean and covariance. A version of principal component analysis for covariance matrices can also be developed. By employing these tools from differential geometry, the objective of the paper is to propose a new framework for modelling covariance dynamics from a differential geometry perspective, and develop multivariate volatility models under the framework. Our models preserve symmetry and positive definiteness of the covariance matrix as it evolves, without imposing any ad hoc restrictions. We demonstrate benefits of the framework via simulation studies and compare our models with existing ones in empirical studies. Both simulation and empirical studies suggest that our models perform well and possess properties desired for covariance dynamics.

The remainder of the paper is organized as follows. Section 2 reviews the geometry of $P(n)$, including its Riemannian structure, evaluation of geodesics in a given direction and also between two arbitrary points, and its natural extension to the computation of intrinsic sample mean and covariance on $P(n)$. Section 3 constructs differential equations on $P(n)$, with a focus on linear and quadratic equations—these are chosen for their ability to model diverse correlation phenomena observed in various multivariate time series data, without sacrificing analytic and computational tractability. From the differential equations, we develop a geometrically well-defined multivariate volatility model. We also propose a variation of the model by employing the principal component analysis. In Sect. 4, the characteristics of our model are illustrated via simulation analysis and its performance is assessed through empirical studies. We conclude in Sect. 5 with a discussion of open problems and future extensions of this geometric approach to time-varying volatility.

2 GEOMETRY OF P(N)

In this section, geometric properties of the covariance space are briefly introduced. More complete description can be found in Fletcher et al. (2003), Fletcher and Joshi (2004), Moakher (2005), Lenglet et al. (2006).

The covariance space $P(n)$ is defined as

$$P(n) = \{\boldsymbol{P} \in \mathbb{R}^{n \times n} \mid \boldsymbol{P} = \boldsymbol{P}^{\top}, \boldsymbol{P} > 0\} \tag{1}$$

$P(n)$ is a differentiable manifold whose tangent space at a point $P \in P(n)$ can be identified with $n \times n$ symmetric matrices, $S(n)$. A basis for $S(n)$ can be constructed in the usual way, i.e., the basis element $E_{ij} \in S(n)$, where $i \geq j$, is a symmetric matrix whose ij and ji elements are one, and the remaining elements zero.

A Riemannian structure can be constructed via the Riemannian metric given by $\langle \boldsymbol{X},\boldsymbol{Y} \rangle_{\boldsymbol{P}} = \mathrm{tr}(\boldsymbol{P}^{-1}\boldsymbol{X}\boldsymbol{P}^{-1}\boldsymbol{Y})$, $X,Y \in S(n)$. In terms of this metric, the length of a curve $\boldsymbol{P}(t) \in P(n)$, $a \leq t \leq b$, is given by

$$L(\boldsymbol{P}) = \int_a^b \sqrt{\mathrm{tr}\left((\boldsymbol{P}^{-1}(t)\dot{\boldsymbol{P}}(t))^2\right)}\, dt \tag{2}$$

This notion of length is invariant not only under reparametrization of $[a, b]$, but also under congruent transformations of the form $\boldsymbol{MPM}^{\top}$, where M is any fixed element in the general linear group, $GL(n)$. Using the fact that $P(n)$ is a complete space (i.e., the geodesics are well-defined for all t), the minimal geodesic $\boldsymbol{\gamma}(t):[0,1] \rightarrow [\boldsymbol{A},\boldsymbol{B}]$ connecting two points $\boldsymbol{A},\boldsymbol{B} \in P(n)$ is given by

$$\gamma(t) = \boldsymbol{G}(\boldsymbol{G}^{-1}\boldsymbol{B}\boldsymbol{G}^{-\top})^{t}\boldsymbol{G}^{\top} \quad , \tag{3}$$

where $\boldsymbol{G}\boldsymbol{G}^{\top} = \boldsymbol{A}$, $\boldsymbol{G} \in \boldsymbol{GL}^{+}(n)$, the identity component of $GL(n)$, i.e., a subgroup of $GL(n)$ with positive determinants. The tangent vector of the geodesic at $\boldsymbol{A}$ is defined by the Riemannian log map

$$\mathrm{Log}_{A}(\boldsymbol{B}) = \boldsymbol{G} \log(\boldsymbol{G}^{-1}\boldsymbol{B}\boldsymbol{G}^{-\top})\boldsymbol{G}^{\top} \tag{4}$$

The inverse of the Riemannian log map, the Riemannian exponential map is also defined. Given an element $\boldsymbol{X}\in S(n)$, the minimal geodesic emanating from some $\boldsymbol{A}\in P(n)$ in the direction of the tangent vector $\boldsymbol{X}$ can be computed as follows:

$$\operatorname{Exp}_A(\boldsymbol{X}) = \boldsymbol{G}\exp(\boldsymbol{G}^{-1}\boldsymbol{X}\boldsymbol{G}^{-\top})\boldsymbol{G}^{\top} \tag{5}$$

Defining the distance between $\boldsymbol{A}$ and $\boldsymbol{B}$ in the usual way by the length of the above minimal geodesic, we have

$$d(\boldsymbol{A},\boldsymbol{B}) = \|\operatorname{Log}_A \boldsymbol{B}\| = \left(\sum_{i=1}^{n}(\log\lambda_i)^2\right)^{1/2}, \tag{6}$$

where $\|\cdot\|$ is the Frobenius norm, and $\lambda_1,\ldots,\lambda_n$ are the eigenvalues of the matrix $\boldsymbol{AB}^{-1}$. Since $\boldsymbol{AB}^{-1}$ is symmetric positive-definite, the eigenvalues of $\boldsymbol{AB}^{-1}$ are all positive, and $\log\lambda_i$ is well defined for each i. Note also that $d(\boldsymbol{A},\gamma(t))=td(\boldsymbol{A},\boldsymbol{B})$.

With the above metric structure on $P(n)$, we now discuss sample means and covariances on $P(n)$. A widely used formula for the sample mean of N symmetric positive-definite matrices $\{\boldsymbol{P}_1,\ldots,\boldsymbol{P}_N\}$ is the arithmetic mean $\frac{1}{N}\sum_{i=1}^{N}\boldsymbol{P}_i$. While the arithmetic mean clearly lies in $P(n)$, it has a number of undesirable properties: e.g., the arithmetic mean of matrices with an equal determinant can have a larger determinant. We therefore focus on the intrinsic mean, defined as

$$\underset{\bar{\boldsymbol{P}}\in P(n)}{\arg\min}\sum_{i=1}^{N} d(\bar{\boldsymbol{P}},\boldsymbol{P}_i)^2 \tag{7}$$

For the case of two points, the intrinsic mean is simply the midpoint of the minimal geodesic. For arbitrary N, the above intrinsic mean is unique on $P(n)$. Fletcher and Joshi (2004) provides a simple steepest descent algorithm for numerically obtaining the intrinsic mean, using the gradient of (7) given by $\sum_{i=1}^{N}\log(\bar{\boldsymbol{P}}\boldsymbol{P}_i^{-1})$.

An intrinsic notion of sample covariance on $P(n)$ can also be defined via the Riemannian structure of $P(n)$. Given N elements, $\boldsymbol{P}_1,\ldots,\boldsymbol{P}_N$, and the intrinsic mean $\bar{\boldsymbol{P}}$, the covariance matrix relative to $\bar{\boldsymbol{P}}$ is defined by

$$\Sigma_{\bar{\boldsymbol{P}}} = \frac{1}{N}\sum_{i=1}^{N}\boldsymbol{X}_i\boldsymbol{X}_i^{\top}, \tag{8}$$

where X_i is the tangent vector at $\bar{P}$ of the geodesic connecting $\bar{P}$ and PPiPPi, i.e., $X_i = \text{Log}_{\bar{P}}(P_i)$. Based on these mean and covariance formulas, a generalized normal distribution on $P(n)$ can be constructed by taking the curvature into account; see Lenglet et al. (2006) for details.

The usual principal component analysis defined on vector spaces is no longer valid on $P(n)$. Instead, principal geodesic analysis (PGA), a version of principal component analysis on $P(n)$, can be defined by utilizing the notions of geodesic and intrinsic mean. PGA allows us to identify the principal axes of variation of covariance matrices, which then can be used to develop a parsimonious covariance dynamics model. Details of PGA is provided in "Appendix".

3 GEOMETRICALLY WELL DEFINED VOLATILITY MODELS

Assume that an n-variable system, r_t is governed by the following dynamics equation:

$$\boldsymbol{r}_t = \boldsymbol{\mu} + \boldsymbol{e}_t, \quad \boldsymbol{e}_t \sim N(\boldsymbol{0}, \boldsymbol{H}_t), \tag{9}$$

where $\boldsymbol{H}_t \in P(n)$ is the covariance matrix of e_t. We assume a constant mean and focus on the covariance dynamics; generalization to a time-varying mean model should be straightforward. The covariance matrix of the shock is assumed to have dynamics of the form

$$d\boldsymbol{H}_t = \boldsymbol{F}_t\, dt, \tag{10}$$

where F_t is a time-varying $n \times n$ symmetric matrix which depends on the information set at t. As $d\boldsymbol{H}_t$ is the differential of $\boldsymbol{H}_t$, it is defined in the tangent space $S(n)$, and it suffices for $\boldsymbol{F}_t$ to be symmetric.

$P(n)$ is a Riemannian symmetric space that is geodesically complete, and as such the minimal geodesics provide a natural way of discretizing general differential equations on $P(n)$. Using the Riemannian exponential map defined in (5), $\boldsymbol{H}_t$ can be obtained by the formula

$$\boldsymbol{H}_t = \text{Exp}_{\boldsymbol{H}_{t-1}}(\boldsymbol{F}_t) \tag{11}$$

Another class of dynamics we consider for $\boldsymbol{H}_t$ is based on the assumption that $\boldsymbol{H}_t$ is mapped by $\boldsymbol{F}_t$ from a constant covariance matrix H_∞, i.e.,

$$\boldsymbol{H}_t = \text{Exp}_{\boldsymbol{H}_\infty}(\boldsymbol{F}_t) \tag{12}$$

In the event that H_∞ is an identity matrix, this model nests the matrix exponential GARCH model by Kawakatsu (2006). In (11), F_t is the tangent vector of the geodesic connecting H_{t-1} and $\boldsymbol{H}_t$, and it is the tangent vector of the geodesic connecting H_∞ and H_t in (12). Various specifications of F_t can be considered. The remainder of the section is devoted to developing standard models for F_t.

3.1 Geometric GARCH models

We assume that F_t depends only on the current and past covariances and shocks. In this case, a natural extension of the univariate GARCH model would yield the equation

$$\boldsymbol{F}_t = \sum_{p=1}^{P}\left(\boldsymbol{A}_p\boldsymbol{H}_{t-p} + \boldsymbol{H}_{t-p}\boldsymbol{A}_p^\top\right) + \sum_{q=1}^{Q}\left(\boldsymbol{B}_q\boldsymbol{e}_{t-q}\boldsymbol{e}_{t-q}^\top + \boldsymbol{e}_{t-q}\boldsymbol{e}_{t-q}^\top\boldsymbol{B}_q^\top\right) \tag{13}$$

We can also construct a quadratic form

$$\boldsymbol{F}_t = \sum_{p=1}^{P}\left(\boldsymbol{A}_p\boldsymbol{H}_{t-p} + \boldsymbol{H}_{t-p}\boldsymbol{A}_p^\top + \boldsymbol{H}_{t-p}\boldsymbol{S}_p\boldsymbol{H}_{t-p}^\top\right) + \sum_{q=1}^{Q}\left(\boldsymbol{B}_q\boldsymbol{e}_{t-q}\boldsymbol{e}_{t-q}^\top + \boldsymbol{e}_{t-q}\boldsymbol{e}_{t-q}^\top\boldsymbol{B}_q^\top\right) \tag{14}$$

In Eqs. (13) and (14), $\boldsymbol{A}_p, \boldsymbol{S}_p$ and $\boldsymbol{B}_q$ are $n \times n$ matrices. In the event that $\boldsymbol{B}_q$ are zero, we have a standard matrix Riccati equation in H, with well-known methods for obtaining solutions:

$$\dot{\boldsymbol{H}} = \boldsymbol{A}\boldsymbol{H} + \boldsymbol{H}\boldsymbol{A}^\top + \boldsymbol{H}\boldsymbol{S}\boldsymbol{H}^\top \tag{15}$$

The functional form of F_t can be easily extended to accommodate phenomena observed in the market, e.g., asymmetric volatility effect as reported in Kroner and Ng (1998). Asymmetric effect of shocks on F_t is modeled by augmenting Eq. (13) with terms of $\boldsymbol{\eta}_t = |\boldsymbol{e}_t| - \boldsymbol{e}_t$:

$$\begin{aligned}\boldsymbol{F}_t = {} & \sum_{p=1}^{P}\left(\boldsymbol{A}_p\boldsymbol{H}_{t-p} + \boldsymbol{H}_{t-p}\boldsymbol{A}_p^\top\right) + \sum_{q=1}^{Q}\left(\boldsymbol{B}_q\boldsymbol{e}_{t-q}\boldsymbol{e}_{t-q}^\top + \boldsymbol{e}_{t-q}\boldsymbol{e}_{t-q}^\top\boldsymbol{B}_q^\top\right) \\ & + \sum_{q=1}^{Q}\left(\boldsymbol{C}_q\boldsymbol{\eta}_{t-q}\boldsymbol{\eta}_{t-q}^\top + \boldsymbol{\eta}_{t-q}\boldsymbol{\eta}_{t-q}^\top\boldsymbol{C}_r^\top\right).\end{aligned} \tag{16}$$

$\boldsymbol{F}_t$ could be more generally specified by transforming it to a vector as in Kawakatsu (2006), but the above specification is sufficient to capture the essential features of covariance dynamics.

The proposed model is geometrically well-defined: the covariance matrix remains in $P(n)$ as it evolves without further restrictions on the dynamics, as long as the differential $d\boldsymbol{H}_t$ is an element of $S(n)$. We call this specification of time-varying volatilities Geometric GARCH, or simply GGARCH model.

3.2 PCA based specifications

Another formulation of F_t we consider is based on the principal component analysis. The PCA can be applied in two ways depending on the class of the covariance dynamics. For the covariance dynamics defined in (11), the usual PCA can be applied to the tangent vectors of the geodesics connecting H_{t-1} and H_t. In conjunction with the dynamics equation in (12), the PGA can be applied to the geodesics connecting H_∞, and H_t. In either case, if the variation of the covariance matrix can be approximated by $K<n(n+1)/2$ principal directions, $\{V_k\}$, F_t can be written in the form

$$\boldsymbol{F}_t \approx \sum_{k=1}^{K} \alpha_{kt} \boldsymbol{V}_k, \tag{17}$$

where α_{kt} are time-varying factor loadings. We specify α_{kt} as a scalar function of the lagged terms of the covariance matrix and residuals, $\{\boldsymbol{H}_{t-p}, \boldsymbol{e}_{t-q}\}$ In particular, the following functional form is considered.

$$\alpha_{kt}(\boldsymbol{H}_{t-p}, \boldsymbol{e}_{t-q}) = \operatorname{tr}\left(\sum_{p=1}^{P} \boldsymbol{A}_{k,p}\boldsymbol{H}_{t-p} + \sum_{q=1}^{Q} \boldsymbol{B}_{k,q}\boldsymbol{e}_{t-q}\boldsymbol{e}_{t-q}^{\top} + \sum_{q=1}^{Q} \boldsymbol{C}_{k,q}\boldsymbol{\eta}_{t-q}\boldsymbol{\eta}_{t-q}^{\top}\right) \tag{18}$$

where A_p, B_q, and C_q are $n \times n$ upper triangular matrices. Another method worth consideration is to use the distance function. For example, the following specification can be considered.

$$\alpha_{kt}(\boldsymbol{H}_{t-p}, \boldsymbol{e}_{t-q}) = d\left(\boldsymbol{H}_{t-1}, \sum_{p=1}^{P} a_{k,p}\boldsymbol{H}_{t-p} + \sum_{q=1}^{Q} b_{k,q}\boldsymbol{e}_{t-q}\boldsymbol{e}_{t-q}^{\top} + \sum_{q=1}^{Q} c_{k,q}\boldsymbol{\eta}_{t-q}\boldsymbol{\eta}_{t-q}^{\top}\right) \tag{19}$$

where $a_{k,p}$, $b_{k,q}$, and $c_{k,q}$ are nonnegative scalar coefficients.[1] By defining the factor loadings as the distance between the current covariance matrix and a function of the residuals, this specification has a better geometric interpretation. If K is sufficiently small, the PCA-based specifications ensure a parsimonious representation of the covariance dynamics.

Although the principal directions can, in principle, be obtained by applying PCA or PGA to a series of covariance matrices, covariance matrices are unobservable and need to be estimated. A good proxy would be a realized covariance obtained from high frequency data. If high frequency data is not available, sample covariance matrix calculated from a sub-sample could be used.

3.3 Parsimonious representations

The GGARCH models involve $O(n^2)$ number of parameters in the most general case and unless some restrictions are imposed, the models will be numerically untenable for a large dimensional system. An obvious and popular practice is to restrict the coefficient matrices to diagonal matrices or scalars. Table 1 compares the number of parameters in each GGARCH specification and in some existing models. Note that, while BEKK and DCC models have the number of parameters of $O(n^2)$ regardless of the degree of restriction, diagonal or more restricted versions of the GGARCH model have parameter numbers of $O(n)$ or a constant. This is because the covariance dynamics is defined in the tangent space and there is no constant term in the dynamics equation. A constant term in the tangent space would imply that the covariance matrix tends to move toward a certain direction regardless of its current position, which is not intuitive. Empirical studies also support this: while a constant matrix plays a dominant role in BEKK and DCC models, it turns out to be insignificant in GGARCH models. This gives our model a huge benefit when applied to large dimensional systems.

Table 1 Multivariate GARCH models and their number of parameters

Model	Parameters	Description
GGARCH PCA DIST	$3K$	α_{kt} is defined as in (19)
GGARCH PCA DIAG	$3nK$	In (18), coefficient matrices are diagonal
GGARCH PCA FULL	$(1.5n^2 + n)K$	In (18), coefficient matrices are upper triangular
GGARCH SCALAR	3	In (16), coefficient matrices are scalar
GGARCH DIAG	$3n$	In (16), coefficient matrices are diagonal
GGARCH LINEAR	$1.5n^2 + 1.5n$	In (16), coefficient matrices are symmetric and the matrix product is the element-wise product
GGARCH FULL	$3n^2$	In (16), coefficient matrices are arbitrary $n \times n$ matrices
BEKK SCALAR	$0.5n^2 + 0.5n + 3$	BEKK model with scalar coefficients
BEKK DIAGONAL	$0.5n^2 + 3.5n$	BEKK model with diagonal coefficients
BEKK FULL	$3.5n^2 + 0.5n$	BEKK model with arbitrary $n \times n$ coefficients
DCC SCALAR	$n^2 + 4n + 3$	DCC model

All models are assumed to have an asymmetric component and be a function of only the first lags

4 EMPIRICAL STUDIES

In this section, we illustrate the characteristics of the proposed models via simulations. They are then applied to a set of real market data and compared with widely used models, i.e., BEKK and DCC. Although the case studies introduced in this section are far from exhaustive, they convey some important lessons about our model and shed light on the future directions of research.

4.1 Properties of the covariance space

Riemannian exponential map serves as a building block of our covariance dynamics models and it is important to understand its properties in order to correctly specify the tangent vector F_t. For illustration purposes, we consider a simple bivariate system; this facilitates visualization and is still sufficient to understand the basic properties.

4.1.1 The shortest path

Consider two covariance matrices

$$\boldsymbol{H}_0 = \begin{bmatrix} 1 & 0 \\ 0 & 1 \end{bmatrix}, \quad \text{and} \quad \boldsymbol{H}_1 = \begin{bmatrix} 1 & 0.5 \\ 0.5 & 1 \end{bmatrix},$$

and let

$$\boldsymbol{H}(t) = \begin{bmatrix} h_{11} & h_{12} \\ h_{12} & h_{22} \end{bmatrix}, \quad 0 \le t \le 1,$$

be a covariance matrix on the geodesic connecting H_0 and H_1. The geodesic is visualized in Fig. 1. In the figure, the first panel plots the trajectory of h_{11} and h_{12}, and the second panel plots the distance between $H(t)$ and $H(t+0.1)$ for $t=0,\dots,0.9$, with $H(0)= H_0$ and $\boldsymbol{H}(1)=H_1$. The solid lines represent the geodesic and the dotted lines represent the linear interpolation. Interestingly, the trajectory of the variance term is convex and that of the covariance term is slightly concave, contrary to the naive linear interpolation that yields straight lines for both variance and covariance terms. This is because, under the Riemmanian metric, the distance between two covariance matrices increases exponentially as one matrix approaches singularity, i.e., perfect correlation, whereas $H(t)$ moves at a constant speed as shown in Fig. 1b: Convex h_{11} and concave h_{12} implies that the correlation increases more rapidly at the beginning and more slowly later compared to the case of linear interpolation. This results in the equal distance between two adjacent points along the geodesic, contrary to the increasing distance between two adjacent points along the linear interpolation line. This is a desired property of covariance dynamics since, for example, an increase in correlation from 0.0 to 0.5 must be more likely than an increase from 0.5 to 1.0. In fact, for $H_1 = \begin{bmatrix} & h0.5 \\ 0.5h & \end{bmatrix}$, the minimum distance between H_0 and H_1 is achieved when h is near 1.275, whereas it would be achieved when h=1 under a conventional metric such as Frobenius norm. This suggests that an increase in covariance is likely to involve an increase in variance more under the Riemannian metric compared to an Euclidean metric, which is consistent with the fact that the correlations are much more persistent than the variances. This is also supported by the empirical analysis demonstrated later in this section, where the first principal component of covariance matrix variation turns out to be simultaneous changes of the variance and covariance.

4.1.2 Riemmanian exponential map

We examine the movement of a covariance matrix when it is repeatedly subject to a tangent vector. This is to assess the effect of repeated shocks in one direction. Suppose the covariance matrix has the initial value

$$\boldsymbol{H}_0 = \begin{bmatrix} 1.0 & 0.5 \\ 0.5 & 1.0 \end{bmatrix}$$

The covariance matrix is repeatedly updated by the Riemannian exponential map H_t= Exp $H_{t-1}(F)$ where F has one of three values,

$$F_1 = \begin{bmatrix} 0.05 & 0.00 \\ 0.00 & 0.00 \end{bmatrix}, \ F_2 = \begin{bmatrix} 0.05 & 0.05 \\ 0.05 & 0.05 \end{bmatrix}, \ \text{or } F_3 = \begin{bmatrix} 0.05 & -0.05 \\ -0.05 & 0.05 \end{bmatrix}$$

$\boldsymbol{F}_1$ implies a shock only on the first variable, while $\boldsymbol{F}_2$ ($\boldsymbol{F}_3$) implies shocks of an equal size in the same (opposite) direction. The results are displayed in Figs. 2, 3, and 4. In each figure, σ_1, σ_2, and ρ respectively denote the standard deviations of the two variables and their correlation coefficient. When $\boldsymbol{H}_t$ is subject to $\boldsymbol{F}_1$, σ_1 increases while ρ decreases. This is intuitive as independent shocks would imply reduced correlation. Note that σ_1 increases at a reduced pace when it is high. This is desirable because when the variance is high, the same shock would have a smaller impact. As $\boldsymbol{F}_2$ implies shocks in the same direction, the correlation increases when $\boldsymbol{H}_t$ is subject to $\boldsymbol{F}_2$, and for the same reason, it decreases when $\boldsymbol{H}_t$ is subject to $\boldsymbol{F}_3$. However, the correlation decreases more rapidly, as we would expect from the actual movements of covariance matrices.

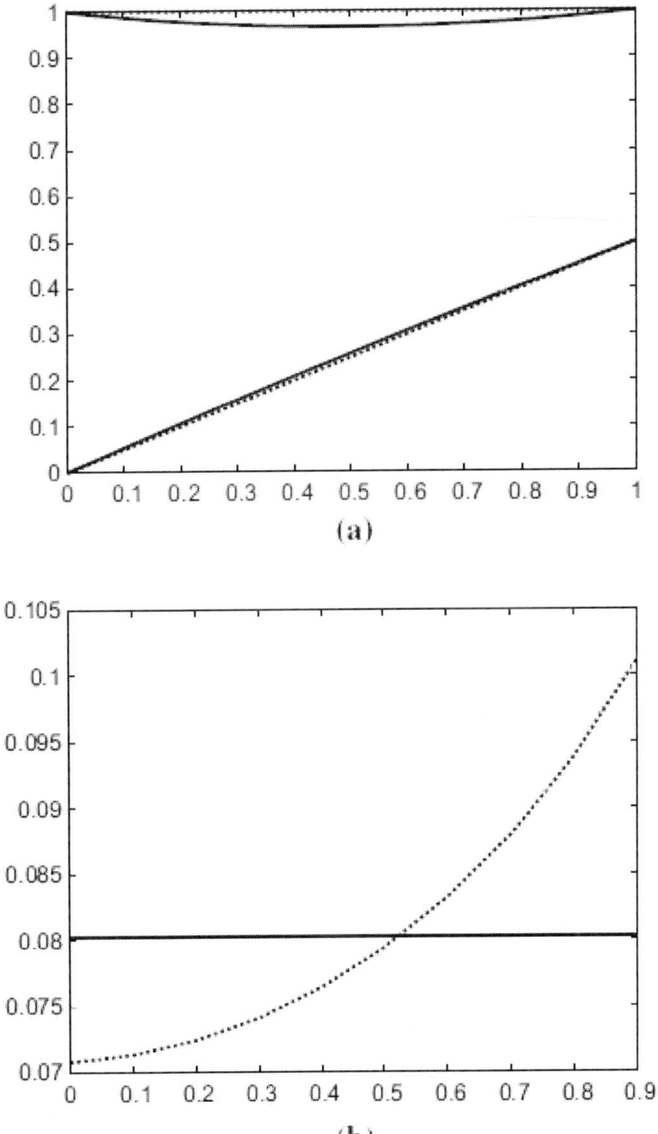

Figure 1 Geodesic between $\boldsymbol{H}_0$ and $\boldsymbol{H}_1$. In **a**, the *lines at the top* are the trajectories of h_{11} and the *lines at the bottom* are the trajectories of h_{12}. *Solid lines* represent the geodesic and *dotted lines* represent the linear interpolation. **a** Trajectory of $\boldsymbol{H}(t)$. **b** $d(\boldsymbol{H}(t),\boldsymbol{H}(t+0.1))$

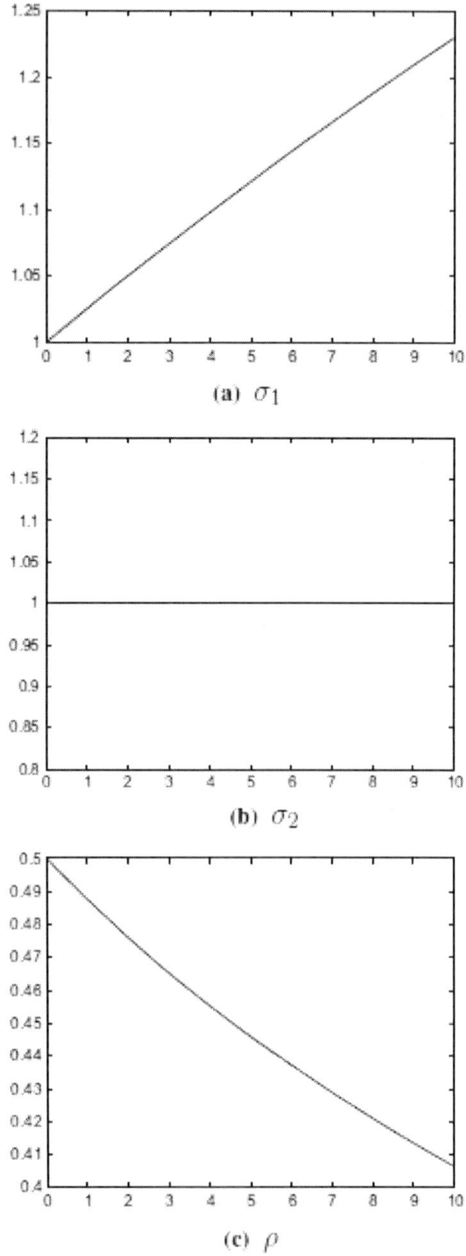

Figure 2 Movement of $\boldsymbol{H}_t$ subject to $\boldsymbol{F}_1$. σ_1, σ_2, and ρ respectively denote the SD of the two variables and their correlation coefficient

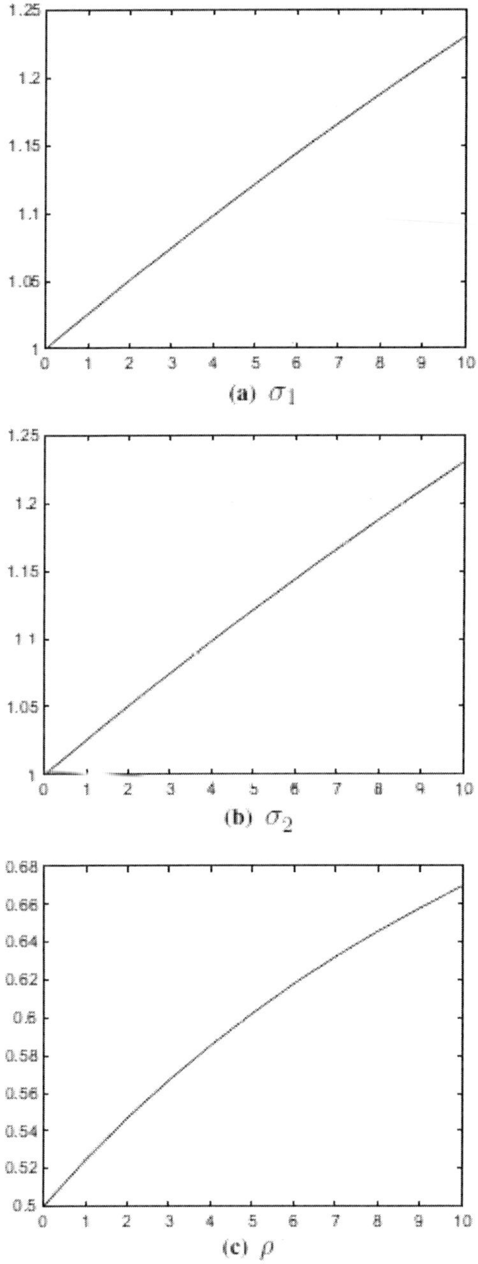

Figure 3 Movement of $\boldsymbol{H}_t$ subject to $\boldsymbol{F}_2$. σ_1, σ_2, and ρ respectively denote the SD of the two variables and their correlation coefficient

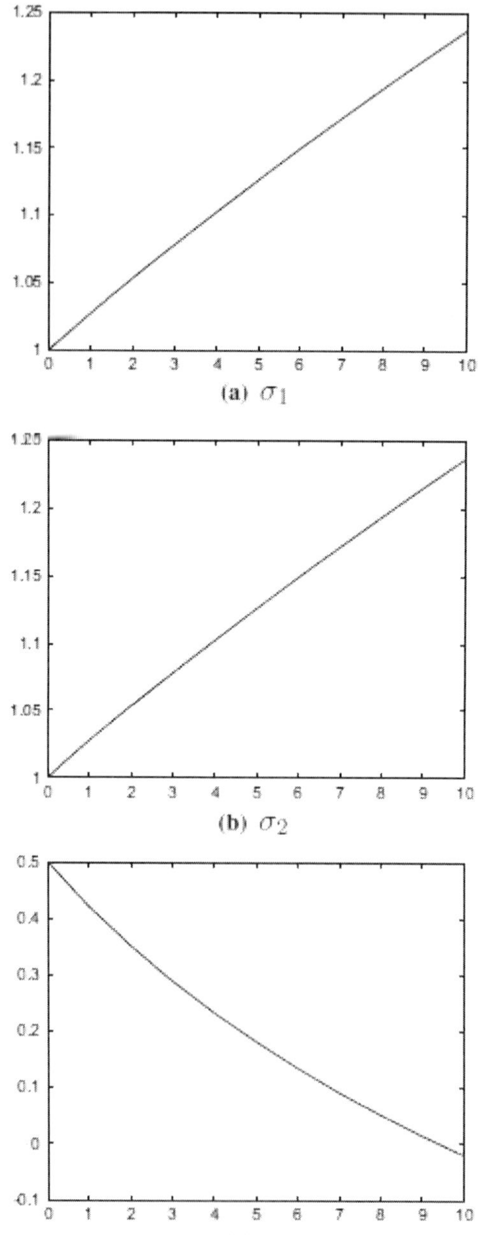

Figure 4 Movement of $\boldsymbol{H}_t$ subject to $\boldsymbol{F}_3$. σ_1, σ_2, and ρ respectively denote the SD of the two variables and their correlation coefficient

While Riemmanian exponential map carries desired properties of covariance dynamics, the behavior of the covariance matrix is not entirely clear due to the nonlinearity of the matrix exponential. This makes specifying the tangent vector and calibrating the model tricky: e.g., one may want to make correlations more persistent than variances. Although the GGARCH model does not allow independent modelling of variances and correlations, one workaround is to incorporate the GGARCH model into the DCC framework, i.e., estimate variances using a univariate GARCH model and estimate correlations using the GGARCH model.

We have investigated various characteristics of the covariance space and related them to some of the features observed in the market. Based on the promising simulation results, we believe that, by correctly specifying the tangent vector, our model can be successfully employed in many financial application domains in which volatilities are best modeled as time-varying.

4.2 Global market correlation

In this section, we apply the GGARCH models to a set of real market data and compare them with the BEKK and DCC models. We choose two stock market indexes, S&P500 and FTSE100, both obtained from Thomson Reuters database. Daily index values are collected from October 1, 2003 to September 30, 2013, from which daily log returns are calculated. The sample mean and covariance matrix of the index returns are reported in Table 2. The two markets are highly correlated during the sample period, and have a similar level of mean and variance.

Table 2 Sample statistics of S&P500 and FTSE100 daily log returns

	Mean	Covariance		Correlation	
		S&P500	FTSE100	S&P500	FTSE100
S&P500	1.924E−04	1.600E−04	0.872E−04	1.000	0.578
FTSE100	1.680E−04	0.872E−04	1.425E−04	0.578	1.000

Sample period is from October 1, 2003 to September 30, 2013

4.2.1 Test models

We consider six GGARCH specifications and compare them with two BEKK specifications (scalar and diagonal coefficient matrices) and a DCC specification. The most general form of the BEKK model was also considered but excluded as the estimation results were too sensitive to initial values and therefore unreliable. In all specifications, the covariance matrix is assumed to be a function of only the first lags of the covariance matrix and the residuals. As our primary interest lies in the covariance dynamics, we assume a constant mean return for simplicity.

$$\boldsymbol{r}_t = \boldsymbol{\mu} + \boldsymbol{e}_t, \quad \boldsymbol{e}_t \sim N(\boldsymbol{0}, \boldsymbol{H}_t), \tag{20}$$

$$\boldsymbol{H}_t = \operatorname{Exp}_{\boldsymbol{H}_{t-1}}(\boldsymbol{F}_t) \tag{21}$$

The six GGARCH specifications are described below. For the details of the BEKK and DCC models, please refer to Engle and Kroner (1995) and Engle (2002), respectively.

- GGARCH SCALAR

$$\boldsymbol{F}_t = \boldsymbol{A}\boldsymbol{H}_{t-1} + \boldsymbol{B}\boldsymbol{e}_{t-1}\boldsymbol{e}_{t-1}^\top + \boldsymbol{C}\boldsymbol{\eta}_{t-1}\boldsymbol{\eta}_{t-1}^\top, \tag{22}$$

where ***A***, ***B***, and ***C*** are scalar.

- GGARCH DIAG

$$F_t = AH_{t-1} + H_{t-1}A^\top + Be_{t-1}e_{t-1}^\top + e_{t-1}e_{t-1}^\top B^\top + C\eta_{t-1}\eta_{t-1}^\top + \eta_{t-1}\eta_{t-1}^\top C^\top \tag{23}$$

where ***A***, ***B***, and ***C*** are diagonal.

- GGARCH LINEAR

$$\boldsymbol{F}_t = \boldsymbol{A} \otimes \boldsymbol{H}_{t-1} + \boldsymbol{B} \otimes \boldsymbol{e}_{t-1}\boldsymbol{e}_{t-1}^\top + \boldsymbol{C} \otimes \boldsymbol{\eta}_{t-1}\boldsymbol{\eta}_{t-1}^\top, \tag{24}$$

where ***A***, ***B***, and ***C*** are symmetric, and $\otimes$ is the element-wise matrix product.

- GGARCH FULL

$$F_t = AH_{t-1} + H_{t-1}A^\top + Be_{t-1}e_{t-1}^\top + e_{t-1}e_{t-1}^\top B^\top + C\eta_{t-1}\eta_{t-1}^\top + \eta_{t-1}\eta_{t-1}^\top C^\top, \tag{25}$$

where ***A***, ***B***, and ***C*** are arbitrary $n \times n$ matrices.

- GGARCH PCA DIAG

$$\boldsymbol{F}_t = \sum_{k=1}^{K} \alpha_{kt} \boldsymbol{V}_k$$

$$\alpha_{kt} = \operatorname{tr}(\boldsymbol{A}_k\boldsymbol{H}_{t-1} + \boldsymbol{B}_k\boldsymbol{e}_{t-1}\boldsymbol{e}_{t-1}^\top + \boldsymbol{C}_k\boldsymbol{\eta}_{t-1}\boldsymbol{\eta}_{t-1}^\top), \tag{26}$$

where $\boldsymbol{A}_k$, $\boldsymbol{B}_k$, and $\boldsymbol{C}_k$ are diagonal matrices.

- GGARCH PCA FULL

$$F_t = \sum_{k=1}^{K} \alpha_{kt} V_k$$

$$\alpha_{kt} = \mathrm{tr}\left(A_k H_{t-1} + B_k e_{t-1} e_{t-1}^{\top} + C_k \eta_{t-1} \eta_{t-1}^{\top}\right), \quad (27)$$

where $\boldsymbol{A}_k$, $\boldsymbol{B}_k$, and $\boldsymbol{C}_k$ are upper triangular matrices.

All the models are estimated via Quasi-Maximum Likelihood Estimation (QMLE).[2]

The PCA-based GGARCH models require a time series of covariance matrices in order to apply PCA prior to model estimation. We generate a covariance matrix time series by calculating sample covariance from the minimum size (two, in our case) subsample at each time t. We tested larger subsamples but the results were qualitatively similar. PCA results are reported in Table 3. The first component explains almost 80% of the change and the first two components combined together explain more than 95% of the change. Based on this, we choose K=2. Brief inspection of the eigenvectors reveals an interesting observation: The first component is related to simultaneous changes of the variances and covariance, the second component is related to independent changes of the variances, and finally the third component is related to an independent change of the covariance. The first component implies that the principal variation of the covariance matrix is due to simultaneous changes of the variances while the correlation remaining still.

Table 3 PCA results of the tangent vectors of the geodesics connecting H_{t-1} and $\boldsymbol{H}_t$

	1st component	2nd component	3rd component
Eigenvalue	1.703E−06	0.375E−06	0.092E−6
	(78.479%)	(17.281%)	(4.240%)
	0.6470	0.6938	0.3163
Eigenvector	0.4972	−0.0693	−0.8649
	0.5781	−0.7168	0.3898

4.2.2 Estimation and diagnosis

Estimation and diagnosis results are reported in Tables 4, 5, 6 and 7. The models are ranked in each table based on their performance: ***, **, and * represent the best three performing models in descending order.

Table 4 Log-likelihood as a result of QMLE

	Log-likelihood
GGARCH SCALAR	1.7353E+04
GGARCH DIAG	1.7356E+04
GGARCH LINEAR	1.7409E+04
GGARCH FULL	1.7371E+04
GGARCH PCA DIAG	1.7404E+04
GGARCH PCA FULL	1.7478E+04
BEKK SCALAR*	1.7479E+04
BEKK DIAGONAL**	1.7509E+04
DCC SCALAR***	1.7554E+04

Table 5 Ljung–Box autocorrelation test results

	Q	*p* value
GGARCH SCALAR	76.8857	0.0000
GGARCH DIAG	71.4855	0.0004
GGARCH LINEAR	37.5462	0.0646
GGARCH FULL	65.4737	0.0012
GGARCH PCA DIAG	70.6275	0.0000
GGARCH PCA FULL	41.8565	0.0060
BEKK SCALAR*	21.7602	0.1427
BEKK DIAGONAL**	17.2584	0.2642
DCC SCALAR***	14.5448	0.4504

Table 6 Minimum variance portfolio

	σ_r	ρ	w_1	w_2
GGARCH SCALAR***	0.0109	0.577	0.5090	0.4910
GGARCH DIAG	0.0110	0.576	0.5076	0.4924
GGARCH LINEAR	0.0110	0.543	0.5046	0.4954
GGARCH FULL	0.0110	0.558	0.5168	0.4832
GGARCH PCA DIAG	0.0110	0.612	0.5055	0.4945
GGARCH PCA FULL	0.0110	0.578	0.4929	0.5071
BEKK SCALAR	0.0110	0.560	0.5215	0.4785
BEKK DIAGONAL	0.0110	0.546	0.5341	0.4659
DCC SCALAR	0.0110	0.520	0.5315	0.4685

Table 7 Value-at-risk test

	20:80			50:50			80:20		
	0.05	0.01	0.001	0.05	0.01	0.001	0.05	0.01	0.001
GGARCH SCALAR*	0.0518	0.0180	0.0042	0.0541	0.0161	0.0050	0.0525	0.0188	0.0046
GGARCH DIAG	0.0525	0.0165	0.0042	0.0541	0.0165	0.0050	0.0521	0.0196	0.0050
GGARCH LINEAR	0.0548	0.0184	0.0046	0.0567	0.0180	0.0058	0.0579	0.0215	0.0050
GGARCH FULL	0.0529	0.0180	0.0042	0.0533	0.0176	0.0050	0.0560	0.0188	0.0042
GGARCH PCA DIAG***	0.0514	0.0153	0.0035	0.0479	0.0138	0.0038	0.0502	0.0150	0.0023
GGARCH PCA FULL**	0.0548	0.0192	0.0046	0.0521	0.0207	0.0035	0.0506	0.0188	0.0035
BEKK SCALAR	0.0537	0.0176	0.0031	0.0556	0.0192	0.0038	0.0541	0.0199	0.0050
BEKK DIAGONAL	0.0571	0.0207	0.0042	0.0579	0.0215	0.0046	0.0560	0.0219	0.0050
DCC SCALAR	0.0602	0.0211	0.0042	0.0590	0.0226	0.0050	0.0575	0.0226	0.0054

The first row indicates portfolio composition between S&P500 and FTSE100, and the second row indicates probability level. The figures are probability of loss exceeding VaR

Table 4 compares the log-likelihood value of each model. The gain in the log-likelihood value from increased model flexibility seems trivial. For example, the difference in the log-likelihood values of GGARCH SCALAR and GGARCH LINEAR is mere 0.3%, even though the former has only 3 parameters, whereas the latter has 9. This means that a model selection criteria such as Bayesian Information Criteria (BIC) would favor the scalar version within each model family. The difference between model families is also small. Still, the DCC and BEKK models achieve higher log-likelihood values than the GGARCH models.

Table 5 reports the results of the autocorrelation test proposed by Ledoit et al. (2003). If a model is correctly specified, normalized residuals should be serially uncorrelated. The test results are in favor of the DCC and BEKK models. The null hypothesis of zero autocorrelation cannot be rejected in these models. Among other models, only the GGARCH LINEAR passes the test at 5% level.

A minimum variance portfolio composed of the two market indexes is obtained from the covariance estimate of each model. The portfolio is rebalanced every 22 business days and held until the next rebalancing day. The results are reported in Table 6. It seems difficult to distinguish the models based on the portfolio optimization results. Portfolio composition is similar across the models and the standard deviation of the portfolio return is almost indistinguishable. This is mainly due to the high correlation and the similar sizes of the variances of the two market returns, as reported in Table 2. Nevertheless, subtle differences among the models can still be observed: The GGARCH family allocate consistently lower weights on S&P500 compared to the BEKK and DCC models. Considering the fact that the volatility of S&P500 is higher overall, and the objective is to minimize variance, the allocations by the GGARCH family seem more plausible.

Finally, we calculate Value-at-Risk (VaR) of three portfolios at three different probability levels. Three portfolios are constructed by mixing the two market indexes at 20:80, 50:50, and 80:20 ratios, and VaR is computed at 95, 99, and 99.9% probability levels. Assuming normality of the returns, VaR at time t is given by

$$VaR_{\alpha,t} = z_{\alpha}\sqrt{\boldsymbol{w}^{\top}\boldsymbol{H}_t\boldsymbol{w}}, \tag{28}$$

where α is the probability level, z_{α} is the z-value at α, and w is the portfolio weights. VaR is computed everyday and the probability of loss exceeding VaR is obtained by comparing the VaR with the actual loss of the portfolio over the sample period. The results are reported in Table 7. The values in the table are the probability of loss exceeding VaR. If VaR is correctly estimated, the probabilityt will converge to the probability level α. The models are ranked based on the overall performance across portfolios and probability levels. It turns out that all the models underestimate VaR, except the GGARCH PCA DIAG for 50:50 portfolio at 95% level. Nevertheless, the GGARCH family appear to estimate VaR more accurately, and PCA-based GGARCH models, in particular, perform best.

We have evaluated the proposed models using various diagnostic tools. The results are mixed and do not consistently support any particular model. This could be because the model specifications considered here are not correct, or more fundamentally, our geometric framework has limitations in modelling covariance dynamics. It will remain inconclusive until more exhaustive research is undertaken. Nevertheless, our models demonstrate clear benefits over the existing models: While the BEKK and DCC models perform better in the in-sample tests, GGARCH models are better performers in the out-of-sample tests, i.e., asset allocation and risk measurement, which are more important in practical applications. Remarkably, this is achieved with fewer model parameters.

5 CONCLUSION

In this paper, we have proposed a new framework for addressing covariance dynamics from a geometric perspective using differential geometry tools developed by Fletcher et al. (2003), Fletcher et al. (2003), and Moakher (2005) among others. Our framework preserves the geometric structure of the covariance matrix without any arbitrary restrictions by respecting the inherent geometric features of the covariance matrix. It also possesses the desired nonlinear nature of the covariance dynamics observed in the market. Based on our geometric framework, we derived two types of covariance dynamics models, a general geometric GARCH (GGARCH) model, and a PCA-based geometric GARCH (GGARCH PCA) model. The major benefits of our model is (1) it allows us to specify the dynamics in an intuitive manner; (2) the nonlinear nature of the covariance dynamics is naturally implied; (3) a parsimonious specification can be easily achieved. The last feature is particularly important as it makes the model suitable for large dimensional systems.

Simulation studies reveal potential benefits of our model by showing that it captures many well-known properties of the covariance dynamics. Empirical studies, however, show mixed results and do not fully support our model against the widely recognized BEKK and DCC models. Nevertheless, it is encouraging that our model outperforms these models when forecasting matters, as evidenced in the asset allocation and VaR estimation tests. Different specifications of the dynamics within our framework could improve the overall performance.

We have primarily focused on introducing a new framework and the underlying mathematical concepts from a theoretical perspective, and demonstrated its potential benefits through case studies. More comprehensive empirical studies and comparison analysis are in order: numerous areas of application can be further considered, e.g., credit risk modelling, asset pricing, and portfolio optimization among others. Particularly intriguing is the possibility of a conditional CAPM model within the context of our framework. Econometric theories in the covariance space, e.g., conditions for stationarity, have yet to be established. Extensions of the fundamental models proposed in this paper can also be considered: different specifications of the tangent vector is of particular interest. Finally, a multivariate stochastic volatility model can be specified in the same framework by adding a diffusion term to the continuous-time dynamics. We expect that the geometric framework proposed here can serve as a building block for developing various covariance dynamics models, and more broadly contribute to building a new paradigm for economic modelling.

FOOTNOTES

[1] $\boldsymbol{H}_{t-1}$ on the right hand side of (19) needs to be replaced by $\boldsymbol{H}_{\infty}$ under the covariance dynamics specified in (12).

[2] BEKK and DCC models are estimated using MFE Matlab toolbox developed by Kevin Sheppard. The toolbox can be downloaded from http://www.kevinsheppard.com/wiki/MFE_Toolbox.

Appendix: Principal geodesic analysis

Principal component analysis (PCA) finds a linear subspace in which the variability of the data is best described. Similarly, principal geodesic analysis (PGA) seeks a submanifold that best represents the variability of the data in a Riemannian manifold. Fletcher and his colleagues investigate the PGA on a Lie group (Fletcher et al. 2003) and also on the space $P(n)$ (Fletcher and Joshi 2004). They first generalize three important concepts in PCA: variance, subspace, and projection, and develop an algorithm for PGA. For completeness, we briefly summarize the process; a more concrete derivation can be found in one of the references cited above.

First, the variance of a set of data on a Riemannian manifold is defined as the expected value of the squared Riemannian distance from the mean:

$$\sigma^2 = E\left[d(\bar{\boldsymbol{P}}, \boldsymbol{P})^2\right] \tag{29}$$

Given a sample of N data, the sample variance is defined as

$$s^2 = \frac{1}{N}\sum_{i=1}^{N} d(\bar{\boldsymbol{P}}, \boldsymbol{P}_i)^2 \qquad (30)$$

Secondly, the manifold counterpart of the linear subspace in the Euclidean space is defined. A submanifold H of a manifold M is said to be *geodesic at* $P \in H$, if all geodesics of H passing through P are also geodesics of M. Since a submanifold geodesic at P preserves the distance from P, submanifolds geodesic at the mean can be regarded as the generalization of the linear subspaces.

We now define the projection operator to a manifold. Fletcher and Joshi (2004) define the projection operator, $\pi_H: M \to H$, as

$$\underset{Q \in H}{\operatorname{argmin}}\, d(\boldsymbol{P}, \boldsymbol{Q})^2 \qquad (31)$$

Projection onto a geodesic submanifold at $\bar{\boldsymbol{P}}$ can be approximated in the tangent space to the mean, $T_{\bar{P}}M$. If $\boldsymbol{V}_1,\dots,\boldsymbol{V}_K$ form an orthonormal basis for $T_{\bar{P}}H$, then the projection operator can be approximated by the formula

$$\operatorname{Log}_{\bar{P}}(\pi_H(\boldsymbol{P})) \approx \sum_{k=1}^{K} \langle \boldsymbol{V}_k, \operatorname{Log}_{\bar{P}}(\boldsymbol{P}) \rangle_{\bar{P}} \boldsymbol{V}_k \qquad (32)$$

With the concepts defined above, it can now be shown that PGA can be performed by applying PCA to the tangent space of the manifold, $T_\mu M$. The algorithm for PGA on $P(n)$ is summarized below.

- Given $\boldsymbol{P}_1,\dots,\boldsymbol{P}_N \in P(n)$,
- Calculate the intrinsic mean of $\{P_i\}$ and denote it $\bar{\boldsymbol{P}}$.
- Compute the tangent vectors at $\bar{\boldsymbol{P}}$ of the geodesics connecting $\bar{\boldsymbol{P}}$ and $\boldsymbol{P}_i$:

$$\boldsymbol{X}_i = \operatorname{Log}_{\bar{P}}(\boldsymbol{P}_i).$$

- Apply PCA to $\{\boldsymbol{X}_i\}$, i.e., find eigenvectors and eigenvalues, $\{\boldsymbol{v}_k, \boldsymbol{\lambda}_k\}$ of

$$\boldsymbol{S} = \frac{1}{N}\sum_{i=1}^{N} \boldsymbol{x}_i \boldsymbol{x}_i^{\top}$$

where $\boldsymbol{x}_i = vech(\boldsymbol{X}_i)$ $\boldsymbol{V}_k = vech^{-1}(\boldsymbol{v}_k) \in S(n)$ are the principal directions and λ_k are the corresponding variances.

Each observation $\boldsymbol{P}_i$ can be reproduced by the formula

$$P_i = \mathrm{Exp}_{\bar{P}}\left(\sum_{k=1}^{n(n+1)/2} \alpha_k V_k\right),$$

where $\alpha_k = v_k^\top x_i$.

REFERENCES

1. Bauwens L, Laurent S, Rombouts V (2006) Multivariate GARCH models: a survey. J Appl Econom 21:79–109
2. Bollerslev T, Engle R, Wooldridge J (1988) A capital asset pricing model with time varying covariances. J Polit Econ 96:116–131
3. De Santis G, Gerard B (1998) How big is the premium for currency risk? J Financ Econ 49:375–412
4. Engle R (1982) Autoregressive conditional heteroscedasticity with estimates of the variance of United Kingdom inflation. Econometrica 50:987–1006
5. Engle R (2002) Dynamic conditional correlation: a simple class of multivariate generalized autoregressive conditional heteroskedasticity models. J Bus Econ Stat 20:339–350
6. Engle R, Kroner F (1995) Multivariate simultaneous generalized arch. Econom Theory 11:122–150
7. Fletcher P, Joshi S (2004) Principal geodesic analysis on symmetric spaces: statistics of diffusion tensors. In: ECCV 2004 workshop on computer vision approaches to medical image analysis (CVAMIA). Springer, vol LNCS 3117, pp 87–98
8. Fletcher P, Lu C, Joshi S (2003) Statistics of shape via principal geodesic analysis on lie groups. In: IEEE conference on computer vision and pattern recognition (CVPR), pp 95–101
9. Hafner C, Herwartz H (1998) Time-varying market price of risk in the CAPM—approaches, empirical evidence and implications. Finance 19:93–112
10. Karolyi G (1995) A multivariate GARCH model of international transmission of stock returns and volatility: the case of the United States and Canada. J Bus Econ Stat 13:11–25
11. Kawakatsu H (2006) Matrix exponential GARCH. J Econom 134:95–128
12. Kearney C, Patton A (2000) Multivariate GARCH modelling of exchange rate volatility transmission in the European monetary system. Financ Rev 41:29–48
13. Kroner F, Ng V (1998) Modelling asymmetric comovements of asset returns. Rev Financ Stud 11:817–844
14. Ledoit O, Santa-Clara P, Wolf M (2003) Flexible multivariate GARCH modeling with an application to international stock markets. Rev Econ Stat 85:735–747
15. Lenglet C, Rousson M, Deriche R, Faugeras O (2006) Statistics on the manifold of multivariate normal distributions: theory and application to diffusion tensor mri processing. J Math Imaging Vis 25(3):423–444
16. Lien D, Tse Y (2002) Some recent developments in futures hedging. J Econ Surv 16:357–396
17. Moakher M (2005) A differential geometric approach to the geometric mean of symmetric positive-definite matrices. SIAM J Matrix Anal Appl 26(3):735–747
18. Nelson D (1991) Conditional heteroskedasticity in asset returns: a new approach. Econometrica 59:347–370

19. Silvennoinen A, Teräsvirta T (2009) Handbook of financial time series. Springer, chap Multivariate GARCH models, pp 201–229
20. Tse Y, Tsui A (2002) A multivariate GARCH model with time-varying correlations. J Bus Econ Stat 20:351–362
21. Wang K, Chen M (2007) The dynamics in the spot, futures, and call options with basis asymmetries: an intraday analysis in a generalized multivariate GARCH-M MSKST framework. Rev Quant Financ Acc 29(4):371–394

CHAPTER 11

A Skewness-Adjusted Binomial Model for Pricing Futures Options—The Importance of the Mean and Carrying-Cost Parameters

Stafford Johnson[1], Amit Sen[2], Brian Balyeat[1]

[1]*Department of Finance, Xavier University, Cincinnati, USA*

[2]*Department of Economics, Xavier University, Cincinnati, USA*

ABSTRACT

In this paper, we extend the Johnson, Pawlukiwicz, and Mehta [1] skewness-adjusted binomial model to the pricing of futures options and examine in some detail the asymptotic properties of the skewness model as it applies to futures and spot options.The resulting skewness-adjusted futures options model shows that for a large number of subperiods, the price of futures options depends not only on the volatility and mean but also on the risk-free rate, asset-yield, and other carrying-cost parameters when skewness exists.

Keywords: Options Pricing; Black-Scholes; Skewness

1. INTRODUCTION

One of the interesting, as well as subtle, features of the Black-Scholes (B-S) [2] model and the binomial option pricing model (BOPM) with a large number of subperiods (n) is that the models depend only on the variance. In these models, the mean is not important in determining the value of spot options and the mean and net carry cost are not important for futures options. These implications of the model, however, depend on the assumption that the logarithmic return of the underlying security is normally distributed. Studies by Johnson, Zuber, and Gandar [3] and [4] have shown that in periods of increasing stock prices or rates, the logarithmic return of stock indexes and interest rates are often characterized by a positive mean and significant negative skewness, and in periods of decreasing prices or rates, the logarithmic returns are often characterized by a negative mean and significant positive skewness. Moreover, several earlier empirical studies have reported that the B-S model consistently underprices options in the presence of skewness; see for instance, Stein and Stein [5], Wiggins [6], and Heston [7].

Jarrow and Rudd [8] and Corrado and Tie Su [9] have extended the B-S model to account for cases in which there is skewness in the underlying security's return distribution. Similarly, Câmara and Chung [10] and Johnson, Pawlukiewicz, and Mehta (JPM) [1] have extended the Cox, Ross, and Rubinstein (CRR) [11] and Rendleman and Bartter (RB) [12] binomial option pricing model to include skewness. In their paper, JPM also show that skewness changes the asymptotic properties of the up (u) and down (d) parameters, elevating the relative importance of the mean in valuing options. This property of their skewness model suggests that when distributions of logarithmic returns are characterized by skewness, the observed pricing biases associated with the B-S model may be due to not only the omission of skewness, but also the mean.

Today, the derivative market for non-stock options (indices, currencies, debt securities, and commodities) is dominated more by options on futures contracts than options on spot securities. The purpose of this paper is to extend the JPM skewness-adjusted binomial model to the pricing of futures options. In addition, given that one of the features of the JPM skewness model for spot options is that skewness changes the asymptotic properties of the u and d parameters, this paper examines in some detail the asymptotic properties of the skewness-adjusted binomial model as it applies to both futures and spot options. Our results show that the skewness model for futures options has similar asymptotic properties as the model for spot options. However, in the case of futures options, the presence of skewness elevates the importance of the mean, as well as the risk-free rate, the asset yield, and other parameters that are defined by the carrying-cost model.

2. BINOMIAL FUTURES OPTIONS PRICING MODEL

The standard binomial option model values futures options recursively by determining the futures option's intrinsic values at expiration and then using the singleperiod binomial model at each node to price the futures option equal to the value of its replicating portfolio:

$$C_0^* = \frac{1}{e^{R(t/n)}}\left[\sum_{j=0}^{n}\frac{n!}{(n-j)!j!}p^j(1-p)^{(n-j)}\left[\text{Max}\left(u^{fj}d^{f(n-j)}f_0\right)-X,0\right]\right] \tag{1}$$

$$P_0^* = \frac{1}{e^{R(t/n)}}\left[\sum_{j=0}^{n}\frac{n!}{(n-j)!j!}p^j(1-p)^{(n-j)}\left[\text{Max}(X-\left(u^{fj}d^{f(n-j)}f_0\right),0\right]\right] \tag{2}$$

where:

C_0 = call price

P_0 = put price

R = annual risk-free rate

t = time to expiration expressed as a proportion of a year

X = exercise price

f_0 = current futures price

u^f = the futures up parameter

d^f = the futures down parameter

n = number of periods to the option's expiration

t/n = length on the binomial period = time to expiration as a proportion of a year (t) divided by number of periods to the option's expiration

p = risk-neutral probability

For the case of an option on a financial futures contract (e.g., index, currency, or debt security) in which the underlying security is adjusted to reflect a continuous asset yield (e.g., dividend yield, foreign risk-free rate, or coupon rate), the equilibrium futures price as determined by the carrying-cost model is:

$$f_0 = S_0 e^{(R-\psi)n_f \Delta t} \tag{3}$$

where:

S_0 = current spot price

ψ = annual asset yield

$\Delta t = t/n$ = length of binomial steps as a proportion of a year

n_f = number of discrete binomial periods of length Δt to the futures' expiration

The risk-neutral probability, p, for options on financial futures defined in terms of the up and down parameters for the underlying spot ($u = S_u/S_0$ and $d = S_d/S_0$, where $S_u = uS_0$ and $S_d = dS_0$) is:

$$p = \frac{e^{(R-\psi)(t/n)} - d}{u - d} \tag{4}$$

If the carrying-cost model holds, then the up and down parameters for the futures price ($u^f = f_u/f_0$ and $d^f = f_d/f_0$, where $f_u = u^f$ and $f_d = df_0$) are given as:

$$u^f = \frac{f_u}{f_0} = \frac{uS_0 e^{(R-\psi)(n_f-1)(t/n)}}{S_0 e^{(R-\psi)n_f(t/n)}} = \frac{u}{e^{(R-\psi)(t/n)}}$$

$$u = u^f e^{(R-\psi)(t/n)} \tag{5}$$

$$d^f = \frac{f_d}{f_0} = \frac{dS_0 e^{(R-\psi)(n_f - 1)(t/n)}}{S_0 e^{(R-\psi)n_f(t/n)}} = \frac{d}{e^{(R-\psi)(t/n)}}$$

$$d = d^f e^{(R-\psi)(t/n)} \tag{6}$$

Substituting Equations (5) and (6) into Equation (4), the risk-neutral probability for futures call and put option prices can be alternatively defined in terms of the futures up and down parameters (u^f and d^f) :

$$p = \frac{e^{(R-\psi)(t/n)} - d}{u - d}$$

$$p = \frac{e^{(R-\psi)(t/n)} - d^f c^{(R-\psi)(t/n)}}{u^f e^{(R-\psi)(t/n)} - d^f e^{(R-\psi)(t/n)}}$$

$$p = \frac{1 - d^f}{u^f - d^f} = p^f \tag{7}$$

2.1. Skewness-Adjusted Formulas for u^f and d^f

In their seminal 1979 paper, Cox, Ross, and Rubinstein [11] derive the formulas for estimating the u and d parameters of the BOPM for a spot option. They do this by setting the equations for the expected value and variance of the logarithmic return of the underlying security equal to their empirical values. The resulting equations are then solved simultaneously for u and d under the assumption that the probability of the underlying security increasing in one period (q) is 0.5. By treating q as an unknown, JPM extend the CRR binomial model to include skewness. Specifically, in the JPM skewness-adjusted model, the u, d, and q values that define a binomial process for a spot security are found by setting the equations for the expected value, variance, and skewness equal to their respective empirical values and then solving the resulting equation system simultaneously for u, d, and q. That is:

$$E(g_n) = \sum_{j=0}^{n} p_{nj}\, g_{nj} = n\left[q \ln u + (1-q)\ln d\right] = t\,\mu_S$$

$$V(g_n) = nq(1-q)\left[\ln(u/d)\right]^2 = t\,V_S$$

$$Sk(g_n) = n\left[q(1-q)^3 - q^3(1-q)\right]\left[\ln(u/d)\right]^3 = t\,\delta_S, \tag{8}$$

where:

g = logarithmic return of the underlying spot price = $ln(S_1/S_0)$

j = number of increases in n periods

p_{nj} = probability of j increases in n periods, where:

μ_S, V_S, σ_S, d_S = the annualized empirical values of the mean, variance, standard deviation, and skewness of the spot price's logarithmic return t = time to expiration as a proportion of a year

n = number of periods

The values of u, d, and q that satisfy this system of equations are:

$$u = e^{\sigma_S \sqrt{((1-q)/q)(t/n)} + \mu_S (t/n)} \tag{9}$$

$$d = e^{-\sigma_S \sqrt{(q/(1-q))(t/n)} + \mu_S (t/n)} \tag{10}$$

$$q = \frac{1}{2} \pm \frac{1}{2}\left[\frac{4(tV_S)^3}{n\delta_S^2} + 1\right]^{-1/2}, \quad - \text{ if } \delta_S > 0, + \text{ if } \delta_S < 0. \tag{11}$$

For futures options, the formulas for estimating u^f, d^f, and q are found by setting the equations for the population moments for the futures price's logarithmic return equal to their respective empirical values, and then solving the resulting equation system simultaneously for u^f, d^f, and q:

$$n\left[q\ln u^f+(1-q)\ln d^f\right]=t\,\mu_f$$

$$nq(1-q)\left[\ln\left(u^f/d^f\right)\right]^2=tV_f$$

$$n\left[q(1-q)^3-q^3(1-q)\right]\left[\ln\left(u^f/d^f\right)\right]^3=t\,\delta_f, \tag{12}$$

where m_f, V_f, σ_f, and d_f are respectively the annualized empirical values of the mean, the variance, the standard deviation, and skewness of the futures price's logarithmic return, $ln(f_1/f_0)$, and t is the time to expiration as a proportion of a year.

The values of u^f, d^f, and q that satisfy this system of equations are:

$$u^f=e^{\sigma_f\sqrt{((1-q)/q)(t/n)}+\mu_f(t/n)} \tag{13}$$

$$u^f=e^{\sigma_f\sqrt{((1-q)/q)(t/n))}+\mu_f(t/n)} \tag{14}$$

$$q=\frac{1}{2}\pm\frac{1}{2}\left[\frac{4(tV_f)^3}{n(t\delta_f)^2}+1\right]^{-1/2},\quad -\text{ if }\delta_f>0,\ +\text{ if }\delta_f<0 \tag{15}$$

2.2. Relations between Futures and Spot Mean and Volatility

The relationships between the spot and futures parameters follow directly from the futures and spot relation defined by the carrying-cost model given in equation (3). Specifically, from equation (3), one can solve the relationship between the spot and futures moments as follows:

$$S_0 = f_0 e^{-(R-\psi)n_f \Delta t}$$
$$S_1 = f_1 e^{-(R-\psi)(n_f - 1)\Delta t}$$
$$\frac{S_1}{S_0} = \frac{f_1 e^{-(R-\psi)(n_f - 1)\Delta t}}{f_0 e^{-(R-\psi)n_f \Delta t}}$$
$$\ln\left(\frac{S_1}{S_0}\right) = \ln\left(\frac{f_1 e^{-(R-\psi)(n_f - 1)\Delta t}}{f_0 e^{-(R-\psi)n_f \Delta t}}\right)$$
$$\ln\left(\frac{S_1}{S_0}\right) = \ln\left(\frac{f_1}{f_0}\right) + \ln\left(e^{(R-\psi)\Delta t}\right)$$
$$\ln\left(\frac{S_1}{S_0}\right) = \ln\left(\frac{f_1}{f_0}\right) + (R - \psi)\Delta t \tag{16}$$

Taking the expected value of both sides of Equation (16) results in the mean over a period of time Δt:

$$E\left[\ln\left(\frac{S_1}{S_0}\right)\right] = E\left[\ln\left(\frac{f_1}{f_0}\right) + (R - \psi)\Delta t\right]$$
$$\mu_S^* = \mu_F^* + (R-\psi)\Delta t$$

where μ_S^* and μ_F^* are respectively the periodic means of a spot option and a futures option for a period of length $\Delta t(= t/n)$. Multiplying both sides by $1/\Delta t$ results in the annualized means (μ_S and μ_F):

$$\mu_S^*(1/\Delta t) = \mu_f^*(1/\Delta t) + (R - \psi)\Delta t$$
$$\mu_S = \mu_f + (R - \psi)$$

Taking the variance of both sides of (16) and annualizing we obtain:

$$Var\left[\ln\left(\frac{S_1}{S_0}\right)\right] = Var\left[\ln\left(\frac{f_1}{f_0}\right) + (R - \psi)\Delta t\right]$$

$$\sigma_S^2 = \sigma_F^2$$

$$\sigma_S = \sigma_F$$

A similar proof shows that $\delta_S = \delta_F$.

Thus, the relationships between the mean, standard deviation, and skewness on the futures and spot logarithmic returns are:

$\mu f = \mu s - (R - \psi)$ (17)

$\sigma f = \sigma s$ (18)

$\delta f = \delta_s$ (19)

Substituting (17) and (18) into equations (13) and (14), u^f and d^f can be expressed in terms of the spot mean and variability, the risk-free rate, and the asset yield:

$$\begin{aligned} u^f &= e^{\sigma_f \sqrt{((1-q)/q)(t/n)} \; + \; \mu_f (t/n)} \\ &= e^{\sigma_S \sqrt{((1-q)/q)(t/n)} \; + \; (\mu_S - (R - \psi))(t/n)} \end{aligned} \quad (20)$$

$$\begin{aligned} d^f &= e^{-\sigma_f \sqrt{((1-q)/q)(t/n)} \; + \; \mu_f (t/n)} \\ &= e^{-\sigma_S \sqrt{(q/(1-q))(t/n)} \; + \; (\mu_S - (R - \psi))(t/n)} \end{aligned} \quad (21)$$

The difference between the futures up and down parameters (u^f and d^f respectively), and the spot up and down parameter (u and d respectively (see Equations (9) and 10)) is the net cost of carry term ($R - \psi$). If $R > \psi$, then the futures market is normal with the futures price (equation (3)) exceeding the spot.

In this market, $u > u^f$ and $d > d^f$. On the other hand, if $R < \psi$, then the futures market is inverted with the futures price less than the spot. In this market, $u^f > u$ and $d^f > d$. Finally, if the net carry cost is zero, then $R = \psi$ and the futures market is neutral; in this market, the futures up and down parameters will be equal to the spot up and down parameters, that is, $u^f = u$ and $d^f = d$. The relations between the parameters are illustrated in **Table 1**. The table shows the futures and spot up and down parameters calculated for a number of scenarios: positive mean and negative skewness cases (increasing price case) for normal, inverted, and neutral futures market; negative mean and positive skewness cases (decreasing price case) for normal, inverted, and neutral futures market; and zero mean and skewness case (stable price case) for normal, inverted, and neutral futures market. Panel A in the table shows the inputs for each scenario and Panel B gives the corresponding parameters.

3. BINOMIAL FUTURES OPTIONS PRICING MODEL

3.1. Decreasing Exchange-Rate Case

Several empirical studies have shown that periods of increasing security prices are often characterized by a positive mean and significant negative skewness in the security price's logarithmic return. As an example of a decreasing price scenario, suppose the current US dollar/British pound exchange rate is S_0 = \$1.60/BP, and there is a market expectation of a dollar appreciation over the next year such that the expected distribution of logarithmic returns for the exchange rate has the following annualized mean, variance, and skewness: $\mu_S = -0.17597$, $V_S = 0.019555$, and $\delta_s = 0.0008602$. Given these empirical moment values, consider the pricing of call and put options on a British pound (BP) futures contract each with X = \$1.60/BP and a time to expiration of 270 days, using a three-period binomial model. In pricing the options, assume the following:

1) The spot \$/BP exchange rate at time 0 is

E_0 = \$1.60/BP

2) The annual risk-free rate paid on US dollars is $R = 0.06$

3) The annual risk-free rate paid on British pounds is $\psi = 0.04$

4) The futures contract on the BP expires in one year

5) Carrying-cost model holds

6) Options on the BP futures options expire in 270 days

7) 360-day year The length of the binomial period in years is $\Delta t = t/n = (270/360)/3 = 0.25$, with the call and put options expiring in $(n_{option})\Delta t = (3)(0.25) = 0.75$ years, and the BP futures expiring in $n_f\Delta t = (4)(0.75) = 1$ year. The up and down parameters for the spot rate are $u = 1.052636$ and $d = 0.9091$, the up and down parameters for the futures contract are $u^f = 1.047388$ and $d^f = 0.90457$, $q = 0.35$, the risk-neutral probability is $p^f = 0.66822$, and the equilibrium futures price is \$1.63232/BP:

$$q = \frac{1}{2} \pm \frac{1}{2}\left[\frac{4\left(t\sigma_e^2\right)^3}{n\left(t\delta_e\right)^2} + 1\right]^{-1/2}$$

$$= \frac{1}{2} - \frac{1}{2}\left[\frac{4\left((0.75)(0.019556)\right)^3}{(3)\left((0.75)(0.00086002)\right)^2} + 1\right]^{-1/2}$$

$$= 0.35$$

$$u^f = e^{\sqrt{V_S((1-q)/q)(t/n)} + ((\mu_S - (R-\psi))(t/n)}$$

$$= e^{\sqrt{0.019555((1-0.35)/0.35))(0.75/3)} + (-0.17597 - (0.06-0.04))(0.75/3)}$$

$$= 1.047388$$

$$d^f = e^{-\sqrt{V_S(q/(1-q))(t/n)} + (\mu_S - (R-\psi))(t/n)}$$

$$= e^{-\sqrt{0.019555(0.35/(1-0.35))(0.75/3)} + (-0.17597 - (0.06-0.04))(0.75/3)}$$

$$= 0.90457$$

$$p^f = \frac{1-d^f}{u^f - d^f} = \frac{1-0.90457}{1.047388-0.90457}$$

$$= 0.66822$$

$$f_0 = S_0 e^{(R-\psi)n_f \Delta t}$$

$$f_0 = \$1.60e^{(.06-.04)4(.25)} = \$1.63232$$

The negative mean and positive skewness in this case yield u^f, d^f, and q values that reflect a decreasing exchange-rate scenario in which the proportional decrease in the futures rate each period is 10.03%, exceeding in absolute value the proportional increase of 4.63%, and the probability of the decrease in each quarterly period ($\Delta t = 0.25$) is $1 - q = 0.65$:

Table 1. u and d and u^f and d^f values for different scenarios.

Panel: A							
Scenario	Market	R	ψ	t/n	Annualized Mean	Annualized Variance	Annualized Skewness
Positive mean, Zero Skewness.	Normal	0.06	0.04	0.25	0.0880	0.0215	0.0000
Positive mean, Negative Skewness.	Normal	0.06	0.04	0.25	0.1760	0.0196	−0.0009
Positive mean. Negative Skewness.	Normal	0.06	0.04	0.25	0.2640	0.0138	−0.0012
Positive mean. Zero Skewness.	Inverted	0.04	0.06	0.25	0.0880	0.0215	0.0000
Positive mean, Negative Skewness.	Inverted	0.04	0.06	0.25	0.1760	0.0196	−0.0009
Positive mean. Negative Skewness.	Inverted	0.04	0.06	0.25	0.2640	0.0138	−0.0012
Positive mean. Zero Skewness	Stable	0.06	0.06	0.25	0.0880	0.0215	0.0000
Positive mean. Negative Skewness	Stable	0.06	0.06	0.25	0.1760	0.0196	−0.0009
Positive mean. Negative Skewness	Stable	0.06	0.06	0.25	0.2640	0.0138	−0.0012
Negative mean. Zero Skewness.	Normal	0.06	0.04	0.25	−0.0880	0.0215	0.0000
Negative mean, Positive Skewness.	Normal	0.06	0.04	0.25	−0.1760	0.0196	0.0009
Negative mean. Positive Skewness.	Normal	0.06	0.04	0.25	−0.2639	0.0138	0.0012
Negative mean. Zero Skewness.	Inverted	0.04	0.06	0.25	−0.0880	0.0215	0.0000
Negative mean. Positive Skewness.	Inverted	0.04	0.04	0.25	−0.1760	0.0196	0.0009
Negative mean. Positive Skewness.	Inverted	0.04	0.06	0.25	−0.2639	0.0138	0.0012
Negative mean. Zero Skewness	Stable	0.06	0.06	0.25	−0.0880	0.0215	0.0000
Negative mean. Positive Skewness	Stable	0.06	0.06	0.25	−0.1760	0.0196	0.0009
Negative mean. Positive Skewness	Stable	0.06	0.06	0.25	−0.2639	0.0138	0.0012
Zero Mean. Zero Skewness.	Normal	0.06	0.04	0.25	0.0000	0.0363	0.0000
Zero Mean. Zero Skewness.	Inverted	0.04	0.06	0.25	0.0000	0.0363	0.0000
Zero Mean. Zero Skewness	Stable	0.06	0.06	0.25	0.0000	0.0363	0.0000
Panel B:							
Scenario	Market	q		Spot u	Spot d	Futures u^f	Futures d^f
Positive mean. Zero Skewness.	Normal	0.5000		1.1000	0.9500	1.0945	0.9453
Positive mean. Negative Skewness.	Normal	0.6500		1.1000	0.9500	1.0945	0.9453
Positive mean. Negative Skewness.	Normal	0.8000		1.1000	0.9500	1.0945	0.9453
Positive mean. Zero Skewness.	Inverted	0.5000		1.1000	0.9500	1.1055	0.9458
Positive mean. Negative Skewness.	Inverted	0.6500		1.1000	0.9500	1.1055	0.9458
Positive mean. Negative Skewness.	Inverted	0.8000		1.1000	0.9500	1.1055	0.9458
Positive mean. Zero Skewness	Stable	0.5000		1.1000	0.9500	1.1000	0.9500
Positive mean. Negative Skewness	Stable	0.6500		1.1000	0.9500	1.1000	0.9500
Positive mean. Negative Skewness	Stable	0.8000		1.1000	0.9500	1.1000	0.9500
Negative mean. Zero Skewness.	Normal	0.5000		1.0526	0.9091	1.0474	0.9046
Negative mean. Positive Skewness.	Normal	0.3500		1.0526	0.9091	1.0474	0.9046
Negative mean. Positive Skewness.	Normal	0.2000		1.0526	0.9091	1.0474	0.9046
Negative mean. Zero Skewness.	Inverted	0.5000		1.0526	0.9091	1.0579	0.9137
Negative mean. Positive Skewness.	Inverted	0.3500		1.0526	0.9091	1.0526	0.9091
Negative mean. Positive Skewness.	Inverted	0.2000		1.0526	0.9091	1.0579	0.9137
Negative mean. Zero Skewness	Stable	0.5000		1.0526	0.9091	1.0526	0.9091
Negative mean. Positive Skewness	Stable	0.3500		1.0526	0.9091	1.0526	0.9091
Negative mean. Positive Skewness	Stable	0.2000		1.0526	0.9091	1.0526	0.9091
Zero Mean. Zero Skewness.	Normal	0.5000		1.1000	0.9091	1.0945	0.9046
Zero Mean. Zero Skewness.	Inverted	0.5000		1.1000	0.9091	1.1055	0.9137
Zero Mean. Zero Skewness	Stable	0.5000		1.1000	0.9091	1.1000	0.9091

$$\left|\ln d^f\right| > \left|\ln u^f\right|$$

$$\left|\ln 0.90457\right| > \left|\ln 1.047388\right|$$

$$\left|10.03\%\right| > \left|4.63\%\right|$$

The binomial tree for the underlying spot $/BP exchange rate, BP futures contract, European and American futures calls, and European and American futures puts are shown in Exhibit 1. In the three-period option case, the binomial model prices the European futures call at $0.0857 and the European futures put at $0.0553. As shown in the exhibit, there is an early exercise advantage for the American futures call at the upper node in period 2, and an early exercise advantage for the American futures put at the lower node in period 2. As a result, both the American futures put and call options are price slightly higher than their European counterparts.

If the up and down parameters are not adjusted for skewness, then $q = 0.5$ and the skewness-adjusted equations for the up and down parameters for the spot and futures simplify to the CRR/RB formulas:

$$u = e^{\sigma_S \sqrt{t/n} + \mu_S (t/n)}$$

$$d = e^{-\sigma_S \sqrt{t/n} + \mu_S (t/n)}$$

$$u^f = e^{\sigma_f \sqrt{t/n} + \mu_f (t/n)}$$

$$= e^{\sigma_S \sqrt{(t/n)} + (\mu_S - (R - \psi))(t/n)}$$

$$d^f = e^{-\sigma_f \sqrt{t/n} ++ \mu_f (t/n)}$$

$$= e^{-\sigma_S \sqrt{(t/n)} + (\mu_S - (R - \psi))(t/n)}$$

In this example, the up and down parameters for the spot rate would be u = 1.026268 and d = 0.892335, the up and down parameters on the futures contract would be u^f = 1.021149 and d^f = 0.887884, and the risk-neutral probability would be p^f = 0.8412978. In this case, in which skewness is assumed to exist but is excluded in the estimates of the up and down parameters the binomial model prices the European call at $0.0786, 8.28% less than the skewness-adjusted model, and the European put at $0.0477, 13.74% less than the skewness model. Additionally, there would not be an early exercise advantage

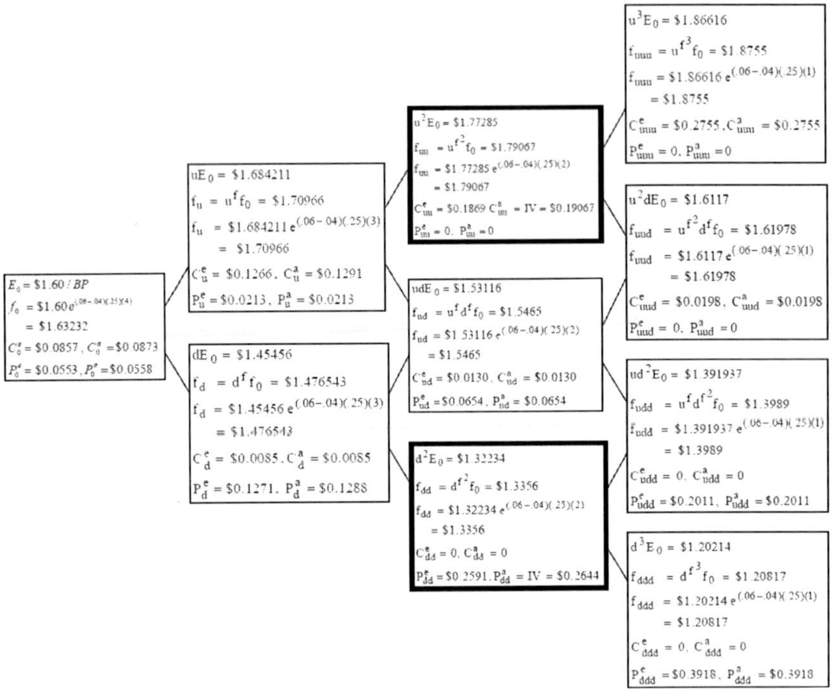

Exhibit 1. Binomial call and put values under decreasing exchange rate case: skewness model. for the American futures call and there would be only one early exercise advantage for the American futures put. As a result, the American futures put option is priced slightly higher than its European counterpart.

3.2. Increasing Exchange Rate Case

Periods of increasing security prices are often characterized by a positive mean and negative skewness in the security price's logarithmic return. To illustrate option pricing under this scenario, suppose the market expects the $/BP exchange rate to increase over the next year such that the expected distribution of the exchange rate's logarithmic return has the following estimated annualized moments: $\mu_S = 0.17597$, $V_S = 0.019555$, and $\delta_S = -0.0008602$. In this increasing exchange rate case, the up and down parameters for the spot rate are $u = 1.10$ and $d = 0.95$, the up and down parameters on the futures contract are $u^f = 1.0945$ and $d^f = 0.94526$, $q = 0.65$, and the risk-neutral probability is $p^f = 0.36675$. The positive mean and negative skewness yield u^f, d^f, and q values that reflect an increasing exchange-rate period in which the proportional increase in the futures rate each period is 9.03%, exceeding in absolute value the proportional decrease of 5.63%, and the probability of an increase in each quarterly period (Δt = 0.25) is q = 0.65:

$$\left|\ln u^f\right| > \left|\ln d^f\right|$$

$$\left|\ln 1.0945\right| > \left|\ln 0.94526\right|$$

$$\left|9.03\%\right| > \left|5.63\%\right|$$

In a three-period option case, the binomial model for the increasing case would price the European futures call at $0.0862 and the European futures put at $0.0553. The American futures put and call options are priced slightly higher than the European option, given early exercise advantages for both the call and put options.

If the up and down parameters for this increasing case were not adjusted for skewness, then q would be equal to 0.5, and the up and down parameters for the spot rate would be u = 1.120667 and d = 0.974407, the up and down parameters on the futures contract would be u^f = 1.115078 and d^f = 0.969547, and the risk-neutral probability would be p^f = 0.20925. In this case in which skewness is assumed to exist but is excluded in the estimates of the up and down parameters, the binomial model prices the European call at $0.084, 2.55% less than the skewness-adjusted model, and the European put at $0.0531, 3.98% less than the skewness model.

4. PROPERTIES OF THE SKEWNESS MODEL

In the case of spot options, the existence of skewness affects the relative contribution of the mean to the values of the up and down parameters and the asymptotic properties, with the skewness-adjusted up and down parameters having different asymptotic properties than the CRR/RB parameters. In the case of futures options, these same properties also hold. In addition, with futures options, the impact skewness has on elevating the importance of the mean depends on the carrying cost value ($R - \psi$) and whether the futures market is normal, inverted or neutral.

4.1. Relative Importance of the Mean Term

In the case of a positive mean, the mean becomes more important in determining the value of the spot up parameter value, the greater the negative skewness (or equivalently the more q exceeds 0.5). Similarly, for futures options, the $\mu - (R - \psi)$ term becomes more important in determining the futures up parameter value (u^f), the greater the negative skewness. By contrast, in the case of a negative mean, the mean becomes more important in determining the value of d, and the $\mu - (R - \psi)$ term becomes more important in determining the value of the futures down parameter (d^f), the greater the positive skewness (or equivalently the more $1 - q$ exceeds 0.5).

To see the impact skewness has on increasing the importance of the mean for spot options and the $\mu - (R - \psi)$ term for futures options, consider the previous three period increasing exchange rate case in which $u = 1.10$ and $d = 0.95$ for the spot exchange rate. If skewness were zero, then q would be equal to 0.5, the expected quarterly mean would be equal to 0.020084 (=[(0.5)ln(1.10) + (0.5)ln(0.95)]) and the annualized mean would be 0.088034. The quarterly variance would be equal to 0.0053731 (= 0.5[ln (1.10) - 0.020084]^2 + 0.5[ln (0.95) - 0.020084]^2), and the annualized variance would be 0.021493. If these were the actual empirical values of μ_S and V_S, then ln(u) would be equal to 0.0953 (u = 1.10), and the mean would contribute 23% to the value of u and the variance would contribute 77% to the value of u:

$$\ln u = \mu_S\left(t/n\right) + \left[V_S\left(\left(1-q\right)/q\right)\left(t/n\right)\right]^{1/2}$$

$$1 = \frac{\mu_S\left(t/n\right)}{\ln u} + \frac{\left[V_S\left(\left(1-q\right)/q\right)\left(t/n\right)\right]^{1/2}}{\ln u}$$

$$1 = \frac{0.088034\left(0.75/3\right)}{\ln\left(1.10\right)}$$

$$+\frac{\left[0.021492\left(\left(1-0.5\right)/0.5\right)\left(0.75/3\right)\right]^{1/2}}{\ln\left(1.10\right)}$$

$$1 = 0.23 + 0.77$$

For the futures option, the ln(u^f) would be equal to 0.0903095 (u^f = 1.094513) given the net cost of carry of 0.02 (=$R - \psi$ = 0.06 - 0.04). The mean minus the net cost of carry ($R - \psi$) would, in turn, contribute 19% to the value of u^f and the variance would contribute 81% to the value of u^f:

$$\ln u^f = \left[\mu_S - (R - \psi)\right](t/n) + \left[V_S\left((1-q)/q\right)(t/n)\right]^{1/2}$$

$$1 = \frac{\left[\mu_S - (R-\psi)\right](t/n)}{\ln u^f} + \frac{\left[V_S\left((1-q)/q\right)(t/n)\right]^{1/2}}{\ln u^f}$$

$$1 = \frac{\left[0.088034 - (0.06 - 0.04)\right](0.75/3)}{\ln(1.094513)} + \frac{\left[0.021492\left((1-0.5)/0.5\right)(0.75/3)\right]^{1/2}}{\ln(1.094513)}$$

$$1 = 0.19 + 0.81$$

If negative skewness were present such that q = 0.65, then the quarterly mean would be equal 0.044 (= [(0.65) ln(1.10) + (0.35) ln(0.95)]) and the annualized mean would be 0.176. The quarterly variance would, in turn, be equal to 0.0389986 (= 0.65 [ln(1.10) – 0.044]^2 + 0.35 [ln(0.95) – 0.044]^2) and the annualized variance would be 0.019558. Given the mean and variance values and a q = 0.65, the implied skewness would be δ = –0.0008602. If these were the actual empirical values of mean, variance, and skewness, then ln(u) would be equal to the nonskewed value of 0.0953 (u = 1.10), but with q = 0.65, the contribution of the mean to the value of u would be 46% and the contribution of the variance to the value of u would be 54%:

$$\ln u = \mu_S(t/n) + \left[V_S((1-q)/q)(t/n)\right]^{1/2}$$

$$1 = \frac{\mu_S(t/n)}{\ln u} + \frac{\left[V_S((1-q)/q)(t/n)\right]^{1/2}}{\ln u}$$

$$1 = \frac{0.175996(0.75/3)}{\ln(1.10)}$$

$$+ \frac{\left[0.019558((1-0.65)/0.65)(0.75/3)\right]^{1/2}}{\ln(1.10)}$$

$$1 = 0.46 + 0.54$$

For the futures options, the $\ln(u^f)$ would likewise be equal to its non-skewed value of 0.0903095 (u^f= 1.094513), but the contribution of the mean and the net cost of carry ($R - \psi$) term to the value of u^f would be 43%, and the contribution of the variance to the value of u^f would be 57%:

$$\ln u^f = \left[\mu_S - (R-\psi)\right](t/n)$$

$$+ \left[V_S((1-q)/q)(t/n)\right]^{1/2}$$

$$1 = \frac{\left[\mu_S - (R-\psi)\right](t/n)}{\ln u^f}$$

$$+ \frac{\left[V_S((1-q)/q)(t/n)\right]^{1/2}}{\ln u^f}$$

$$1 = \frac{\left[0.175996 - (0.06-0.04)\right](0.75/3)}{\ln(1.094513)}$$

$$+ \frac{\left[0.019558((1-0.65)/0.65)(0.75/3)\right]^{1/2}}{\ln(1.094513)}$$

$$1 = 0.43 + 0.57$$

Thus, in the case of a positive mean, negative skewness increases the relative importance of the mean on the up parameters for the spot and futures rates1.

By contrast, in the case of a positive mean, the mean term has an opposite directional impact on d and d^f than the variance term, with its negative impact increasing the greater the negative skewness (or equivalently the more q exceeds 0.5). For example, in the no skewness case for the futures option, $\ln(d^f)$ is equal to −0.056292 ($d^f = 0.945262$). In this case, the $\mu - (R - \psi)$ term would contribute −30% to the value of d^f and the variance term would contribute 130% to the value of d^f. In the skewness case, the $\mu - (R - \psi)$ term would contribute −69% to the value of d^f, whereas the variance would contribute 169% to the value of d^f:

Just the opposite relationships hold in the case of a negative mean with positive skewness. Specifically for a negative mean, the mean becomes more important in determining d and d^f the greater the positive skewness, whereas the mean term has an opposite directional impact on u and u^f than the variance term, with its negative impact on the up parameter increasing the greater the positive skewness. **Table 2** summarizes the relative contributions of the mean and variance terms to the spot and futures up and down parameters for the decreasing exchange-rate case and other scenarios with different levels of skewness and different futures markets. Panel A details the relative contributions for the spot and Panel B details the contributions for the futures.

4.2. Asymptotic Properties

In the CRR/RB model, as the number of subperiods (n) increases, the mean term in the exponent for spot options and the $\mu - (R - \psi)$ term for futures options go to zero faster than the square root term. Thus for large n, the CRR/RB model depends only on the volatility. Moreover, with $\sigma_S = \sigma_f$, the up and down parameters for the futures are equal to the spot, regardless of whether the futures market is normal or inverted:

$$u^f = e^{\sigma_f \sqrt{t/n}} = e^{\sigma_S \sqrt{t/n}} = u$$

$$d^f = e^{-\sigma_f \sqrt{t/n}} = e^{-\sigma_S \sqrt{t/n}} = d$$

Using theses up and down parameters, the BOPM for futures options converges to the seminal Black futures option model [13] as n gets large.

The skewness-adjusted up and down parameters, though, do not have the same asymptotic properties as the CRR/ RB parameters. In the skewness model, equation (20) for u^f includes $a(1 - q)/q$ term, and equation (21) for d^f includes a $q/(1 - q)$ term, both of which change the order of magnitude as n gets large. Specifically, for the case of negative skewness, the $(1 - q)/q$ term can be rewritten as:

Table 2. (a) Relative contributions of the mean and variance terms to up and down parameters; (b) Relative contributions of the mean and variance terms to up and down parameters.

Panel: A		Relative Contributions of			
Scenario	Market	Mean to u	Variance to u	Mean to d	Variance to d
Increasing: Positive mean, Zero Skewness,	Normal	0.2309	0.7691	−0.4291	1.4291
Increasing: Positive mean, Negative Skewness,	Normal	0.4616	0.5384	−0.8578	1.8578
Positive mean, Negative Skewness,	Normal	0.6924	0.3076	−1.2865	2.2865
Increasing: Positive mean, Zero Skewness,	Inverted	0.2309	0.7691	−0.4291	1.4291
Increasing: Positive mean, Negative Skewness,	Inverted	0.4616	0.5384	−0.8578	1.8578
Increasing: Positive mean, Negative Skewness,	Inverted	0.6924	0.3076	−1.2865	2.2865
Increasing: Positive mean, Zero Skewness	Stable	0.2309	0.7691	−0.4291	1.4291
Increasing: Positive mean, Negative Skewness	Stable	0.4616	0.5384	−0.8578	1.8578
Increasing: Positive mean, Negative Skewness	Stable	0.6924	0.3076	−1.2865	2.2865
Decreasing: Negative mean, Zero Skewness,	Normal	−0.4290	1.4290	0.2309	0.7691
Decreasing: Negative mean, Positive Skewness,	Normal	−0.8577	1.8577	0.4616	0.5384
Decreasing: Negative mean, Positive Skewness,	Normal	−1.2864	2.2864	0.6924	0.3076
Decreasing: Negative mean, Zero Skewness,	Inverted	−0.4290	1.4290	0.2309	0.7691
Decreasing: Negative mean, Positive Skewness,	Inverted	−0.8577	1.8577	0.4616	0.5384
Decreasing: Negative mean, Positive Skewness,	Inverted	−1.2864	2.2864	0.6924	0.3076
Decreasing: Negative mean, Zero Skewness	Stable	−0.4290	1.4290	0.2309	0.7691
Decreasing: Negative mean, Positive Skewness	Stable	−0.8577	1.8577	0.4616	0.5384
Decreasing: Negative mean, Positive Skewness	Stable	−1.2864	2.2864	0.6924	0.3076
Stable Market: Zero Mean, Zero Skewness,	Normal	0.0001	0.9999	−0.0001	1.0001
Stable Market: Zero Mean, Zero Skewness,	Inverted	0.0001	0.9999	−0.0001	1.0001
Stable Market: Zero Mean, Zero Skewness	Stable	0.0001	0.9999	−0.0001	1.0001

Panel: B		Relative Contributions of			
Scenario	Market	$\mu-(Rf-\psi)$ to u^f	Variance to u^f	$\mu-(Rf-\psi)$ to d^f	Variance to d^f
Increasing: Positive mean, Zero Skewness,	Normal	0.1883	0.8117	−0.3021	1.3021
Increasing: Positive mean, Negative Skewness,	Normal	0.4318	0.5682	−0.6928	1.6928
Positive mean, Negative Skewness,	Normal	0.6753	0.3247	−1.0834	2.0834
Increasing: Positive mean, Zero Skewness,	Inverted	0.2692	0.7308	−0.5834	1.5834
Increasing: Positive mean, Negative Skewness,	Inverted	0.4855	0.5115	−1.0584	2.0584
Increasing: Positive mean, Negative Skewness,	Inverted	0.7077	0.2923	−1.5335	2.5335
Increasing: Positive mean, Zero Skewness	Stable	0.2309	0.7691	−0.4291	1.4291
Increasing: Positive mean, Negative Skewness	Stable	0.4616	0.5384	−0.8578	1.8578
Increasing: Positive mean, Negative Skewness	Stable	0.6924	0.3076	−1.2865	2.2865
Decreasing: Negative mean, Zero Skewness,	Normal	−0.5833	1.5833	0.2692	0.7308
Decreasing: Negative mean, Positive Skewness,	Normal	−1.0583	2.0583	0.4885	0.5115
Decreasing: Negative mean, Positive Skewness,	Normal	−1.5333	2.5333	0.7077	0.2923
Decreasing: Negative mean, Zero Skewness,	Inverted	−0.3021	1.3021	0.1883	0.8117
Decreasing: Negative mean, Positive Skewness,	Inverted	−0.8577	1.8577	0.4616	0.5384
Decreasing: Negative mean, Positive Skewness,	Inverted	−1.0833	2.0833	0.6753	0.3247
Decreasing: Negative mean, Zero Skewness	Stable	−0.4290	1.4290	0.2309	0.7691
Decreasing: Negative mean, Positive Skewness	Stable	−0.8577	1.8577	0.4616	0.5384
Decreasing: Negative mean, Positive Skewness	Stable	−1.2864	2.2864	0.6924	0.3076
Stable Market: Zero Mean, Zero Skewness,	Normal	−0.0553	1.0553	0.0498	0.9502
Stable Market: Zero Mean, Zero Skewness,	Inverted	0.0499	0.9501	−0.0554	1.0554
Stable Market: Zero Mean, Zero Skewness	Stable	0.0001	0.9999	−0.001	1.0001

$$\frac{1-q}{q}=\frac{4\left(V_f\right)^3}{2n\left(\delta_f\right)^2+4\left(V_f\right)^3+2\sqrt{n\left(\delta_f\right)^2+4\left(V_f\right)^3}\sqrt{\left(\delta_f\right)^2}} \tag{22}$$

The expression $(1 - q)/q$ in equation (22), in turn, is the same order of magnitude as $1/n$. This can be seen by observing that the term $[(1 - q)/q]/[1/n]$ approaches the constant as n gets large. That is:

$$\frac{(1-q)/q}{1/n}=$$

$$\frac{4\left(V_f\right)^3}{2\left(\delta_f\right)^2+\left[4\left(V_f\right)^3\right]/n+2\sqrt{\left(\delta_f\right)^2+\left[4\left(V_f\right)^3/n\right]}\sqrt{\left(\delta_f\right)^2}} \tag{23}$$

and in the limit:

$$\frac{(1-q)/q}{1/n}\rightarrow\frac{\left(V_f\right)^3}{\left(\delta_f\right)^2} \quad \text{as } n\rightarrow\infty \tag{24}$$

With $(1 - q)/q$ having the same order of magnitude as 1/n, the term $[(1 - q)/q*V_f/n]$ 1/2 in Equation (13) approaches a constant multiplied by $[V_f]^{1/2}/n$, as n gets large. As a result, for the case of large n, the first term in the exponent in equation (13) approximates a constant divided by n, which is in the same form as the second term, $\mu_f/n = (\mu_S - (R - \psi))/n$.

Consequently, both terms in the exponent of equation (13) for u^f contribute equally, even when n is large. Thus, as n gets large, u^f depends not only on the variance and skewness, but also on the mean, risk-free rate, and asset yield. By contrast, equation (14) for d^f is defined in terms of $q/(1 - q)$. In this case, as n gets large, the variance term in the exponent for d^f approaches the constant δ_f/V_f and the μ_f/n term approaches zero. Thus, for the case of negative skewness with a large n, d^f depends on the variance and skewness, but not on the mean, risk-free rate and dividend yield, whereas u^f depends on all three parameters. Just the opposite asymptotic relationships occur when skewness is positive. In this case, the mean, variance, skewness, risk-free rate, and dividend yield determine d^f (even when n is large), whereas just the variance and skewness determine u^f.

Note, similar asymptotic relations also hold for the spot up and down parameters, with the spot and futures asymptotic relations being equivalent when the net cost of carry is zero $(R - \psi = 0)$.

To illustrate the relative importance of the mean for the case of large n, consider the preceding increasing exchange-rate case (option expiring in 0.75 year, R = 0.06 and ψ = 0.04) where the estimated annualized mean, variance, and skewness were μ_S = 0.17597, V_S = 0.0019555, and δ_S= −0.0008602. If we subdivide the option period (t = 0.75) into 3 subperiods (n = 3), then for the spot exchange rate the mean term would contribute 46.164% to the value of u and the variance term would contribute 53.836% to the value of u; for the futures, the $\mu - (R - \psi)$ term would contribute 43.18% to the value of u^f and the variance term would contribute 56.82% to the value of u^f:

Spot:

$$\mu_S (t/n)/\ln(u) = 0.17597(0.75/3)/\ln(1.10)$$
$$= 0.46164$$

Futures :

$$[\mu_S - (R-\psi)](t/n)/\ln(u^f)$$
$$= [0.17597 - (0.06 - 0.04)] * (0.75/3)/\ln(1.094513)$$
$$= 0.4318$$

If we subdivide the option period into nine monthly subperiods (n = 9; t/n = 0.75/9 = 0.08333; u = 1.03940, d = 0.9481; u^f = 1.037673, d^f = 0.9465), the mean term would contribute 37.95% and the variance term would contribute 62.05% for the spot rate, and for the futures, the mean term would contribute 35.15% and the variance term would contribute 64.85%:

Spot:

$$\mu_S (t/n)/\ln(u) = 0.17599586(0.75/9)/\ln(1.0394)$$
$$= 0.3795$$

Futures:

$$[\mu_S - (R-\psi)](t/n)/\ln(u^f)$$
$$= [0.17599586 - (0.06 - 0.04)] * (0.75/9)/\ln(1.037673)$$
$$= 0.3515$$

For larger n, the contribution of the mean is approximately the same: when n = 39 (weekly periods; t/n = 0.75/39 = 0.01923), the relative contribution of the mean would be 31.59% for the spot rate and 29.04% for the futures; when n = 270 (daily; t/n = 0.75/270 = 0.0028), the contributions would be

28.91% for the spot and 26.49% for the futures; when n = 1000, the contributions would be 28.51% and 26.11% for the spot and futures, respecttively; when n = 1000000, the contributions would be 28.35% and 25.97%.

By contrast, if skewness were zero (the CRR/RB model with $\mu_S = 0.17597$, $V_S = 0.019558$, and $\delta_S = 0$), the relative contributions of the mean term for the spot rate and futures would be 38.62% and 35.80% for $n = 3$ (($n = 3$; t/n = 0.75/9 = 0.08333; u = 1.120667, d = 0.974407; u^f = 1.115078, d^f = 0.969547):

Spot:

$$\mu_S(t/n)/\ln(u) = 0.17599586\,(0.75/3)/\ln(1.120667)$$
$$= 0.3862$$

Futures:

$$[\mu_S - (R - \psi)](t/n)/\ln(u^f)$$
$$= [0.17599586 - (0.06 - 0.04)] * (0.75/3)/\ln(1.115078)$$
$$= 0.3580$$

When n = 9, the mean contribution for the spot and futures, respectively, would be 26.64% and 24.36%; when n = 39, 14.86% and 14.40%; when n = 270, 6.22% and 5.55%; when n = 1000, 3.33% and 2.96%; and when n = 1000000, 0.1% and 0.1%.

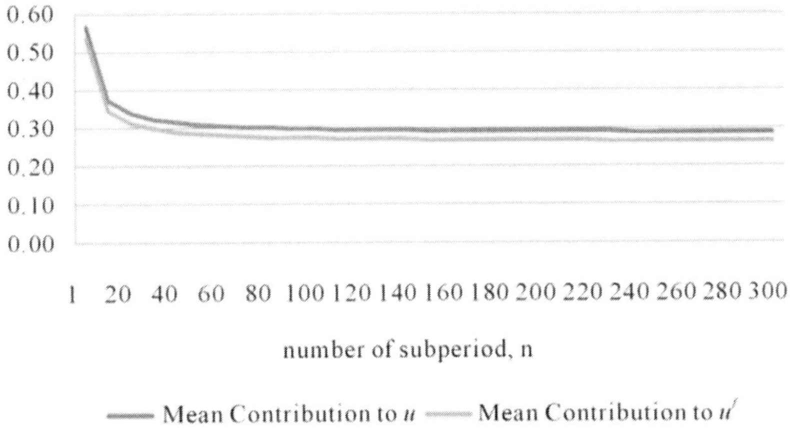

Figure 1. Mean contribution to u and u^f, $\mu_S = 0.175996$, $V_S = 0.019558$, $\delta_S = -0.0086019$.

Figure 1shows graphically the relationship between the number of subperiods and the mean's contribution to the up parameter for the skewness-adjusted increasing exchange-rate case.

The graph in **Figure 1** highlights the asymptotic relation, showing that as n increases, the mean's contribution to both the sopt and futures up parameters decreases asymptotically with the asymptote occurring at approximately $n^* = 30$ where the minimum mean contribution is 30%2. **Figure 2** summarizes the relationship between the number of subperiods and the mean's contribution to the up and down parameters for the CRR/RB case in which skewness is zero.

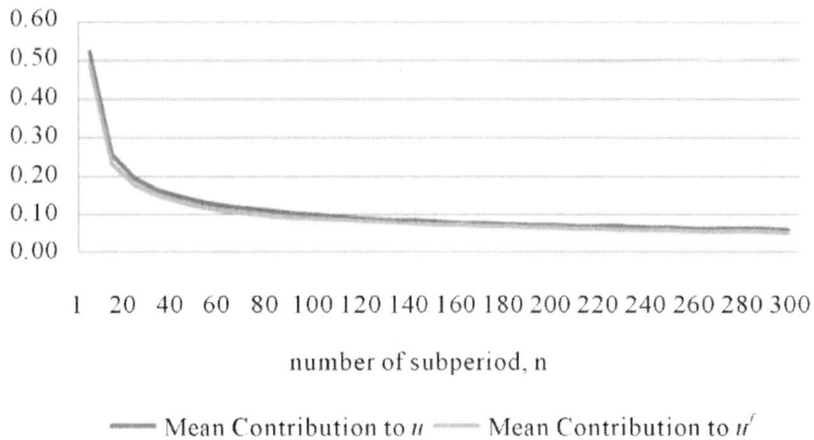

Figure 2. Mean contribution to u and u^f, $\mu_S = 0.175996$, $V_S = 0.019558$, $\delta_S = 0$.

Figure 2, in turn, shows a similar asymptotic relation between the mean's contribution and n for the CRR/RB model as the skewness case, but with the minimum mean contribution being close to zero.

Figure 3 shows graphically the relationship between the number of subperiods and the mean's contribution to the down parameters for the skewness-adjusted decreasing exchange-rate case characterized by a negative mean and positive skewness: $\mu_S = -0.17597$; $V_S = 0.019555$; $\delta_S = 0.0008602$. The graph highlights the asymptotic relation, showing that as n increases, the mean's contribution to the down parameters decreases asymptotically with the asymptote occurring at approximately $n^* = 30$ where the minimum mean contribution is 30%.

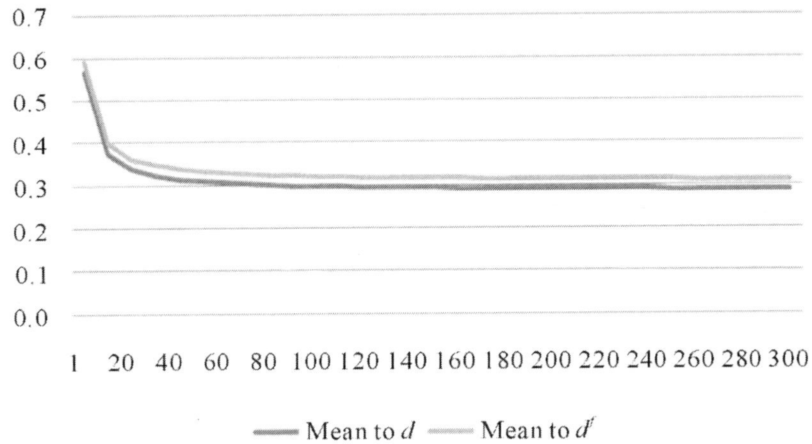

Figure 3. Mean contribution to d and d^f, $\mu_S = -0.17597$, $V_S = 0.019555$, $\delta_S = 0.00086019$.

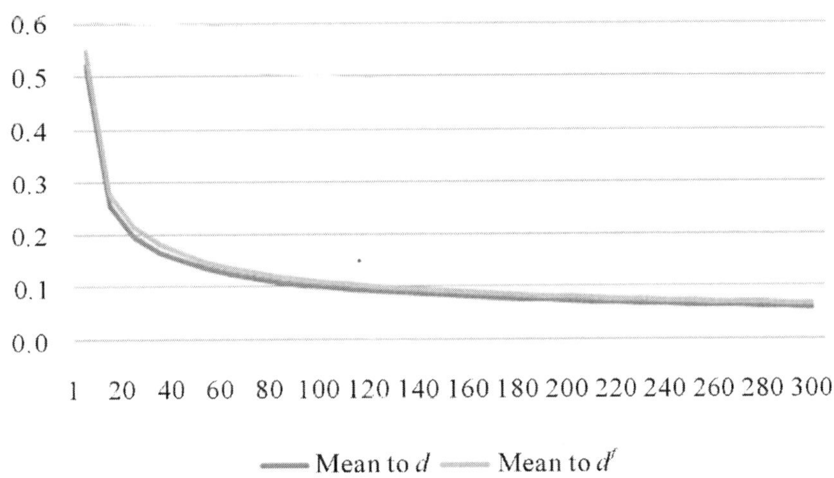

Figure 4. Mean contribution to d and d^f, $\mu_S = -0.17597$, $V_S = 0.019555$, $\delta_S = 0$.

Figure 4 shows a similar asymptotic relation between the mean's contribution to the down parameter and n for the CRR/RB case, but with the minimum contribution being close to zero.

5. DIFFERENCES IN FUTURES OPTION PRICES BETWEEN THE CRR/RB MODEL AND THE SKEWNESS-ADJUSTED MODEL

Values for European futures call options on the British pound obtained using the skewness-adjusted model and the CRR/RB Model are presented in Table 3, and values for the European futures put options on the British pound are shown in Table 4. The call and put futures options each have exercise prices of $1.60/BP and expire in 0.75 years, and the British pound futures contract expires in one year, with the futures price assumed to be equal to its carrying-coat value. The binomial model used for pricing is subdivided into 60 periods of length 6 days. The tables show three futures markets: a normal futures market where the annualized risk-free rate is assumed to be 6% on US dollars and 4% on British pounds, an inverted market where the dollar rate is 4% and British pound rate is 6%, and a neutral market where each rate is equal to 6%. Finally, the tables show two exchange-rate scenarios:

1) An increasing exchange-rate case characterized by a positive mean and a negative skewness: $\mu_S = 0.17599586$, $V_S = 0.019558$, $\delta_S = -0.000860192$.

2) A decreasing exchange-rate case characterized by a negative mean and a positive skewness: $\mu_S = 0.17599586$, $V_S = 0.019558$, $\delta_S = -0$.

A comparison of the futures option values obtained using the skewness-adjusted model with the CRR/RB model illustrates the pricing differences that occur under increasing or decreasing exchange-rate cases characterized by skewness. In general, for both scenarios, the CRR/RB model prices the American and European futures call and the American and European futures puts less than the skewness model, with the greatest pricing differences occurring for out-of-the-money options. Specifically, the simulations show underpricing of the CRR/RB model, ranging from approximately 5% for in-the-money options to approximately 35% for out-of-the-money options. For the futures call option, in both the increasing and decreasing cases, the CRR/RB model underprices more for the inverted futures market case (range 30.56% - 10.11% (increasing); 39.43% - 5.93% (decreasing)) than the normal futures market case (21.59% - 5.63% (increasing); 34.79% - 4.10% (decreasing)). For puts, the CRR/RB model underprices more for the normal market case (range 5.05% - 38.36% (increasing); 9.14% - 30.80% (decreasing)) than the inverted market case (4.67% - 33.06% (increasing); 4.85% - 21.25% (decreasing))[3].

Finally, Table 5 compares futures call and put prices with corresponding spot call and put prices for normal, inverted, and neutral futures market under a stable exchange-rate scenario in which the mean and skewness are zero. As shown, with zero skewness, the skewness model and the CRR/RB model for spot and futures options are the same. The simulations, in turn, also show that this is the only case in which European futures and spot options are equal.

Table 3. Comparison of skewness-adjusted futures option model with crr/rb model for normal, inverted and neutral futures markets and increasing and decreasing exchange-rate trends: european futures call options.

Panel A: Normal					
Rate Trend	Spot $/BP	Futures Price	Skewness Model European	CRR/RB Model European	% Difference CRR/RB-Skew
Increase	$1.45	$1.4793	$0.0352	$0.0276	−21.59%
	1.55	1.5813	0.0765	0.0646	−15.56%
	1.65	1.6833	0.1349	0.1217	−9.79%
	1.75	1.7854	0.2078	0.1961	−5.63%
Decrease	$1.45	$1.4793	$0.0411	$0.0268	−34.79%
	1.55	1.5813	0.0797	0.0637	−20.08%
	1.65	1.6833	0.1346	0.1213	−9.88%
	1.75	1.7854	0.2048	0.1964	4.10%

Panel B: Inverted					
Rate Trend	Spot $/BP	Futures Price	Skewness Model European	CRR/RB Model European	% Difference CRR/RB-Skew
Increase	$1.45	$1.4213	$0.0216	$0.0150	−30.56%
	1.55	1.5193	0.0524	0.0397	−24.24%
	1.65	1.6173	0.0993	0.0829	−16.52%
	1.75	1.7153	0.1612	0.1449	−10.11%
Decrease	$1.45	$1.4213	$0.0246	$0.0149	−39.43%
	1.55	1.5193	0.0527	0.0400	−24.10%
	1.65	1.6173	0.0961	0.0838	−12.80%
	1.75	1.7153	0.1552	0.1460	−5.93%

Panel C: Neutral					
Rate Trend	Spot $/BP	Futures Price	Skewness Model European	CRR/RB Model European	% Difference CRR/RB-Skew
Increase	$1.45	$1.45	$0.0276	$0.0203	−26.54%
	1.55	1.55	0.0632	0.0508	−19.62%
	1.65	1.65	0.1154	0.1006	−12.82%
	1.75	1.75	0.1823	0.1682	−7.73%
Decrease	$1.45	$1.45	$0.0319	$0.0201	−36.99%
	1.55	1.55	0.0649	0.0506	−22.03%
	1.65	1.65	0.1137	0.1008	−11.35%
	1.75	1.75	0.1779	0.1691	−4.95%

Table 4. Comparison of skewness-adjusted futures option model with crr/rb model for normal, inverted and neutral futures markets and increasing and decreasing exchange-rate trends: european futures put options.

Panel A: Normal					
Rate Trend	Spot $/BP	Futures Price	Skewness Model European	CRR/RB Model European	% Difference CRR/RB-Skew
Increase	$1.45	$1.4793	$0.1506	$0.1430	−5.05%
	1.55	1.5813	0.0944	0.0824	−12.71%
	1.65	1.6833	0.0533	0.0420	−24.05%
	1.75	1.7854	0.0305	0.0188	−38.36%
Decrease	$1.45	$1.4793	$0.1565	$0.1422	−9.14%
	1.55	1.5813	0.0975	0.0816	−16.31%
	1.65	1.6833	0.0549	0.0417	−24.04%
	1.75	1.7854	0.0276	0.0191	−30.80%
Panel B: Inverted					
Rate Trend	Spot $/BP	Futures Price	Skewness Model European	CRR/RB Model European	% Difference CRR/RB-Skew
Increase	$1.45	$1.4213	$0.1950	$0.1859	−4.67%
	1.55	1.5193	0.1307	0.1180	−9.72%
	1.65	1.6173	0.0825	0.0661	−19.88%
	1.75	1.7153	0.0493	0.0330	−33.06%
Decrease	$1.45	$1.4213	$0.1980	$0.1884	−4.85%
	1.55	1.5193	0.1311	0.1184	−9.69%
	1.65	1.6173	0.0793	0.0670	−15.51%
	1.75	1.7153	0.0433	0.0341	−21.25%
Panel C: Neutral					
Rate Trend	Spot $/BP	Futures Price	Skewness Model European	CRR/RB Model European	% Difference CRR/RB-Skew
Increase	$1.45	$1.45	0.1710	0.1637	−4.27
	1.55	1.55	0.1110	0.0986	−11.17
	1.65	1.65	0.0676	0.0528	−21.89
	1.75	1.75	0.0389	0.0248	−36.25
Decrease	$1.45	$1.45	0.1753	0.1635	−6.73
	1.55	1.55	0.1127	0.0984	−12.69
	1.65	1.65	0.0659	0.0530	−19.58
	1.75	1.75	0.0345	0.0257	−25.51

Table 5. Comparison of european spot and future option prices for normal, inverted and neutral futures markets and a stable exchange-rate market.

	Spot $/BP	Skewness and CRR/RB Spot Option Price European	Futures Price	Skewness and CRR/RB Futures Option Price European	%Difference Futures – Spot Option Price
Panel A: Normal					
Put	$1.45	0.0258	$1.48	$0.0275	6.59%
	1.55	0.0610	1.58	0.0648	6.23%
	1.65	0.1165	1.68	0.1221	4.81%
	1.75	0.1896	1.79	0.1965	3.64%
Call	$1.45	0.1482	$1.48	$0.1429	−3.58%
	1.55	0.0864	1.58	0.0827	−4.28%
	1.65	0.0448	1.68	0.0425	−5.13%
	1.75	0.0209	1.79	0.0192	−8.14%
Panel B: Inverted					
	Spot $/BP	Skewness and CRR/RB Spot Option Price European	Futures Price	Skewness and CRR/RB Futures Option Price European	%Difference Futures – Spot Option Price
Put	$1.45	$0.0166	$1.42	$0.0155	−6.63%
	1.55	0.0429	1.52	0.0405	5.50%
	1.65	0.0883	1.62	0.0841	−4.76%
	1.75	0.1522	1.72	0.1462	−3.94%
Call	$1.45	$0.1831	$1.42	$0.1889	3.17%
	1.55	0.1138	1.52	0.1188	4.39%
	1.65	0.0636	1.62	0.0673	5.82%
	1.75	0.0320	1.72	0.0343	7.19%
Panel C: Neutral					
	Spot $/BP	Skewness and CRR/RB Spot Option Price European	Futures Price	Skewness and CRR/RB Futures Option Price European	%Difference Futures – Spot Option Price
Put	$1.45	$0.0206	$1.45	$0.0206	0.00%
	1.55	0.0510	1.55	0.0510	0.00%
	1.65	0.1010	1.65	0.1010	0.00%
	1.75	0.1692	1.75	0.1692	0.00%
Call	$1.45	$0.1640	$1.45	$0.1640	0.00%
	1.55	0.0988	1.55	0.0988	0.00%
	1.65	0.0532	1.65	0.0532	0.00%
	1.75	0.0258	1.75	0.0258	0.00%

6. CONCLUSIONS

A subtle feature of the B-S model and the BOPM for large n is that these models depend only on the variance. The mean is not important in determining the value of spot options, and the mean and net carry cost are not important for futures options. This feature is a consequence of the assumption that the logarithmic return of the underlying security is normally distributed. In this paper, we show that in cases where skewness exists, the skewness-adjusted up and down parameters for spot options depend more on the mean than non-skewness-adjusted parameters, and that the skewness-adjusted up and down parameters for futures options depend more on the $\mu - (R - \psi)$ term than non-skewness-adjusted parameters. Furthermore, the skewness-adjusted up and down parameters also have different asymptotic properties such that for large n, the mean for spot prices and the $\mu - (R - \psi)$ term for futures maintain their relative importance.

Thus, the presence of skewness serves to augment the relative importance of the mean for spot options and the $\mu - (R - \psi)$ term for futures options. Using simulations, we show that when there is an expected increasing price trend characterized by a positive logarithmic mean and negative skewness or an expected decreasing price trend characterized by a negative logarithmic mean and positive skewness, the CRR/RB model for large n when compared to the skewness-adjusted futures options model underprices futures options between 4% and 30%, with the larger underpricing occurring for out-of-the money options.

Stafford Johnson, Amit Sen, Brian Balyeat Moreover, with the relative contribution of the mean and carrying cost values for a large number of subperiods being at least 30% for one of the up or down parameters in the skewness model and minimal in the CRR/RB, the underpricing can be explained by the change in asymptotic properties resulting from skewness that elevate the importance of the mean and carrying cost parameters. Finally, it should be noted that the BOPM for futures options converges to the seminal Black futures option model as n gets large given a normal distribution for the logarithmic return. The Black futures model, in turn, differs from the B-S Merton model used for pricing spot options by the exclusion of the risk-free rate and the asset yield in the equations for d_1 and d_2. This difference, though, only holds given the assumption of normality. If skewness exists, then the risk-free rate and asset yield, as well as the mean, become important in pricing futures options for the discrete binomial model, as well as the Black futures option model.

REFERENCES

1. R. S. Johnson, J. E. Pawlukiewicz and J. Mehta, "Binomial Option Pricing with Skewed Asset Returns," Review of Quantitative Finance and Accounting, Vol. 9, No. 1, 1997, pp. 89-101. doi:10.1023/A:1008283011490
2. F. Black and M. Scholes, "The Pricing of Options and Corporate Liabilities," Journal of Political Economy, Vol. 81, No. 3, 1973, pp. 637-659. doi:10.1086/260062

3. R. S. Johnson, R. A. Zuber and J. M. Gandar, "Binomial pricing of Fixed-Income Securities for Increasing and Decreasing Interest Rate Cases," Applied Financial Economics, Vol. 16, No. 14, 2006, pp. 1029-1046. doi:10.1080/09603100500426473
4. R. S. Johnson, R. A. Zuber and J. M. Gandar, "Pricing Stock Options under Expected Increasing and Decreasing Price Cases," Quarterly Journal of Business and Economics, Vol. 46, No. 4, 2007, pp. 63-90.
5. E. M. Stein and J. C. Stein, "Stock Price Distributions with Stochastic Volatility: An Analytical Approach," Review of Financial Studies, Vol. 4, No. 4, 1991, pp. 727-752. doi:10.1093/rfs/4.4.727
6. J. B. Wiggins, "Option Values under Stochastic Volatility: Theory and Empirical Estimates," Journal of Financial Economics, Vol. 19, No. 2, 1987, pp. 351-372. doi:10.1016/0304-405X(87)90009-2
7. S. L. Heston, "A Closed Form Solution for Options and Stochastic Volatility with Applications to Bond and Currency Options," Review of Financial Studies, Vol. 6, No. 2, 1993, pp. 327-344. doi:10.1093/rfs/6.2.327
8. R. Jarrow and A. Rudd, "Approximate Option Valuation for Arbitrary Stochastic Processes," Journal of Financial Economics, Vol. 10, No. 3, 1982, pp. 347-369. doi:10.1016/0304-405X(82)90007-1
9. C. J. Corrado and T. Su, "Skewness and Kurtosis in S&P 500 Index Returns Implied by Option Prices," The Journal of Financial Research, Vol. 19, No. 2, 1996, pp. 175- 192.
10. A. Câmara and S. Chung, "Option Pricing for the Transformed Binomial Class," The Journal of Futures Markets, 2006, Vol. 26, No. 8, pp. 759-787. doi:10.1002/fut.20218
11. J. C. Cox, S. A. Ross and M. Rubinstein, "Option Pricing: A Simplified Approach," Journal of Financial Economics, Vol. 7, No. 3, 1979, pp. 229-263. doi:10.1016/0304-405X(79)90015-1
12. R. J. Rendleman and B. J. Bartter, "Two-State Option Pricing," Journal of Finance, Vol. 34, No. 5, 1979, pp. 1093- 1110. doi:10.2307/2327237
13. F. Black, "The Pricing of Commodity Contracts," Journal of Financial Economics, Vol. 3, No. 1-2, 1976, pp. 167-179. doi:10.1016/0304-405X(76)90024-6

NOTES

[1]Note that in this case, $\mu > (R - \psi)$ and the futures market is normal ($R - \psi > 0$) with the futures price exceeding the spot price. As a result, negative skewness increases the mean's impact on the up parameter (46%) for the spot option more than the up parameter for the futures option (43%). In contrast, if the market were inverted ($R - \psi < 0$), then negative skewness would have decreased the impact of the mean on the futures up parameter more than the spot parameter.

[2]It should be noted that since all the parameters contribute to either lnu or lnd as n gets large, the n value in which the skewness-adjusted model approaches a continuous one depends on the relative values of μ_S, V_S, and δ_S. For the case of $\delta_S < 0$, the term $(1 - q)/q$ approaches a constant divided by n in the limit. The critical value, n^*, can therefore be found by solving for the n that makes $(1 - q)/q$ (Equation (20)) equal to a large proportion (e.g., 0.99) of the limit (Equation (22)). Defining the proportion as $1 - \varepsilon$, where ε is equal to the proportion of error (e.g., $\varepsilon = 0.01$), the n^* that is equal to $1 - \varepsilon$ of the limit is:

$$n^* = \frac{(V_e)^3 (1-\varepsilon)\left(1+\sqrt{1+\varepsilon}\right)}{(\delta_e)^2 \varepsilon}$$

[3]The pricing differences between the CRR/RB model and the skewness model are consistent with the aforementioned empirical studies of Stein and Stein [4], Wiggins [5], and Heston [6] who demonstrate that when skewness exists, the B-S model consistently underprices options. Also, as expected, there were no significant difference in the prices for the European call and put options obtained using the Black futures option model (not shown) and the 60-period CRR/RB binomial prices shown in Table 5.

CHAPTER 12

Practical Bayesian support vector regression for financial time series prediction and market condition change detection

T. Law* And J. Shawe-Taylor

Department of Computer Science, University College London, Gower Street, London WC1E 6BT, United Kingdom

ABSTRACT

Support vector regression (SVR) has long been proven to be a successful tool to predict financial time series. The core idea of this study is to outline an automated framework for achieving a faster and easier parameter selection process, and at the same time, generating useful prediction uncertainty estimates in order to effectively tackle flexible real-world financial time series prediction problems. A Bayesian approach to SVR is discussed, and implemented. It is found that the direct implementation of the probabilistic framework of Gao *et al.* returns unsatisfactory results in our experiments. A novel enhancement is proposed by adding a new kernel scaling parameter μ to overcome the difficulties encountered. In addition, the multi-armed bandit Bayesian optimization technique is applied to automate the parameter selection process. Our framework is then tested on financial time series of various asset classes (i.e. equity index, credit default swaps spread, bond yields, and commodity futures) to ensure its flexibility. It is shown that the generalization performance of this parameter selection process can reach or sometimes surpass the computationally expensive cross-validation procedure. An adaptive calibration process is also described to allow practical use of the prediction uncertainty estimates to assess the quality of predictions. It is shown that the machine-learning approach discussed in this study can be developed as a very useful pricing tool, and potentially a market condition change detector. A further extension is possible by taking the prediction uncertainties into consideration when building a financial portfolio.

Keywords: Support vector machines regression, Kernel scaling, Machine learning, Bayesian inference, Multi-armed bandit Bayesian optimization, Gaussian process

1. INTRODUCTION

Support vector regression (SVR) has been one of the most popular Machine-learning algorithms for over a decade. Its ability to predict financial time series has been demonstrated in various studies (Müller *et al.* 1997, Tay and Cao 2001, Cao and Tay 2003, Kim 2003, Lu *et al* 2009, Gündüz and Uhrig-Homburg 2011) with satisfactory empirical results. Although the base assumption of most machine-learning models requires i.i.d. data, there is literature (Mohri and Rostamizadeh 2008, Ralaivola *et al.* 2010) suggesting non-i.i.d. data can be used to train the statistical learning system by increasing sample sizes. This supports SVR as a relevant tool for time series prediction. Comparing to other linear time series models, which require careful designing of inputs, SVR allows flexible mapping of high dimensional features to capture non-linear relationships and at the same time with a regularization technique to reduce over-fitting. However, on the other hand, financial experts in the industry may step back from using this algorithm for two reasons: (1) it is not easy, and sometimes computationally expensive to determine the parameters for the algorithm, (2) it does not provide an estimate of prediction uncertainties.

We propose to investigate extensions to the approach that make up for these two disadvantages by generalizing it within a Bayesian approach. This allows a more efficient parameter selection procedure as well as the derivation of a prediction uncertainty estimate. This is particularly important in the financial context to avoid decisions based on unreliable predictions. A Bayesian approach to SVR (Gao *et al.* 2002) is discussed, and implemented. It is found in our experiments that direct implementation of the probabilistic framework proposed by Gao *et al.* (2002) gives unsatisfactory results. A novel enhancement is proposed by adding a new kernel scaling parameter μ to overcome the difficulties encountered. In addition, because the gradient optimization approach proposed by Gao *et al.* (2002) is found to be impractical, we develop an alternative optimization method. Multi-arm bandit Bayesian optimization has been gaining popularity in recent years (Shahriari *et al.* 2016), mainly for its capability to actively and efficiently search for the global optimum for a complex or even unknown function. We make use of this technique to fully automate the parameter selection process.

We test our framework to predict financial time series of various asset classes to ensure its flexibility. The generalization performance from models selected by different parameter selection approaches (cross-validation, grid search, and Bayesian optimization) are compared. The practicality of the proposed framework is further examined by generating daily predictions, aiming to allow a thorough investigation of the prediction uncertainty estimates derived. A calibration process is established to show how these uncertainty estimates may be utilized to significantly improve predictions as well as to detect market condition changes.

The paper is organized as follows: section 2 gives an overview of SVR. Section 3 discusses the Bayesian probabilistic framework (Gao *et al.* 2002), and the rationale behind our novel enhancement to the probabilistic framework. Section4 covers the theoretical background of the multi-armed bandit Bayesian optimization algorithm. In section 5, the implementation and results of various experiments are explained with the introduction of an adaptive calibration process. Section 6 concludes the findings and discusses possible research extensions.

2. SUPPORT VECTOR REGRESSION

The concepts of support vector machine (SVM) can naturally be applied to handle regression problems. Instead of classifying testing examples into one of the two outcomes (class 1 or class 2), it is used to address target variables with real values. The ε-Support Vector Regression (ε-SVR) can be considered similar to the approach of applying a linear regression in the feature space. In contrast to the squared loss function used in the least squares regression, the error function for ε-SVR is the ε-insensitive loss function. This dictates that the errors smaller than ε are ignored. Figure 1 demonstrates this intuition. This method leads to sparsity similar to SVMs meaning that the form of the regression depends only on the support vectors.

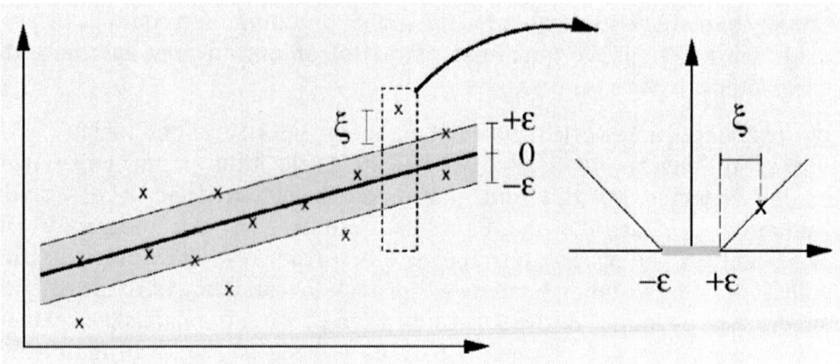

Figure 1. The loss function of ε -SVR.

Consider we have training examples $x_1, \ldots, x_n \in \mathbb{R}$, and a target variable $y_i \in \mathbb{R}$ for each $\boldsymbol{x}_i$. The ε -insensitive loss function has the form

$$|y - f(\boldsymbol{x})|_\varepsilon := \max\{0, |y - f(\boldsymbol{x})| - \varepsilon\}$$

To estimate the linear regression $f(x)=w\cdot x+b$ with this loss function, we minimize

$$\frac{1}{2}\|\boldsymbol{w}\|^2 + \frac{C}{n}\sum_{i=1}^{n}|y_i - f(\boldsymbol{x}_i)|_\varepsilon \tag{1}$$

While the second term of equation (1) is the ε-insensitive loss function, the first term is considered the regularization constraint to prevent the ε-SVR from over-fitting. The parameter *C* controls the trade-off between the complexity of the model and the error. Introducing the “slack variables” (ξ_i^+, ξ_i^-) into this optimization problem, it becomes

$$\begin{aligned} \min_{\boldsymbol{w},b} \quad & \frac{1}{2}\|\boldsymbol{w}\|^2 + C\sum_{i=1}^{n}(\xi_i^+ + \xi_i^-) \\ \text{subject to} \quad & y_i - \boldsymbol{w}\cdot\boldsymbol{x}_i - b \le \epsilon + \xi_i^+ \\ & \boldsymbol{w}\cdot\boldsymbol{x}_i + b - y_i \le \epsilon + \xi_i^- \\ & \xi_i^+, \xi_i^- \ge 0 \qquad \forall i \end{aligned} \tag{2}$$

This optimization is very similar to the one for SVMs. Smola and Schölkopf provide detailed derivation on how the dual problem can be obtained using the Lagrange multiplier method (Smola and Schölkopf 2004).

$$\max_{\alpha^+,\alpha^-} \quad W(\boldsymbol{\alpha}^+,\boldsymbol{\alpha}^-) = -\epsilon \sum_{i=1}^{n}(\alpha_i^+ + \alpha_i^-) + \sum_{i=1}^{n} y_i(\alpha_i^+ - \alpha_i^-)$$

$$-\frac{1}{2}\sum_{i,j-1}^{n}(\alpha_i^+ - \alpha_i^-)(\alpha_j^+ - \alpha_j^-)K(\boldsymbol{x}_i,\boldsymbol{x}_j)$$

$$\text{subject to} \quad 0 \le \alpha_i^+,\alpha_i^- \le C, \quad \sum_{i=1}^{n}(\alpha_i^+ - \alpha_i^-) = 0 \tag{3}$$

Calculating the partial derivative of the Lagrangian L with respect to w, we can calculate the optimal w using the optimal $\boldsymbol{\alpha}^+, \boldsymbol{\alpha}^-$.

$$\boldsymbol{w} = \sum_{i=1}^{n}(\alpha_i^+ \quad \alpha_i^-)\phi(\boldsymbol{x}_i)$$

For a new data point z, the regression estimate can be obtained using the formula

$$f(z) = \boldsymbol{w}^* \cdot \phi(z) + b^* = \sum_{i-1}^{n}(\alpha_i^{+*} - \alpha_i^{-*})K(\boldsymbol{x}_j, z) + b^*$$

Similar to SVMs, this optimization problem satisfies the KKT conditions. The dual complementarity condition suggests that $\alpha_i^+, \alpha_i^- \ge 0$ only when $|f(x_i) - y| \ge \epsilon$, which equivalently means the corresponding training examples on or outside of the ϵ band. These training examples are the support vectors.

3. PROBABILISTIC FRAMEWORK FOR SVR

The selection of parameters has always been one of the most important tasks when training supervised learning algorithms, and SVR is no exception. In order to achieve good generalization performance, a process is required to fine tune the parameters in order to balance the trade-off between variance and bias. Traditionally, the multi-fold cross-validation process is employed. This process is generally very effective, but usually computationally heavy. In this section, a probabilistic framework for SVR is investigated aiming to accelerate the parameter selection process. Gao *et al.* (2002) demonstrates how MacKay's evidence framework (MacKay 1992) is used to determine the parameters. Also, the approximation of an error bar formula for the SVR predictions is derived.

We propose a new parameter μ in addition to the SVR probabilistic framework of Gao *et al.* (2002). This parameter μ scales the kernel, and is defined to be the *a priori* estimate of the output variance. We show that adding such a parameter does not affect the SVR quadratic programming problem, but may act as a scaling parameter to allow the evidence function to be more flexible when handling data with different ranges. Because error bar estimate derived by Gao *et al.* (2002) is solely from the kernel features, this parameter also adjusts the level of this variance to better match the range of the SVR prediction.

3.1. Bayesian Evidence Framework

In this section, the Bayesian evidence framework for SVR proposed by Gao *et al.* (2002) is discussed. SVR can be interpreted as a maximum a posteriori (MAP) solution to inference problems with Gaussian priors and an appropriate likelihood function based on a probabilistic interpretation. Gao *et al.* (2002) derived the evidence formula, and the error bar approximation under this framework.

In regression problems, the functional dependency $f(\cdot)$ between a set of sampled points $X = \{\mathbf{x1}, \mathbf{x2}, \ldots, \mathbf{xn}\}$ taken from Rd , and target values $Y = \{y1, y2, \ldots, yn\}$ with $yi \in$ R is defined as follows. The training data $D = \{X, Y\}$ are collected by randomly sampling from the model:

$$y_i = f(\boldsymbol{x}_i) + \delta_i \qquad i = 1, 2, \ldots, n$$

where $f(\cdot)$ is the underlying function, and δ_i are independent, identically distributed (i.i.d.) random noise. Regression aims to estimate the function f from the data-set D. In the Bayesian approach, the function $\boldsymbol{f}$ is regarded as the realization of a random field with a known *a priori* probability. Let $\boldsymbol{f} = [f(x_1), f(x_2), \ldots, f(x_n)]^T$. The *a posteriori* probability of $\boldsymbol{f}$ given the training data D can then be derived by Bayes' theorem:

$$P(f|D) = \frac{P(D|f)P(f)}{P(D)} \tag{4}$$

where $P(\boldsymbol{f})$ is the *a priori* probability of the random field, and $P(D|\boldsymbol{f})$ is the conditional probability of the data D given the function values $\boldsymbol{f}$. This conditional probability $P(D|\boldsymbol{f})$, which is also the likelihood, can simply be interpreted as the noise model. $P(D|\boldsymbol{f})$ is evaluated by:

$$P(D|f) = \prod_{i=1}^{n} P(\delta_i) = \prod_{i=1}^{n} P(y_i - f(\boldsymbol{x}_i))$$

Gao *et al.* (2002) suggested that the likelihood function be written as:

$$P(D|f) = [G(C, \varepsilon)]^n \exp\left(-C \sum_{i=1}^{n} L_\varepsilon(y_i - f(\boldsymbol{x}_i))\right) \tag{5}$$

where $G(C, \varepsilon) = \frac{C}{2(\varepsilon C+1)}$ is the corresponding normalizing constant. For ε-SVR, asmentioned in section 2, the ε-insensitive loss function is given by:

$$L_\varepsilon(y_i - f(\boldsymbol{x}_i)) = \begin{cases} 0, & \text{for} |y - f(\boldsymbol{x})| < \varepsilon, \\ |y - f(\boldsymbol{x})| - \varepsilon, & \text{otherwise.} \end{cases}$$

The *a priori* distribution $P(f)$ is assumed to be a multivariate Gaussian with a zero mean and covariance function

$K(\cdot, \cdot)$.

$$P(f) = \frac{1}{\sqrt{\det 2\pi \; K_{X,X}}} \exp\left\{-\frac{1}{2} f^T K_{X,X}^{-1} f\right\} \tag{6}$$

where $K_{X,X} = [K(\mathbf{x}_i, \mathbf{x}_j)]$ is the covariance matrix at the training points X. Following Bayes' theorem given in equation (4), the *a posteriori* distribution is determined by combining equation (5) and (6).

$$P(f|D) = \frac{[G(C, \varepsilon)]^n}{\sqrt{\det 2\pi \; K_{X,X}} P(D)} \times \exp\left\{-C \sum_{i=1}^{n} L_\varepsilon(y_i - f(\boldsymbol{x}_i)) - \frac{1}{2} f^T K_{X,X}^{-1} f\right\} \tag{7}$$

where $G(C, \varepsilon) = \frac{C}{2(\varepsilon C+1)}$, and $P(D)$ is the evidence which is explained later in the section. The MAP estimate of f is computed by maximizing the *a posteriori* distribution provided in equation (7), which is the same as minimizing the risk function as follows:

$$R(f) = C \sum_{i=1}^{n} L_\varepsilon(y_i - f(\boldsymbol{x}_i)) + \frac{1}{2} f^T K_{X,X}^{-1} f \tag{8}$$

This is equivalent to the primal problem for SVR (equation (2)) when $\frac{1}{2} f^T K_{X,X}^{-1} f = \frac{1}{2}\|w\|^2$, which is suggested in the Lagrangian condition. As per the SVR optimization problem, the model that minimizes the risk function is defined as:

$$f^* = \sum_{i=1}^{n} (\alpha_i^+ - \alpha_i^-) K(x_i, x) \tag{9}$$

With respect to the optimal solution f^**, the training dataset X are divided into four parts:

$$\begin{aligned} X_0 &= \{x_i \mid |y_i - f^*(x_i)| < \varepsilon && \text{with } \alpha_i^+ = \alpha_i^- = 0\} \\ X_C &= \{x_i \mid |y_i - f^*(x_i)| > \varepsilon && \text{with } \alpha_i^+ = C, \alpha_i^- = 0, \\ & && \text{or } \alpha_i^+ = 0, \alpha_i^- = C\} \\ X_{M^-} &= \{x_i \mid f^*(x_i) - y_i - \varepsilon = 0 && \text{with } 0 \le \alpha_i^- \le C\} \\ X_{M^+} &= \{x_i \mid y_i - f^*(x_i) - \varepsilon = 0 && \text{with } 0 \le \alpha_i^+ \le C\} \end{aligned} \tag{10}$$

For later derivation, let us define $X_M = X_M+ \cup X_M-$, and the support vectors $X_{sv} = X_M \cup X_C$. Also, denote that $.X_M = X \backslash X_M$.

In order to determine how to use the probabilistic framework to select the best parameters, the evidence $P(D)$ has to be determined. The evidence is simply defined as the marginal likelihood of the data for a particular model, and can be obtained by integrating the *a posteriori* distribution over the parameter space $\boldsymbol{f}$.

$$P(D) = \int P(f) P(D \mid f) \mathrm{d} f \tag{11}$$

However, inserting equation (5) and (6) into (11) leads to analytically intractable integrals. Gao *et al.* (2002) used the Laplace method for approximation, and the detailed proof is provided in their study.

$$\begin{aligned} \ln P(D) \approx & -C \sum_{x_i \in X_C} L_\varepsilon(y_i - f^*(x_i)) \\ & - \frac{1}{2} (\alpha_i^+ - \alpha_i^-)^T K_{SV,SV} (\alpha_i^+ - \alpha_i^-) \\ & + l \ln G(C, \varepsilon) - \frac{1}{2} \ln \det(2\pi K_{X_M, X_M}) \\ & + \sum_{x_i \in X_M} \ln \frac{C}{|\alpha_i|(C - |\alpha_i|)} \end{aligned} \tag{12}$$

To find the 'best' parameters for the SVR problem, the logarithmic evidence ln $P(D)$ from different combinations of parameters are compared. Under this evidence framework, the set of parameters which give the optimal logarithmic evidence is determined to form the 'best' model. The number of support vectors required is implicitly fixed by the parameter selection under the Bayesian framework. In the study of Gao *et al.* (2002), this evidence formula is differentiated with respect to the parameter of interest and set to zero in order to find the optimal parameters. However, we realize that finding the gradient of equation (12) is nontrivial. The count and location of support vectors are implicitly related to the set of parameters selected, which implies that the terms that include the kernel $K_{XM,XM}$, $K_{SV,SV}$ or the loss function $\sum_{\mathbf{x}_i \in X_C} L_\varepsilon(y_i - f^*(\mathbf{x}_i))$ must be taken into account when taking the derivatives, and these are not considered in the original study. Such gradient approach to find the optimal parameters may not be practical, and we suggested using other optimization methods to maximize the log evidence of the model.

3.2. Error bar estimation

Gao *et al.* (2002) also introduced an error bar estimate for the SVR predictions under the evidence framework described in the previous section. It was suggested that the prediction error is made up of two terms. One of them is from the posterior uncertainty of the model $\boldsymbol{f}$, and the other is due to random noise in the data. The prediction model of a new test data point $\mathbf{z}$ is defined to be:

$$t = f(z) + e(z) \tag{13}$$

where t is the target prediction, $\boldsymbol{f}(\mathbf{z})$ is the model output, and $e(\mathbf{z})$ is random noise, which is independent of $\boldsymbol{f}(\mathbf{z})$. The goal is to find the variance of the target t on the new test data point $\mathbf{z}$.

Gao *et al.* (2002) derives the estimate of the variance due to model uncertainty, let $\boldsymbol{f}(X)$ be the trained model with training data X, and the predictive distribution of $\boldsymbol{f}(\mathbf{z})$ corresponding to a new data point $\mathbf{z}$ is defined to be

$$\begin{aligned} P(f(z)|D) &= \frac{1}{P(D)} \int p(D|f(X))P(f(X), f(z))\mathrm{d}f(X) \\ &\propto \int p(D|f(X))P(f(X), f(z))\mathrm{d}f(X) \end{aligned} \tag{14}$$

The covariance matrix K of the training dataset $X = X_M \cup \bar{X}_M$ is defined as

$$K_{X,X} = \begin{pmatrix} K_{X_M,X_M} & K_{X_M,\bar{X}_M} \\ K^T_{X_M,\bar{X}_M} & K_{\bar{X}_M,\bar{X}_M} \end{pmatrix}$$

and the covariance matrix between the training data *X* and the new test data *z* is

$$K_{[X,z],[X,z]} = \begin{pmatrix} K_{X,X} & K_{X,z} \\ K^T_{X,z} & K_{z,z} \end{pmatrix}$$

Similar to the evidence approximation derivation, Gao *et al.* (2002) again suggested to use the Laplace method to approximate the integral in equation (14). While the detailed steps can be found in their study, the predictive distribution has the form

$$P(f(z)|D) \propto \exp\left\{ -\frac{1}{2}\left(f(z) - f(z)^*\right) \times \left(K_{z,z} - K^T_{X_M,z} K^{-1}_{X_M,X_M} K_{X_M,z}\right)^{-1} \left(f(z) - f(z)^*\right)\right\}$$

which has the form of a Gaussian distribution with mean ***f**** (**z**), and variance

$$\sigma_f^2(z) = K_{z,z} - K^T_{X_M,z} K^{-1}_{X_M,X_M} K_{X_M,z} \tag{15}$$

Regarding the variance of the random noise *e*(**z**), Gao *et al.* (2002) made use of the noise model suggested in equation (5), which is interpreted as $P(e(z)) \propto \exp\{-C\, L_\varepsilon(e(z))\}$ with

$$E(e(z)) = 0 \quad \text{and } \mathrm{Var}(e(z)) = \sigma_e^2(z) = \frac{2}{C^2} + \frac{\varepsilon^2(\varepsilon C + 3)}{3(\varepsilon C + 1)} \tag{16}$$

Considering the prediction model suggested earlier in equation (13), the model and noise components are additive. By combining equation (15) and (16), the prediction variance is

$$\begin{aligned}\sigma_t^2(z) &= \sigma_f^2 + \sigma_e^2 \\ &= K_{z,z} - K_{X_M,z}^T K_{X_M,X_M}^{-1} K_{X_M,z} \\ &\quad + \frac{2}{C^2} + \frac{\varepsilon^2(\varepsilon C + 3)}{3(\varepsilon C + 1)}\end{aligned} \tag{17}$$

3.3. New scaling parameter μ

The effort of deriving this evidence framework by Gao *et al.* (2002) is highly appreciated. Not only does it give a very good methodology to choose the 'best' parameters, but also it provides an error bar estimate of the SVRprediction. This is implemented in our study, but the results were rather unsatisfying, which lead to our idea of introducing a new scaling parameter μ to the evidence framework. The problems observed, as well as the rationale behind this new parameter μ is explained in this section.

Let us begin this discussion by cracking open the final logarithmic evidence formula derived by Gao *et al.* (2002) as given in equation (12). It is not difficult to notice that the first part of the formula is equivalent to negative of the the optimal value from the objective function in the SVR optimization problem.

ln *P(D)*

≈ −(optimal objective value of SVR optimization problem)

$$+ \, l \ln \frac{C}{2(\varepsilon C + 1)} - \frac{1}{2} \ln \det(2\pi \, K_{X_M,X_M}) + \sum_{x_i \in X_M} \ln \frac{C}{|\alpha_i|(C - |\alpha_i|)} \tag{18}$$

The level of the optimal objective value heavily depends two main factors: the model structure, and the level of the slack variables ξ_i^+, ξ_i^-. While the model structure is what we want to determine through the evidence framework, the level of slack variables has huge impact on the parameters in order to compute the 'best' model. For example, if the range of the target variable in the training data is wider than the one of the kernel features, which is quite common because inputs are usually normalized, the level of the slack variables is high. In such case, the level of the parameter *C* has to be large in order to bring up the level of the kernel features to match the target variable given the $0 \leq \alpha_i^+, \alpha_i^- \leq C$ constraint in the optimization problem. Therefore, the optimal objective value of the optimization problem is inflated. When observing the remaining terms in the logarithmic evidence formula, there are no other terms to balance such effect. This implies that if the data experiment is set up such that the level of the target variable ismuch higher than the features, the evidence formula may not be able to handle the scale difference. As a result, the evidence formula may be dominated by the objective value of the optimization problem, and therefore not be able to select the 'best' model.

The same scaling effect is also observed in the error bar estimation. Equation (15) provides a very nice formula to estimate the variance of the SVR prediction. However, it is solely derived from the kernel features. This means that if the range of the target variable is different from the one of the kernel features, the prediction error obtained from the formula is only going to be proportional to the estimated variance.

Given the above observations, we introduces a new scaling parameter μ to the evidence framework, aiming to handle such effect caused by the data, with a view to enhance the evidence framework. This scaling parameter is simply a scalar to inflate the kernel such that

$$\widetilde{K} = \mu \cdot K$$

where $\mu > 0$ is a scalar interpreted as *a priori* of the output variance, and K is the kernel matrix in the SVR problem. First of all we show that adding such a parameter results in the same SVR quadratic programming problem. Consider the dual problem of SVR given in equation (3), scaling the kernel K with the parameter μ result in the following:

$$\max_{\alpha^+,\alpha^-} \widetilde{W}(\widetilde{\boldsymbol{\alpha}}^+, \widetilde{\boldsymbol{\alpha}}^-) = -\epsilon \sum_{i=1}^{n} (\widetilde{\alpha}_i^+ + \widetilde{\alpha}_i^-) + \sum_{i=1}^{n} y_i (\widetilde{\alpha}_i^+ - \widetilde{\alpha}_i^-)$$

$$-\frac{1}{2} \sum_{i,j=1}^{n} (\widetilde{\alpha}_i^+ - \widetilde{\alpha}_i^-)(\alpha_j^+ - \widetilde{\alpha}_j^-) \widetilde{K}(\boldsymbol{x}_i, \boldsymbol{x}_j)$$

$$\text{subject to } 0 \leq \widetilde{\alpha}_i^+, \widetilde{\alpha}_i^- \leq \widetilde{C}, \quad \sum_{i=1}^{n} (\widetilde{\alpha}_i^+ - \widetilde{\alpha}_i^-) = 0 \tag{19}$$

Where $\widetilde{\alpha_i} = \frac{\alpha_i}{\mu}$, $\widetilde{C} = \frac{C}{\mu}$, and $\widetilde{W}(\widetilde{\boldsymbol{\alpha}}^+, \widetilde{\boldsymbol{\alpha}}^-) = \frac{1}{\mu} \cdot W(\boldsymbol{\alpha}^+, \boldsymbol{\alpha}^-)$. The new objective function $\widetilde{W}$ is simply a scaled version of the original W, which means that the optimization problem has exactly the same solution.

Adding such a parameter does not change the MAP estimate computation, but has an impact when introduced into the evidence formulation. Consider the logarithmic evidence formula shown in equation (12), introducing μ results in the following.

$$\begin{aligned}\ln P(D) \approx & -\frac{1}{\mu}\Bigg[C \sum_{x_i \in X_C} L_\varepsilon(y_i - f^*(\boldsymbol{x}_i)) \\ & - \frac{1}{2}(\alpha_i^+ - \alpha_i^-)^T K_{SV,SV}(\alpha_i^+ - \alpha_i^-)\Bigg] \\ & + l \ln \frac{C}{2(\varepsilon C + \mu)} - \frac{1}{2}\ln \det(2\pi \mu K_{X_M,X_M}) \\ & + \sum_{x_i \in X_M} \ln \frac{\mu C}{|\alpha_i|(C - |\alpha_i|)}\end{aligned} \tag{20}$$

By simply observing the new evidence function, the first part of the formula, which is the optimal objective value of the SVR optimization as discussed earlier, is scaled by the parameter μ. However, μ is also in the remaining terms, so its impact to the overall evidence formulation is not obvious.Note that equation (20) is only used to demonstrate the impact of μ when comparing to the original evidence function. In practice, the parameters are implicitly altered once the kernel is scaled, and the new logarithmic evidence function for implementation is simply equivalent to the original one with the scaled kernel and parameters as below.

$$\begin{aligned}\ln P(D) \approx & -\widetilde{C} \sum_{x_i \in X_C} L_\varepsilon(y_i - f^*(\boldsymbol{x}_i)) \\ & - \frac{1}{2}(\widetilde{\alpha}_i^+ - \widetilde{\alpha}_i^-)^T \widetilde{K}_{SV,SV}(\widetilde{\alpha}_i^+ - \widetilde{\alpha}_i^-) \\ & + l \ln G(\widetilde{C}, \varepsilon) - \frac{1}{2}\ln \det(2\pi \widetilde{K}_{X_M,X_M}) \\ & + \sum_{x_i \in X_M} \ln \frac{\widetilde{C}}{|\widetilde{\alpha}_i|(\widetilde{C} - |\widetilde{\alpha}_i|)}\end{aligned} \tag{21}$$

Same as the original evidence framework, the parameters with the largest evidence value is chosen to give the 'best' model. On top of the original parameters, the 'best' scaling parameter μ is also determined through maximizing the evidence value. We test this altered evidence framework in our experiment, and the performance is found to be much better than the original evidence derivation. The results of the experiments are shown in section 5. Similarly, this new parameter μ also has its influence on the error bar estimation. Consider introducing μ to the original error bar estimation as shown in equation (17), the new error bar estimate becomes:

$$\begin{aligned}\widetilde{\sigma}_t^2(z) &= \widetilde{\sigma}_f^2 + \widetilde{\sigma}_e^2 \\ &= \mu\left(K_{z,z} - K_{X_M,z}^T K_{X_M,X_M}^{-1} K_{X_M,z}\right) \\ &\quad + \frac{2\mu^2}{C^2} + \frac{\varepsilon^2(\varepsilon C + 3\mu)}{3(\varepsilon C + \mu)}\end{aligned} \tag{22}$$

From simply observing the new error bar estimate in equation (22), the prediction error is scaled by the parameter$\mu$when comparing to the original formulation. Therefore, we interpret μ as the *a priori* variance of the output to scale the kernel features to match the variance of the target variable. Similarly, equation (22) is only provided for the purpose of illustrating the impact of μ when comparing to the original error bar estimate. In practice, the parameters are implicitly altered once the kernel is scaled, and the new error bar formulation for implementation is simply equivalent to original one with the scaled kernel and parameters as follows, and the results are discussed in section 5.

$$\begin{aligned}\widetilde{\sigma}_t^2(z) &= \widetilde{\sigma}_f^2 + \widetilde{\sigma}_e^2 \\ &= \widetilde{K}_{z,z} - \widetilde{K}_{X_M,z}^T \widetilde{K}_{X_M,X_M}^{-1} \widetilde{K}_{X_M,z} \\ &\quad + \frac{2}{\widetilde{C}^2} + \frac{\varepsilon^2(\varepsilon \widetilde{C} + 3)}{3(\varepsilon \widetilde{C} + 1)}\end{aligned} \tag{23}$$

Preprocessing the target variable may be an alternative approach to the problem, but it requires careful reverse engineering of the output to ensure appropriate interpretability.Building the new parameter into the Bayesian framework avoid this extra step, and therefore fit better to the idea of developing an automated flexible prediction framework.

4. BAYESIAN OPTIMIZATION FOR PARAMETERS SELECTION

In section 3.1, we demonstrated that using the gradient method to optimize the log evidence is impractical, and seeking for another optimization method is necessary. Recently, the popularity of Bayesian optimization has grown significantly in the artificial intelligence and machine learning community (Srinivas *et al.* 2010, Shahriari *et al.* 2016). Its capability to actively and efficiently search for the global optimum for a complex or even unknown function greatly enhance the automatic tuning process for many powerful machine learning algorithms (Snoek *et al.* 2012). In this study, we borrow such technique to optimize the log evidence function in order to obtain the 'best' parameters. Bayesian optimization is explained in this section, and are used in our experiments.

4.1. Multi-armed Bandit

Multi-armed bandit (MAB) problem is a problem in which a gambler faces k slot machines, a.k.a. 'one-armed bandits'. Each machine has unknown probability of winnings. The gambler is allowed to play one machine each round, and eventually has to define strategy to maximize the total winnings. The key here is to trade-off between exploration (try a new machine), and exploitation (continue to play with an already observed winning machine). This problem has been widely studied in various areas such as machine learning, operational research and control etc.

Recently, the task of globally optimizing a complex, sometimes unknown function is being formulated in a MAB setting (Srinivas *et al.* 2010). The problem becomes a sequential optimization of an objective function $f: \boldsymbol{D} \rightarrow$ R, considering each input as one of k arms in the data $\boldsymbol{D}$. In each time step t, an input $\boldsymbol{x_t} \in D$ is chosen, and observe the corresponding noisy function value $y_t = f(\boldsymbol{x_t}) + \delta_t$, where δ_t is the noise perturbation. The goal is to maximize the 'sum of rewards' $\sum_{t=1}^{T} f(x_t)$ as rapidly as possible, which is essentially searching for $\boldsymbol{x}^* = \text{argmax}_{\boldsymbol{x} \in D} f(\boldsymbol{x})$.

To efficiently perform the trade-off between exploration and exploitation, the dependencies across arms are assumed and modelled (Dorard *et al.* 2009). Such dependencies allow exploration to be faster. When an arm is explored, knowledge is gained on that arm as well as similar arms. The rewards of arms are assumed to be correlated, which means that the resulting rewards are similar if the arms pulled are similar. The correlations are modelled by assuming that f is a function drawn from a Gaussian Process (GP).

The GP prior distribution is initialized with mean $\mu 0$ and the correlation between arms are described by the GP covariance matrix. Each entry (i, j) of the matrix specifies the covariance between arm i and j. The covariance can be defined using a kernel to measure the similarity between data points.After each inputs point is drawn, the corresponding reward is observed, and the GP posterior distribution is updated. An acquisition function is then used to determined the next point to examine. The algorithm iterates until a near optimal value is reached. A few common acquisition functions are listed in section 4.3. Figure 2 provides a clear overview of the process, and the pseudo code is listed in algorithm 1.

Algorithm 1: Pseudo-code for Bayesian optimization

Input: Input space $\boldsymbol{D}_0$; GP Prior μ_0, Σ_0
for $t = 1, 2, \ldots$ **do**
  Select new $\boldsymbol{x}_{t+1}$ by optimizing acquisition function α
  $$\boldsymbol{x}_{t+1} = \underset{x \in D_t}{\operatorname{argmax}}\, \alpha(\boldsymbol{x}; D_t)$$
  Query objective function to obtain y_{t+1}
  Augmented data $\boldsymbol{D}_{t+1} = \{D_t, (\boldsymbol{x}_{t+1}, y_{t+1})\}$
  Perform Bayesian update to obtain μ_{t+1} and Σ_{t+1}
end

4.2. Gaussian process

GP is a widely used method in machine learning. It defines a distribution over the function f, which maps some input space X toR. For n input data points $x_{1:n}$, the corresponding function values and noisy observations are respectively $f_i := f(\mathbf{x}_i)$, and y_i, with $i = 1, 2, \ldots, n$. In the GP regression setting, the function value $\boldsymbol{f} := f_{1:n}$ is assumed to be jointly Gaussian with mean $\boldsymbol{m}$ and covariance $\boldsymbol{K}$.

$$P(\boldsymbol{f}|\boldsymbol{X}) = \mathcal{N}(\boldsymbol{m}, \boldsymbol{K}) \tag{24}$$

Suppose noisy observations are samples from the distribution over the function perturbed by noise $\delta_i \overset{iid}{\sim} \mathcal{N}(0, \sigma^2)$. The noisy observations $\boldsymbol{y}$ are normally distributed given $\boldsymbol{f}$, and the likelihood is

$$P(\boldsymbol{y}|\boldsymbol{f}) = \prod_{i=1}^{n} \mathcal{N}(f_i, \sigma^2) = \mathcal{N}(\boldsymbol{f}, \sigma^2 \boldsymbol{I}) \tag{25}$$

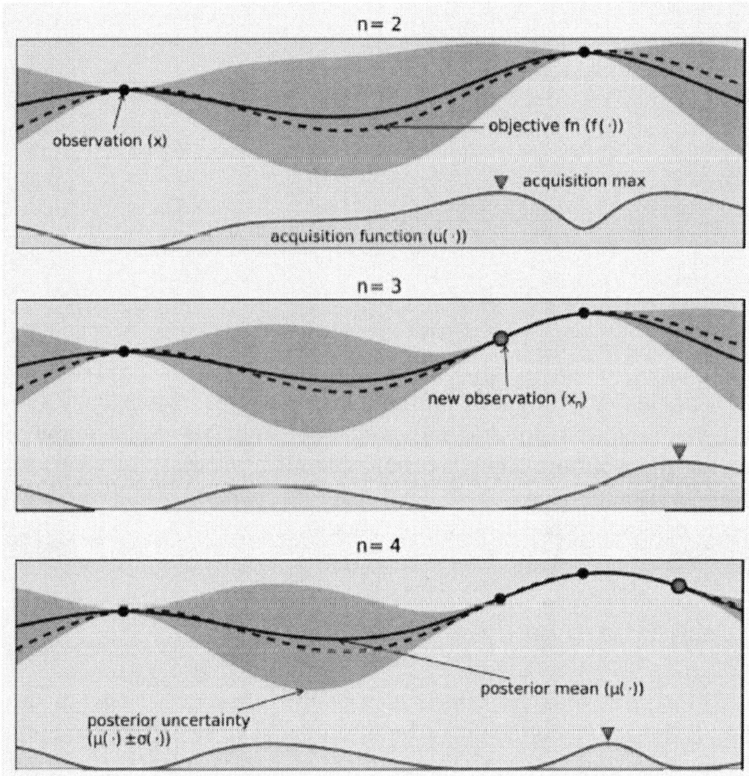

Figure 2. GP Bayesian optimization (Shahriari *et al.* 2016).

Equation (24) represents the prior distribution $p(\boldsymbol{f})$ with mean $m_i := \mu_0(\mathbf{x_0})$ and $K_{i,j} = k(\mathbf{x}_i, \mathbf{x}_j)$, and $K_{i,j}$ is a positive definite kernel to measure the covariance between point I and j. The likelihood is given in equation (25). The posterior distribution is proportional to the product of equations (24) and (25).

$$P(\boldsymbol{f} \mid \boldsymbol{y}, X) \propto P(\boldsymbol{y} \mid \boldsymbol{f})\, P(\boldsymbol{f}, X) \tag{26}$$

Since the function is assumed to be jointly Gaussian, the joint distribution over the observations $\boldsymbol{y}$ and a new test point f^* can be defined as

$$P\begin{pmatrix} \boldsymbol{y} \\ f^* \end{pmatrix} = \mathcal{N}\left(\begin{matrix} \boldsymbol{m} \\ \mu(\boldsymbol{x}_*) \end{matrix}, \begin{bmatrix} K(X,X) + \sigma^2 \boldsymbol{I} & K(X, \boldsymbol{x}_*) \\ K(\boldsymbol{x}_*, X) & K(\boldsymbol{x}_*, \boldsymbol{x}_*) \end{bmatrix} \right) \tag{27}$$

Given the conditioning property of a joint Gaussian distribution, the predictive posterior distribution for the test point f^* is derived to be

$$P(f^*|\boldsymbol{x}_*, \boldsymbol{y}, X) = \int P(f^*|\boldsymbol{f}, \boldsymbol{x}_*, \boldsymbol{y}, X)\, P(\boldsymbol{f}|\boldsymbol{y}, X)\, \mathrm{d}\boldsymbol{f}$$
$$= \mathcal{N}(\mu^*, \Sigma^*)$$

where,

$$\mu^* = \mu(\boldsymbol{x}_*) + K(\boldsymbol{x}_*, X)\left(K(X, X) + \sigma^2 \boldsymbol{I}\right)^{-1}(\boldsymbol{y} - \boldsymbol{m})$$
$$\Sigma^* = K(\boldsymbol{x}_*, \boldsymbol{x}_*) - K(\boldsymbol{x}_*, X)\left(K(X, X) + \sigma^2 \boldsymbol{I}\right)^{-1} K(X, \boldsymbol{x}_*) \tag{28}$$

In the context of Bayesian optimization as described in section 4.1, the process begins by fitting a GP on the function values from a small subset of inputs, and sequentially add points to the input set while updating the posterior distribution. The Bayesian updating process iterates until the function optimum is reached. The order of sequence in selecting input points is decided by the acquisition function that is going to be discussed in section 4.3.

4.3. Acquisition method

The acquisition function, which is also the expected utility in decision theory, is the key to define the order of sequence in selecting input points. It is designed to trade off exploration of the search space and exploitation of the currently known area. The inputs that correspond to the optimum of the acquisition function are usually the location of the next input. Some widely used acquisition function in the current academic and industrial research including but not limited to probability of improvement (PI), expected improvement (EI), upper confidence bound (UCB), and Thompson sampling (TS).

PI is a policy that makes use of the mean μ_n and variance σ_n of the GP posterior distribution to define the highest probability in leading to an improvement over the best current value $f(\boldsymbol{x}_{\text{best}})$, thus, becomes the next input to be selected in the sequence.Φ is the standard normal cumulative distribution function.

$$\alpha_{PI}(\boldsymbol{x}; D_n) = \Phi(\gamma(\boldsymbol{x}))$$
$$\gamma(\boldsymbol{x}) = \frac{\mu_n(\boldsymbol{x}) - f(\boldsymbol{x}_{\text{best}})}{\sigma_n(\boldsymbol{x})}$$

While PI is proven to exploit quite aggressively in some cases (Bergstra and Bengio 2012), EI is a great alternative. Because the random variable $f(\boldsymbol{x})$ is normally distributed, the expectation can be computed analytically (Brochu *et al.* 2010). ø is defined to be the standard normal probability density function.

$$\alpha_{EI}(\boldsymbol{x}; D_n) = \begin{cases} (\mu_n(\boldsymbol{x}) - f(\boldsymbol{x}_{\text{best}}))\,\Phi(\gamma(\boldsymbol{x})) + \sigma_n(\boldsymbol{x})\,\phi(\gamma(\boldsymbol{x})), & \text{if } \sigma_n > 0. \\ 0, & \text{if } \sigma_n = 0. \end{cases}$$

UCB is relatively more recent approach to trade off exploration and exploitation (Srinivas *et al.* 2010). It is constructed to minimize regret for the optimization. κ is a user-defined parameter to balance exploration against exploitation.

$$\alpha_{UCB}(\boldsymbol{x}; D_n) = \mu_n(\boldsymbol{x}) + \kappa\sigma_n(\boldsymbol{x})$$

While conceptually different from other acquisitionmethods mentioned, recently TS has gained wide interest among the MAB community. Multiple theoretical and empirical evaluations have been performed (Chapelle and Li 2011, Scott 2010, Agrawal and Goyal 2013, Kaufmann *et al.* 2012, Russo and Van Roy 2014), and there are claims about its advantage over the other methods. A sample of the reward function is drawn from the posterior distribution, and the arm with the highest simulated reward is selected. This is the method employed in the experiments of this study.

$$\alpha_{TS}(\boldsymbol{x}; D_n) = f^{(n)}(\boldsymbol{x})$$
$$\text{where } f^{(n)} \sim GP(\mu_0, k|D_n)$$

5. EXPERIMENTS

SVR has long been proven to have good prediction performance. The focus of this study is to outline a Bayesian probabilistic framework for the algorithm mainly for two purposes. (1) Multi-fold cross validation process is well known to be an effective method to select parameters for machine learning algorithms, but is usually computationally expensive. Borrowing the MAP approach from Bayesian statistics, it may be possible to achieve similar performance as the original SVR with less computation to select the 'best' parameters. (2) Introducing the probabilistic framework allows assessment of prediction uncertainties. This is extremely important especially in the financial context, where decisions are better based on predictions with some level of confidence.

Before describing the implementation details, it may be useful to provide an overview of the prediction system to bring the proposed methodologies together. Figure 3 presents the complete setup of the prediction process. The SVR model is first trained with the parameters defined (i.e. CV, Grid Search or BayesOpt). Under the Bayesian framework, while making use of the model output, the predictive mean and variance are then computed. The quality of the prediction is evaluated using the corresponding error bar. This, at the same time, is interpreted as an assessment of whether the market condition has changed. This complete process is iterated in a rolling window manner.

5.1. Data

Eight financial time series from various asset classes (i.e. Equity Index, CDS spreads, Bond yields, Commodity Futures) are used in this study to assess the performance and flexibility of the Bayesian SVR framework proposed in this study. The details of each time series are listed in appendix 1. They are gathered from the Reuters Eikon platform at the Thomson-Reuters Laboratory, University College London. Daily quotes are used in the analyses. The time frame used for each time series may vary slightly depending on the availability of data.

5.2. Implementation

In this first set of experiments, the original RBF kernel SVR algorithm is first trained and tested with parameters selected through five-fold cross validation. TheMATLAB package *libsvm* (Chang and Lin 2011) is used for SVR implementation. The results are used as the benchmark performance when comparing to the ones from the Bayesian SVR where parameters are selected through the MAP approach. The following combination of parameters is used for all experiments that involve cross validation procedure to select parameters.

$$C = 2^{(-3,-1,1,3,5,7,9,11,13,15)} \qquad \varepsilon = 2^{(-9,-8,-7,-6,-5,-4,-3)}$$

$$\gamma = 2^{(-8,-7,-6,-5,-4,-3,-2,-1)}$$

The original Bayesian probabilistic framework by Gao *et al.* (2002) is implemented. However, for the theoretical reasons explained in section 3.3, the results are poor.We have therefore not included these results. The same sets of data are then processed using the amended Bayesian probabilistic framework with parameters selected by optimizing the logarithmic evidence formula suggested in equation (21). As mentioned previously, the gradient method to optimize the log evidence is impractical, and seeking for another optimization method was necessary. The easiest optimization method is grid search. Although grid search is still quite computationally expensive, and its performance is highly dependent on the design of the grid, it is very easy to implement and the results are easy to assess. In section 3.3, we introduces the new parameter μ to scale the RBF kernel K.This new parameter is incorporated into the grid search process in maximizing the evidence. Note that because the parameter μ is a factor that scales the kernel, we control the kernel width parameter γ to be a fixed small value in order to avoid the interactive effect of these two parameters within the probabilistic framework. In our experiments, the small γ value chosen works consistently, and the same value is used in all experiments. The parameters C and ε values are the ones considered in the cross validation process.

$$C = 2^{(-3,-1,1,3,5,7,9,11,13,15)} \qquad \varepsilon = 2^{(-9,-8,-7,-6,-5,-4,-3)}$$

$$\mu = 2^{(-3,-1,1,3,5,7,9,11,13,15,17,19)} \qquad \gamma = 2^{(-8)}$$

Lastly, a similar exercise is carried out with the logarithmic evidence formula being optimized with the multi-armed bandit Bayesian optimization approach, aiming to automate and accelerate the parameter selection process, while achieving comparable performance. The MATLAB package *bayesopt* (Martinez-Cantin 2014) is used. TS is used as the acquisition method for our experiments, and observation noise (σ^2 in equation (28)) for the GP is chosen to be a very small number (1e-14) to reflect the fact that the log evidence is a deterministic function. For Bayesian optimization, only the upper and lower bound of the parameter search space is required.

$$C = 2^{(-3:15)} \qquad \varepsilon = 2^{(-9:-3)}$$

$$\mu = 2^{(-3:19)} \qquad \gamma = 2^{(-8)}$$

To reduce the computational cost, algorithms are trained with 2.5 years of data (630 quotes), and tested on the following 0.5 year (125 quotes), assuming 252 trading days in a year. Predictions are performed and assessed in a rolling window manner to cover the entire time series.

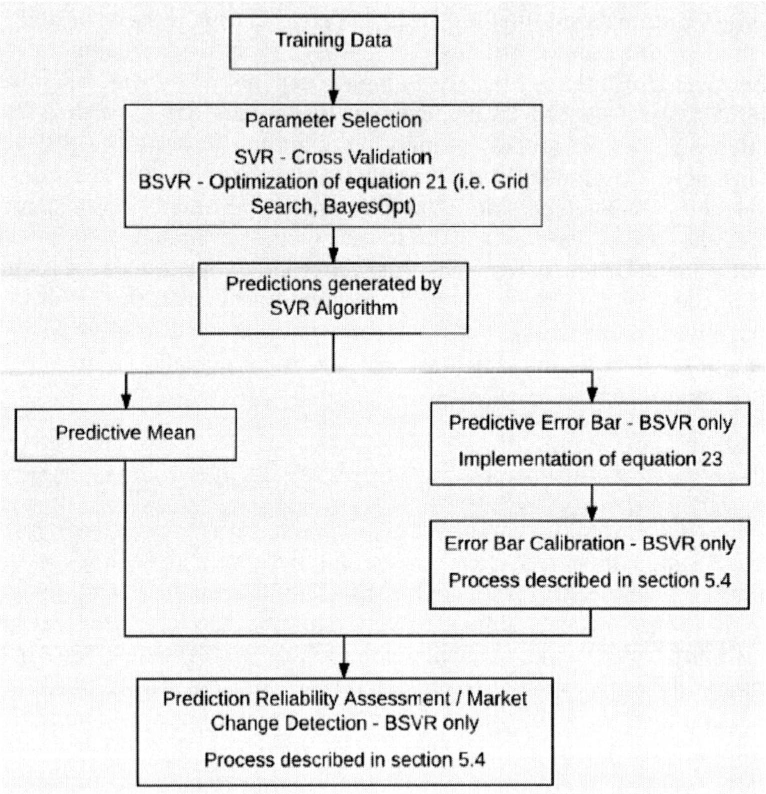

Figure 3. Overview of the proposed methodologies setup.

In the second set of experiments, only Bayesian SVR with Bayesian optimization selected parameters is examined. Because of the reduction in computational cost for the parameters selection process, the algorithm is trained with 3 months of data (63 quotes), and tested daily. Providing the rapid-changing nature of financial time series, this greatly enhances the sensitivity of predictions and their uncertainty measures. The error bar estimates are computed using equation (23). These are closely examined and practical applications are discussed.

In all the above mentioned experiments, quotes from the last 14 days are chosen to capture the patterns in order to predict a one day ahead quote. The same 14-day window was used in another study (Gündüz and Uhrig-Homburg 2011), which we follow the same approach. However, the Bayesian SVR framework proposed is generalizable to use any reasonable features. This study aims to define a flexible framework for the algorithm rather than to provide the best features for predictions. It is also important to mention that each input variable in the training samples are linearly scaled to the range [0, 1] by

$$z_{ij} = \frac{x_{ij} - \min(x_i)}{\max(x_i) - \min(x_i)}$$

where x_{ij} is the j th sample of the i th input variable, and z_{ij} is the corresponding scaled input value. The testing samples are also scaled in the samemanner use the maximum and minimum from the training set.

MeanAbsolute Percentage Errors (MAPE) are used to assess the prediction performance.

$$\text{MAPE}(\%) = \frac{1}{T} \sum_{t=1}^{T} \frac{\left| \hat{y}_t^{pred} - y_t^{obs} \right|}{y_t^{obs}} \tag{29}$$

where $\hat{y}_t^{pred}$ is the predicted quote of the asset on day t, and $t = 1, \ldots, T$ is the number of prediction days. This metric is computed in all experiments to compare prediction performances.

5.3. Results

In the first set of experiments, the original RBF kernel SVR algorithm is first trained and tested with parameters selected through five-fold cross validation. The results are then used as the benchmark performance when comparing to the ones from the Bayesian SVR where parameters are selected through the evidence maximization approach using grid search or Bayesian optimization. MAPEs (equation (29)) are computed to assess the predictive performance of each model for each times series. Algorithms are trained with 2.5 years of data (630 quotes), and test on the following 0.5 year (125 quotes), assuming 252 trading days in a year. Predictions are performed and assessed in a rolling window manner to cover the entire time series.

The bars in figure 4 show the averaged MAPEs for all the rolling windows, and the arm describes the corresponding coefficient of variation (CoV) which is computed by dividing the mean by the standard deviation (SD). CoV acts as the SD standardized by the level of the mean, which allows comparison across different models and assets. The longer CoV arms suggest prediction performance inconsistencies in some of the rolling windows.

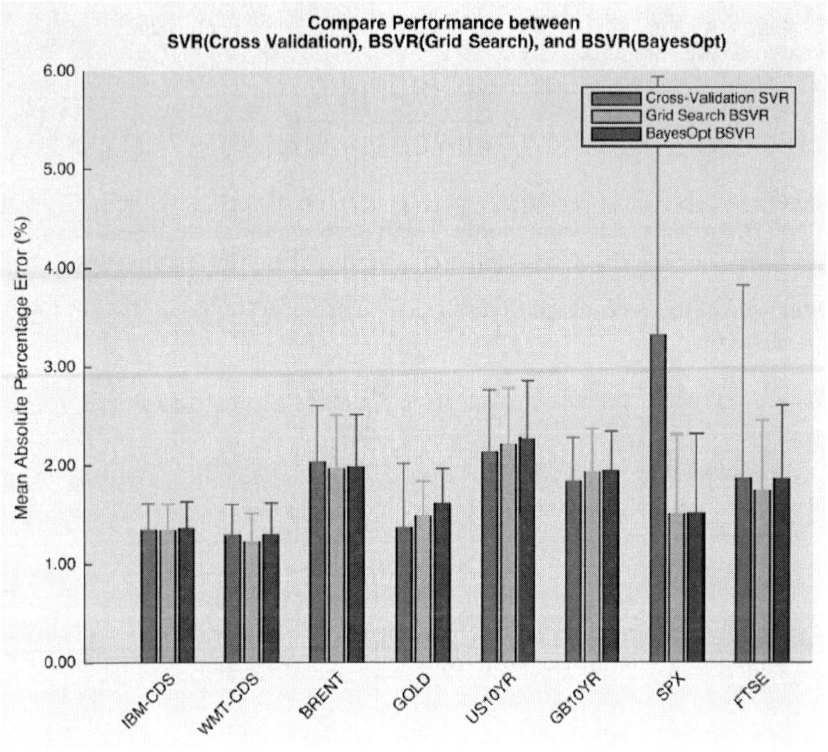

Figure 4. MAPE for SVR(Cross Validation), BSVR(Grid Search), and BSVR(BayesOpt).

Figure 4 shows that the predictive performance varies across assets mainly due to difference in individual behaviour such as volatility level, regime change etc. The MAPE values are provided in appendix 2. It is interesting to observe that with S&P500 (SPX), the original SVR with five-fold cross validation shows inconsistent prediction performance. It may be a sign suggesting, in this case, cross-validation is incapable to generalize the training data distribution to the testing data. It may be fixed by fine tunes such as changing the size of training or testing window, using different number of folds in the cross-validation process etc. This is purposely left unfixed to ensure the consistency in our experiments as well as to demonstrate that evidence maximization approach in Bayesian SVR may sometimes allow better parameter selection than cross validation. Other than SPX, the models generated from the three different parameters selection methods give comparable performance in most of the cases. This is encouraging as it suggests that it may be worthwhile to generalize SVR to Bayesian SVR to gain the additional features such as a faster and easier

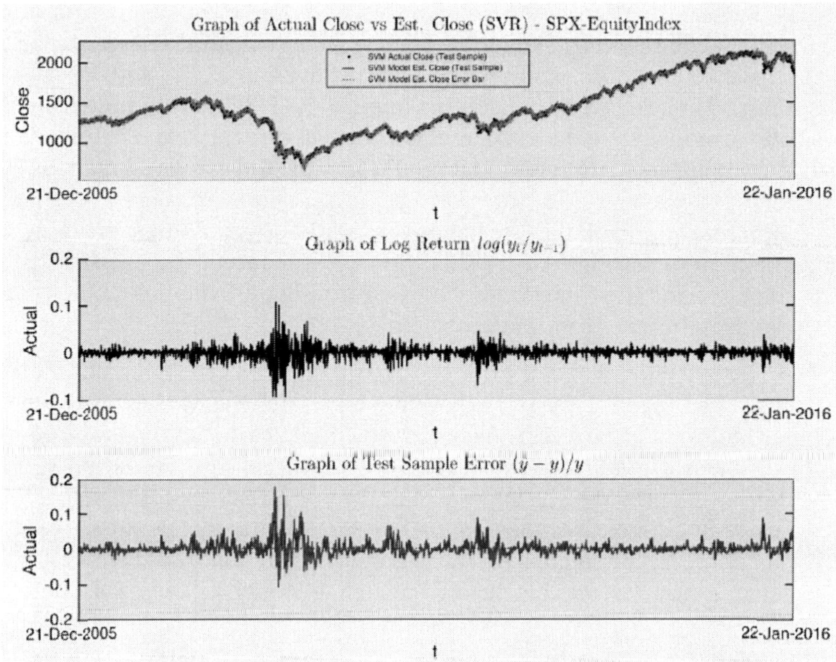

Figure 5. Actual quotes vs predicted quotes (S&P500 Equity Index).

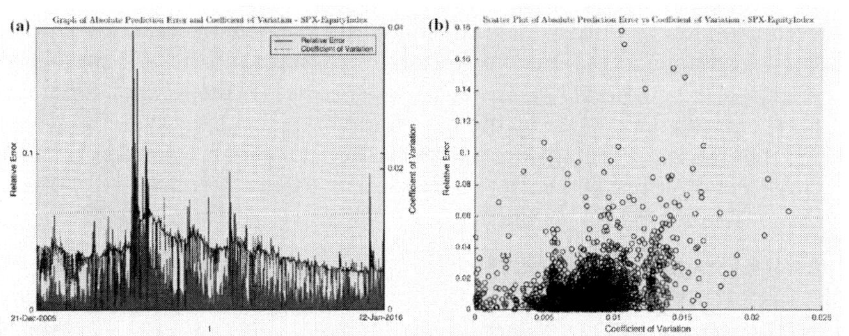

Figure 6. The absolute relative error and the prediction CoV (S&P500 Equity Index).

parameter selection process, as well as an estimate of prediction uncertainty, without sacrificing much predictive performance.

Results from the last experiment provide initiatives to spend more effort in examining further the performance and use of Bayesian SVR. It is shown that Bayesian optimization accelerates the log evidence optimization process while retaining similar performance as using grid search. Hence, in the second set of experiments, only Bayesian SVR with Bayesian optimization selected parameters is applied. S&P500 Equity Index is used as an example to demonstrate the practical use of this Bayesian SVR framework. Since the computational cost for the parameters selection process has been significantly reduced, we are able to increase the number of iterations by training the algorithm with three months of data (63 quotes), and testing daily. This greatly enhances the sensitivity of predictions and their uncertainty measures in better adapting to the rapid-changing nature of financial time series

The S&P500 Equity Index time series is examined. The prediction error drops from roughly 1.5% to 1.2%. While this is quite encouraging, the more important findings are yet to be discussed. In figure 5, the top graph shows the actual vs. predicted quotes as well as the predictive variance. The middle graph provides the log return of the daily asset quotes to display the volatility in different regimes. At the bottom, the graph plots the prediction error of everyday quotes. It is quite obvious that the Bayesian SVR predictions are less reliable during the high volatility regimes. This matches expectation as the data distribution may not be well represented in the training data during high volatility circumstances. Therefore, it is important to take into account prediction uncertainties to assess the quality of predictions.

As mentioned previously, another useful functionality of Bayesian SVR is prediction uncertainty estimation. In the top graph of figure 5, some dotted lines are found. These are the error bar estimates for the SVR predictions provided in equation (23). These are used to assess the level of confidence for the predictions made by the Bayesian SVR algorithm. Given the scale of the figure, the relationship between the error bar estimates and the corresponding predictive performance can hardly be observed. A better demonstration is provided in figure 6.

In figure 6(a), the absolute value of the relative errors ($|Predicted/Actual-1|$) is the vertical bars plotted on the left axis, and the prediction coefficient of variance (σ_{pred}/μ_{pred}) is the dotted line overlaid on the right axis. The coefficient of variance is used instead of the error bar estimates to account for the different level of predictionmean across the time series. It is easy to see that when the coefficient of variance is high, the prediction errors are higher. Similarly, figure 6(b) shows a scatter plot of the absolute value of the relative errors against the coefficient of variance, and the Pearson correlation is 34.06% with p-value close to zero. This suggests a significant relationship between size of the error bars and the prediction quality. In the financial context, when higher coefficient of variance is observed, the quote prediction may not be as reliable, and

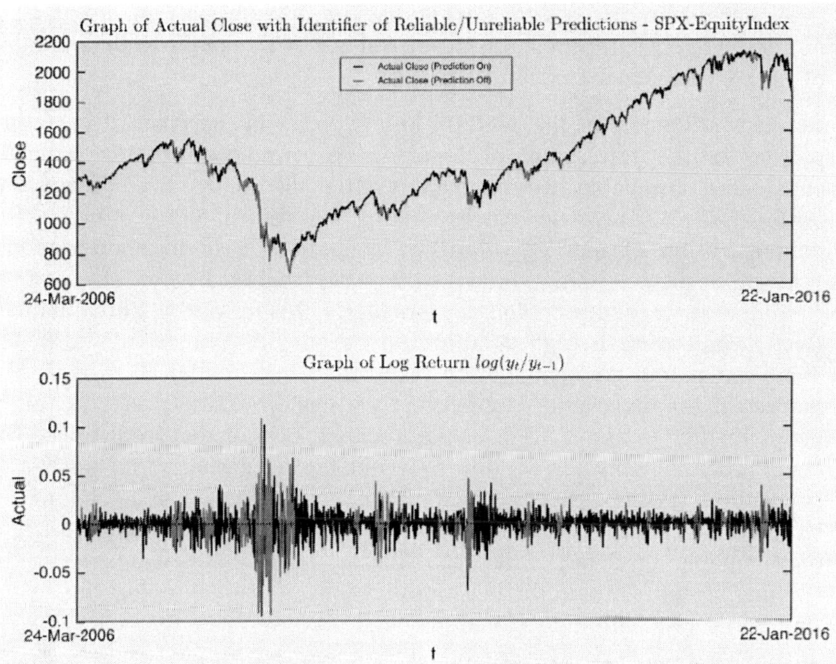

Figure 7. Actual Quotes with identifier of 'Reliable' predictions (S&P500 Equity Index). careful considerations should be made before basing decisions on such predictions.

5.4. Calibration

It is definitely useful to have such prediction uncertainty estimates. However, again because of the rapid changing nature of financial time series, it is not easy to identify one coefficient of variation cut-off value that is suitable to consistently distinguish 'reliable' predictions over time. It is equally difficult to identify such a value across different time series. Therefore, we introduce an automated calibration procedure to hopefully define this cut-off value adaptively in order to best utilize the prediction uncertainty estimates.

We begin by defining a rolling window procedure similar to the one in the previous experiments. In this case, to be consistent, three months of data (63 quotes) are used as training data, and are applied one day forward. The predictions generated in the last three months are assessed. A criterion is arbitrary defined suggesting that only the predictions that have error less than the average daily change in quotes are considered 'reliable'. Among these prediction points, considering that the CoV and prediction error are not perfectively correlated, the cut-off value is set to be the 80th percentile of all CoV values to exclude outliers.

The closing quotes of the S&P500 Equity Index in the past 10 years are plotted in the top graph of figure 7. The prediction points that are considered 'unreliable' are coloured in red, while the rest are in black. The bottom graph shows the corresponding log return of the quotes as a demonstration of market volatility. It is quite easy to spot that the predictions are coloured in red when the training data are not reflecting the current market condition. This is particularly obvious during 2008 financial crisis, when there is high volatility and prediction is extremely difficult. Some red points are also spotted when there are sudden changes in the market. If the 'unreliable' predictions are neglected, the prediction error drops further to about 1.1%, while keeping 77% of the predictions. The calibration not only allows one to avoid basing decisions on unreliable predictions, it can possibly act as a market condition change detection tool. However, it is important to note that the algorithm, in its current form, is only capable to detect when the market condition deviates from the training period. Without further modifications, it is not a tool to cluster the time series into different regimes (i.e. 'high' or 'low' volatility etc.).

There are two user-defined parameters in this calibration procedure. The first parameter is the reliability criteria definition. We define predictions to be considered 'reliable' when error is less than the average daily quote changes in the last three month. However, this can be adjusted based on the user's preference. For example, if this prediction is used as part of a trading strategy, this reliability criteria can be defined as some minimum return of the strategy over a period of time. The second parameter is the quantile used to exclude the outliers in the coefficient of variation estimates. The parameter may vary, but can be easily optimized by setting an objective function to consider other factors such as the number of predictions to be included, the target magnitude of error reduction, etc. While this calibration process is efficient and simple to implement, other clustering methods (i.e. KNN, Naive Bayes, SVM, Logistic Regression etc.) may be considered to differentiate the 'reliable' and 'unreliable' predictions.

6. CONCLUSIONS AND FUTURE EXTENSIONS

With the intention to adopt the powerful SVR in predicting financial time series, this study introduces a framework to generalize the algorithm with a Bayesian approach. This gives a more efficient parameter selection procedure as well as a prediction uncertainty estimate. A Bayesian approach to SVR (Gao *et al.* 2002) is discussed, and implemented. It is found that direct implementation of the probabilistic framework proposed by Gao *et al.* (2002) returns poor results in our experiments.A novel enhancement is proposed by adding a new kernel scaling parameter μ to overcome the difficulties encountered. In addition, the multi-armed bandit Bayesian optimization technique is applied to automate the parameter selection process.

The framework is then tested on financial time series of various asset classes to ensure its flexibility. In the experiments, it is shown that the generalization performance of the model selected through our framework can reach or sometimes surpass the ones from the model selected through the computationally more expensive cross-validation process. While taking advantage of the reduction in computational cost, iterations are increased to generate daily predictions. It is shown with the S&P500 Equity Index that the prediction error has decreased, while generating a sensible prediction uncertainty estimate. An adaptive calibration process is then presented to demonstrate the practical use of the prediction uncertainty estimates to identify 'unreliable' predictions, which greatly enhances the prediction performance. The machine learning approach discussed in this study can be developed as a pricing tool, and possibly as a market condition change detector.

While the framework is quite effective as it is, there are a few directions that may justify further investigations. The multi-armed bandit optimization technique employed at the moment is sequential. Although it is quite efficient, it can be further improved by parallelizing the algorithm. This has been discussed in a few studies (Ginsbourger and Riche 2011 Snoek *et al.* 2012 Desautels *et al.* 2014). Also, as mentioned, the calibration process introduced is efficient and simple to implement. However, it may be interesting to find out if other clustering algorithms give better results, or possibly lead to a regime clustering algorithm. Furthermore, it may be interesting to relax the Gaussian prior distribution assumption of the probabilistic framework proposed by Gao *et al.* (2002) .Agood example is to extend the SVR model to aMulti-Kernel Learning formulation which assumes a non-Gaussian prior distribution. However, this may lead to significant complexity in the reconstruction of the Bayesian evidence approximation. From the financial point of view, the prediction uncertainty estimates may be extended as inputs to build financial portfolios.

FUNDING

This work is supported by the Engineering and Physical Sciences Research Council (EPSRC).

Appendix 1. Financial time series dataset

Asset class	Symbol	Name	Start date	End date
Equity Index	FTSE	FTSE 100	02 September 2005	22 January 2016
Equity Index	SPX	S&P 500	02 September 2005	22 January 2016
Commodity Futures	BRENT	ICE BRENT Crude Oil Front Month Futures	02 September 2005	22 January 2016
Commodity Futures	GOLD	CME GOLD Front Month Futures	02 September 2005	22 January 2016
Bond Yield	GB10YR	UK Gilt 10YR	02 September 2005	22 January 2016
Bond Yield	US10YR	US Treasury 10YR	02 September 2005	22 January 2016
CDS Spread	IBM-CDS	INTERNATIONAL BUSINESS MACHINES CORPORATION 5YR	21 July 2008	21 January 2016
CDS Spread	WMT-CDS	WAL-MART STORES INC. 5YR	21 July 2008	21 January 2016

Appendix 2. MAPE (630 quotes training, 126 quotes testing)

Symbol	SVR (Cross-Validation)	BSVR (Grid Search)	BSVR (BayesOpt)
FTSE	1.88 (3.69)†	1.76 (1.25)	1.88 (1.40)
SPX	3.34 (8.69)	1.50 (1.24)	1.52 (1.23)
BRENT	2.03 (1.16)	1.96 (1.06)	1.98 (1.06)
GOLD	1.37 (0.89)	1.48 (0.54)	1.61 (0.58)
GB10YR	1.85 (0.82)	1.94 (0.86)	1.95 (0.79)
US10YR	2.13 (1.34)	2.21 (1.25)	2.27 (1.35)
IBM-CDS	1.34 (0.36)	1.34 (0.37)	1.36 (0.37)
WMT-CDS	1.29 (0.41)	1.22 (0.36)	1.29 (0.42)

†Average of MAPEs from all rolling windows with standard deviation within parentheses.

REFERENCES

1. Agrawal, S. and Goyal, N., Thompson sampling for contextual Bandits with linear payoffs. *Proceedings of the 30th International Conference on Machine Learning*, Atlanta, GA, 2013, 127–135.
2. Bergstra, J. and Bengio, Y., Random search for hyper-parameter optimization. *J. Mach. Learn. Res.*, 2012, **13**(1), 281–305.
3. Brochu, E., Cora, V. and de Freitas, N., A tutorial on Bayesian optimization of expensive cost functions, with application to active user modeling and hierarchical reinforcement learning. 2010, preprint, arXiv:1012.2599.
4. Cao, L.J. and Tay, F.E., Support vector machine with adaptive parameters in financial time series forecasting. *IEEE Trans. Neural Networks*, 2003, **14**(6), 1506–1518.
5. Chang, C. and Lin, C., LIBSVM: A library for support vector machines. *ACM Trans. Intell. Syst. Technol.*, 2011, **2**(3), 1–27.
6. Chapelle, O. and Li, L., An empirical evaluation of thompson sampling. *Adv. Neural Inf. Proc. Syst* 2011, **24**, 2249–2257.
7. Desautels, T., Krause, A. and Burdick, J., Parallelizing exploration-exploitation tradeoffs with Gaussian process bandit optimization. *J. Mach. Learn. Res.*, 2014, **15**(1), 3873–3923.
8. Dorard, L., Glowacka, D. and Shawe-Taylor, J., Gaussian process modelling of dependencies in multi-armed bandit problems. *Proceedings of the 10th International Symposium on Operations Research*, Nova Gorica, Slovenia, 2009, 77–84.
9. Gao, J., Gunn, S., Harris, C. and Brown, M., A probabilistic framework for SVM regression and error bar estimation. *Mach. Learn.*, 2002, **46**(1), 71–89.

10. Ginsbourger, D. and Riche, R., Dealing with asynchronicity in parallel Gaussian process based global optimization. 4th International Conference of the ERCIM WG on Computing & Statistics (ERCIM'11), London, UK, 2011.
11. Gündüz, Y. and Uhrig-Homburg, M., Predicting credit default swap prices with financial and pure data-driven appraoches. *Quant. Finance*, 2011, **11**(12), 1709–1727.
12. Kaufmann, E., Korda, N. and Munos, R., Thompson sampling: An asymptotically optimal finite-time analysis. *Algorithmic Learn. Theory* 2012, **7568**, 199–213.
13. Kim, K., Financial time series forecasting using support vector machines. *Neurocomputing*, 2003, **55**(1), 307–319.
14. Lu, C., Lee, T. and Chiu, C., Financial time series forecasting using independent component analysis and support vector regression. *Decis. Support Syst.*, 2009, **47**(2), 115–125.
15. MacKay, D., Bayesian interpolation. *Neural Comput.*, 1992, **4**(3), 415–447.
16. Martinez-Cantin, R., BayesOpt: A Bayesian optimization library for nonlinear optimization, experimental design and bandits. *J. Mach. Learn. Res.*, 2014, **15**(1), 3735–3739.
17. Mohri, M. and Rostamizadeh, A., Stability bounds for non-iid processes. *Adv. Neural Inf. Process. Syst.*, 2008, **20**, 1025–1032.
18. Müller, K., Smola, A., Rätsch, G., Sch{\"o}lkopf, B., Kohlmorgen, J., Vapnik, V. Predictin g time series with support vector machines. In *Artificial Neural Networks: ICANN'97*, Vol. 20, edited by W. Gerstner , A. Germond , M. Hasler and J.D. Nicoud, pp. 999–1004, 1997 (Springer: Berlin).
19. Ralaivola, L., Szafranski, M. and Stempfel, G., Chromatic PAC-Bayes bounds for non-iid data: Applications to ranking and stationary-mixing processes. *J. Mach. Learn Res.* 2010, **11**, 1927–1956.
20. Russo, D. and Van Roy, B., Learning to optimize via posterior sampling. *Math. Oper. Res.*, 2014, **39**(4), 1221–1243.
21. Scott, S., A modern Bayesian look at the multi-armed bandit. *Appl. Stochastic Models Bus Ind.*, 2010, **26**(6), 639–658.
22. Shahriari, B., Swersky, K., Wang, Z., Adams, R. and de Freitas, N., Taking the human out of the loop: A review of bayesian optimization. *Proc. IEEE*, 2016, **104**(1), 148–175.
23. Smola, A. and Schölkopf, B., A tutorial on support vector regression. *Statistics Comput.*, 2004, **14**(3), 199–222.
24. Snoek, J., Larochelle, H. and Adams, R., Practical Bayesian optimization of machine learning algorithms. *Adv. Neural Inf. Process. Syst.*, 2012, 2951–2959.
25. Srinivas, N., Krause, A., Kakade, S. and Seeger, M., Gaussian process optimization in the bandit setting: No regret and experimental design. *Proceedings of the 27th International Conference on Machine Learning, Haifa, Israel*, 2010, 1015–1022.
26. Tay, F.E. and Cao, L., Application of support vector machines in financial time series forecasting. *Omega*, 2001, **29**(4), 309–317.

Index